A Concise History of Ornithology

A CONCISE HISTORY OF ORNITHOLOGY

Michael Walters

YALE UNIVERSITY PRESS
NEW HAVEN AND LONDON

The author would like to thank Gerald Baker, Selwyn Tillet and Peter Bircham
for their helpful comments on the manuscript

Photographs supplied by kind permission of the library of the Natural History Museum

Published 2003 in the United Kingdom by Christopher Helm, an imprint of A & C Black Publishers Ltd. and in the United States by Yale University Press

ISBN 0-300-09073-0
Library of Congress Control Number: 2003104752
Produced and designed by Fluke Art, Cornwall
Printed and bound in Great Britain by Bath Press

A catalogue record for this book is available from the British Library.
The paper in this book meets the guidelines for permanence and durability of the Committee on Production Guidelines for Book Longevity of the Council on Library Resources.

10 9 8 7 6 5 4 3 2 1

Contents

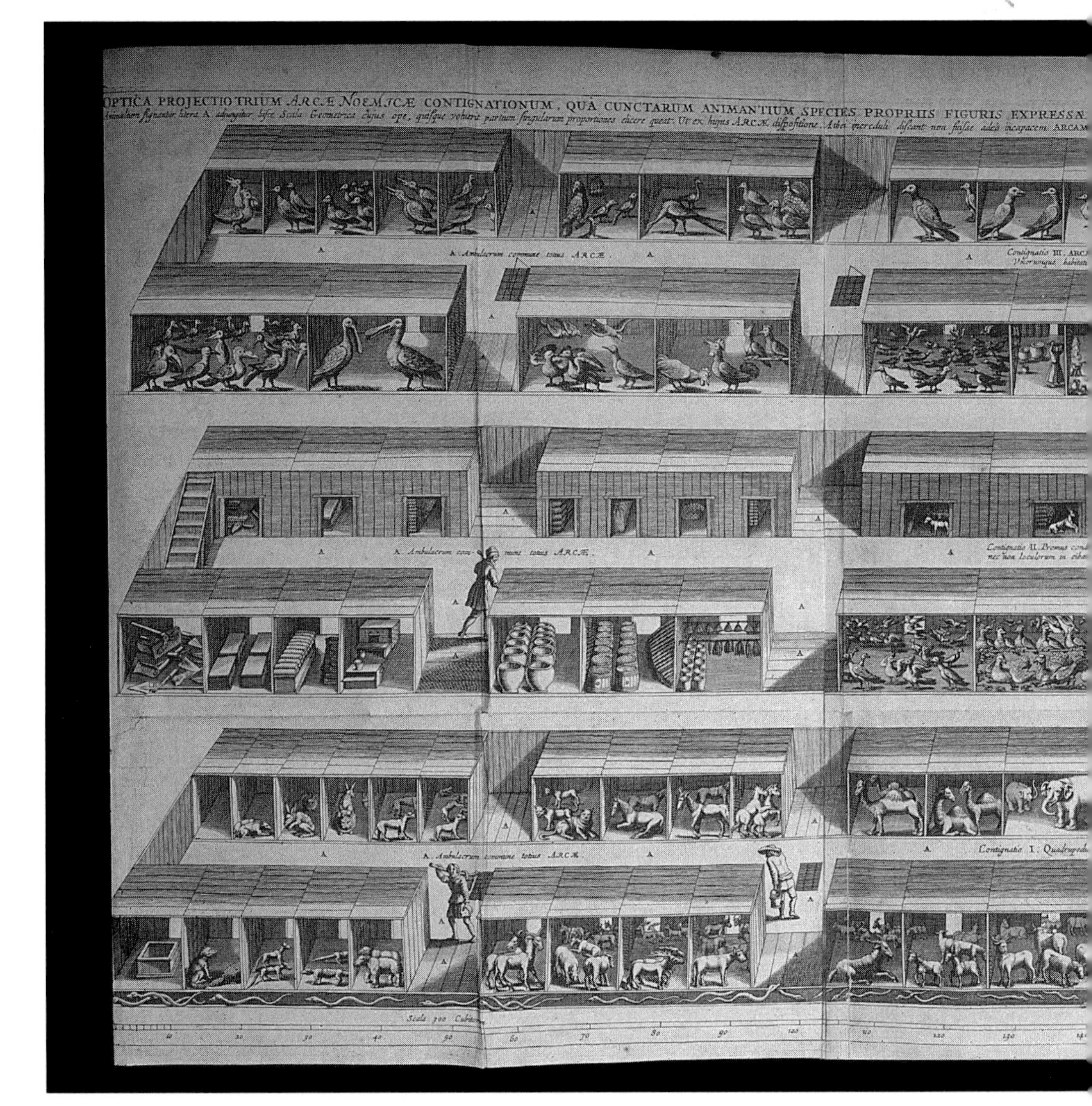

Arca Noë in Tres Libros Digesta, by A. Kircher, published 1675 at Amsterdam

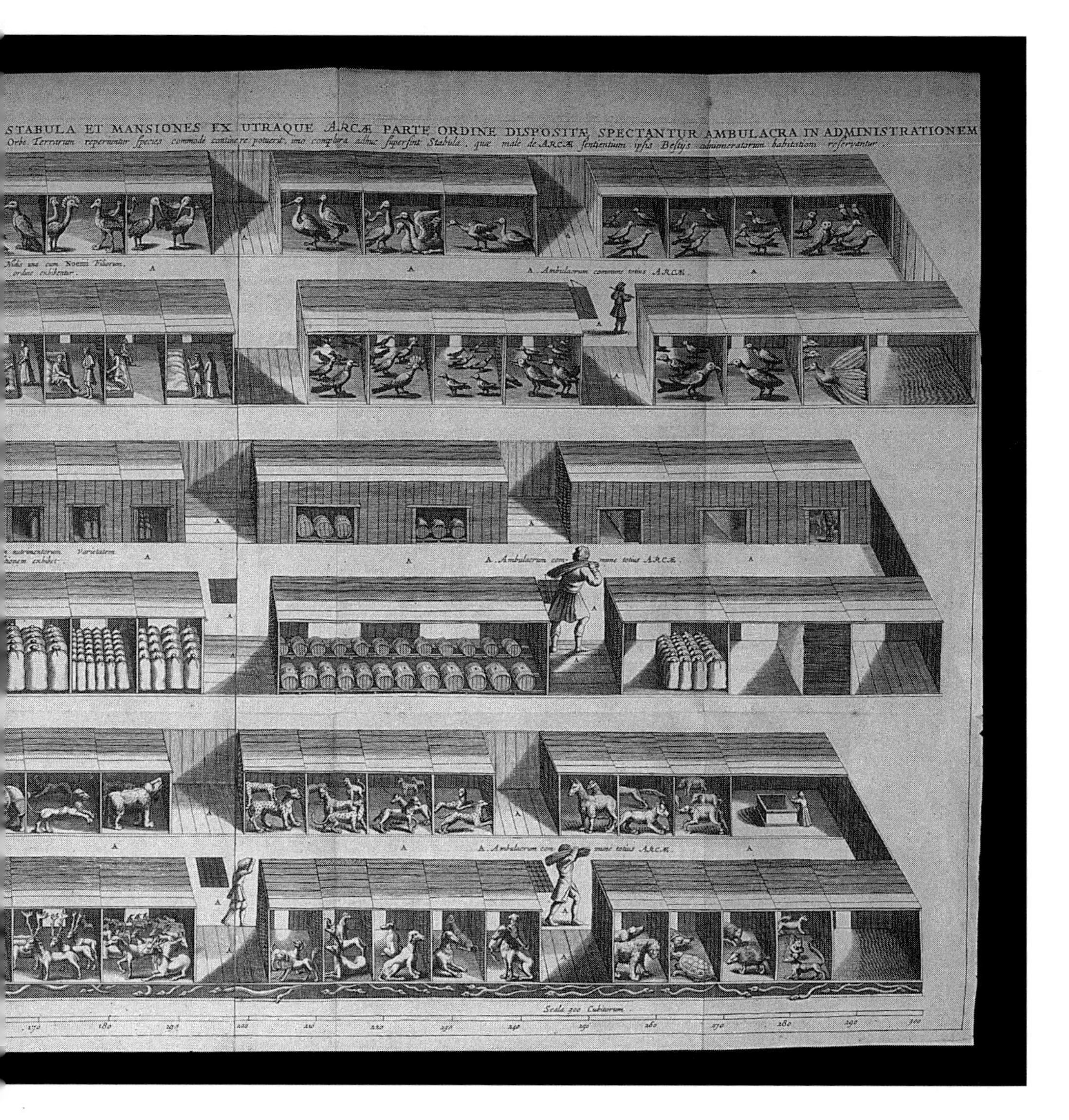
STABULA ET MANSIONES EX UTRAQUE ARCÆ PARTE ORDINE DISPOSITÆ SPECTANTUR AMBULACRA IN ADMINISTRATIONEM
Orbe Terrarum reperiuntur species commode continere potuerit; imo complura adhuc superfint Stabula, quæ male de ARCÆ sentientium ipsis Bestijs adnumeratarum habitationi reservantur.
A. Ambulacrum commune totius ARCÆ.
A. Ambulacrum commune totius ARCÆ.
Scala 300 Cubitorum.
170
180
190
200
210
220
230
240
250
260
270
280
290
300

Preface

The definitive history of ornithology has not been written. Nor, I suspect, will it be for a long time, in contrast to the histories of many branches of the arts, for our subject is as vast as it is neglected. Early attempts by writers like William MacGillivray and James Wilson, author of the article on 'Ornithology' in the eighth edition of the *Encyclopaedia Britannica*, were no more than cursory; the first serious attempt to survey the history of the study being made by Alfred Newton, first in the ninth edition of the *Encyclopaedia Britannica* (1875–89) and subsequently in the introduction to his *Dictionary of Birds* (Newton 1896). Useful though it is, much of Newton's account is basically uncritical, being devoted to a catalogue of the works that had been written, with some brief information about each writer. This, however, is not to belittle its importance - Newton had undoubtedly a greater knowledge of ornithological literature than any other writer of his time. For nearly three-quarters of a century, Newton's account stood as the only one, until Erwin Stresemann published his *Die Entwicklung der Ornithologie* (Stresemann 1951). This was translated into English in 1975 as *Ornithology from Aristotle to the present* (Stresemann 1975) with an Epilogue by Ernst Mayr updating it as far as American ornithology is concerned. Stresemann's was one of the most fertile minds of this century, and his book was a landmark (indeed a starting point for all time) in treatment of the history of the subject. It should be pointed out that both editions are long out of print. But in places Stresemann seems to have been biased, and he omitted a number of significant figures and topics. He was never a kind critic, and a number of the writers he discusses are treated harshly - though not necessarily unjustly. This present book is aimed at the reader who is knowledgeable about birds, but may know nothing of the history, and, with the exception of a handful of names such as Aristotle, Linnaeus and Darwin, may not have heard of any of the people whose lives and work are here discussed. I cannot hope to compete with Stresemann's erudition, and I remain in his debt for a great deal of the information which he presented. But I have omitted much that he included, as being of insufficient interest to the general reader, and added some that he did not cover. With what success, or lack of it, I must leave the reader to judge.

To say that a lot has been written on ornithology would be an understatement of the greatest magnitude. To assess every writer and accurately assign him or her a place in history would be the work of several lifetimes. Here it has been possible to mention only those who seemed the most important in the *development* of the subject. Inevitably, many have had to be omitted, and of many authors and their books which warrant a good deal of study, it has been possible to say only a few words. Perhaps, however, I may have said enough to encourage others to undertake in-depth studies of those many fascinating authors whom space and time have forced me to bypass, and thus help to provide the material for the authoritative study when it comes to be written. Unlike many other branches of science, zoology, and ornithology in particular, had a late start. Unlike botany, in which plants had many implications for medicine, birds were pretty objects with little potential importance, except as food. The comparative lack of documentation in the early centuries may have been partly as a result of a belief that things had always been as they are now, and would always be thus. Documentation was therefore superfluous. Likewise, until the time of Darwin, there was no real need to enquire why certain animals were found in one place and not in others. The answer was clear: God had decreed that this should be so, therefore it was so. Further research was not required. Even in the 18th century, zoogeography was not understood. We may sigh with frustration at the opportunities wasted on the Pacific voyages of Cook, Bougainville and others, when so little serious collecting or observation on islands was made; but in fairness we must realise that the explorers of those days could have had no idea that the fauna of each Pacific Island was not only unique, but fast vanishing. That no-one at that time did understand it is shown by the cavalier treatment shown to those gallant few like the Forsters, and Commerson, who strove to bring back valuable and important information.

A serious interest in birds probably did not begin till the middle of the 17th century when that monumental landmark, Willughby's *Ornithologiae* (Willughby 1676) fired the enthusiasm of at least a handful of men. But for the next century progress was slow, and it was not till the late 18th century, with the publication of books like Brisson's *Ornithologie* (Brisson 1760), Linnaeus's *Systema Naturae* (Linnaeus 1758, 1766) and Buffon's *Histoire Naturelle* (Buffon 1770–83), that study began in earnest. Throughout the 19th century, progress in systematics was astounding, so that by the end of it, Hans Gadow was to draw up a system of classification (Gadow 1892, 1893) which, with modifications, is still in use today. In the 20th century, the emphasis shifted from systematics to questions of behaviour, ecology and conservation. The volume of ornithological literature has become overwhelming, and the number of individuals working on the subject has increased to the extent that a comprehensive survey of the subject in the 20th century is beyond the scope of this book.

CHAPTER 1

EARLY TIMES

Since the dawn of civilisation, birds have fascinated men. Their beauty of plumage, of voice, and the fact that they could travel through the air (a medium denied to him), caused primitive man to regard birds as beings apart; they became woven into his myths and religions. The Hindu god Vishnu had as his heavenly steed the great eagle Garuda. To the North American Indians, the Thunderbird was a creature so vast in size it could swallow whales whole, and its voice was the storm and the tempest. The Egyptians saw the god Horus as a gigantic falcon whose wings stretched from one end of the universe to the other. Birds appear, apparently in semi-mystical guise, in Old Stone Age paintings in the caves at Lascaux. From these scattered scraps of bygone cultures and human thoughts we can learn something of the mind of man and his attitudes to these captivating creatures. In rare instances they can be of use in historical documentation of the birds themselves. We know, for instance, from frescoes in Egyptian tombs, that the Red-breasted Goose (*Branta ruficollis*) formerly wintered in the Nile delta although it no longer does so.

Kien Hexagram

Birds are mentioned in one of the oldest extant books in the world, the *I-Ching*, or Book of Changes, also generally considered to be one of the most incomprehensible. It consists of a series of hexagrams each made up of six horizontal lines which are either entire or with a gap in the middle. These hexagrams were accompanied by a series of commentaries which were used for an ancient form of divination, or fortune-telling. For instance, hexagram no. 53, the KIEN Hexagram, has an accompanying text as follows:

> The first line, divided, shows the wild geese gradually approaching the shore. A young officer (in similar circumstances) will be in a position of danger, and be spoken against; but there will be no error. The second line, divided, shows the geese gradually approaching the large rocks, where they eat and drink joyfully and at ease. There will be good fortune. The third line, undivided, shows them gradually advanced to the dry plains. It suggests also the idea of a husband who goes on an expedition from which he does not return, and

> of a wife who is pregnant, but will not nourish her child. There will be evil. The case symbolised might be advantageous in resisting plunderers. The fourth line, divided, shows the geese gradually advanced to the trees. They may light on the flat branches. There will be no error. The fifth line, undivided, shows the geese gradually advanced to the high mound. It suggests the idea of a wife who for three years does not become pregnant, but in the end the natural issue cannot be prevented. There will be good fortune. The sixth line, undivided, shows the geese gradually advanced to the large heights (beyond). Their feathers can be used as ornaments. There will be good fortune.

The age of the *I-Ching* is uncertain, but it was already an old book when it was selected by Confucius in c. 551–479 BC as one of the four great classics; this fact saved it from the notorious burning of the books by the Emperor Chin-Shih Huang-ti (Qin Shi Huangdi). The passage quoted indicates that the anonymous author or authors were moderately familiar with the habits of geese, but were less interested in the birds themselves than in the lessons their habits had for mankind.

Although little purely ornithological information appeared for many centuries, much of general biological interest has come down to us from ancient writers, and as the history of biology and zoology cannot be entirely divorced from that of ornithology, partial discussion of wider topics is necessary here. Thales of Miletus (650–580 BC) believed all things had their origins in water. The earth, he believed, floated like a disc on a vast sea which surrounded it on all sides. His disciple Anaximander (611–546 BC) further believed that the earth came into being through condensation of water. Originally mud, it then became solid, floating on the vast expanse of ocean like a loaf. Anaximander conceived live beings as having evolved from the primeval mud; animals and plants first, then human beings. The latter originally appeared in the form of fish which subsequently shed their scales and emerged from the water to live on land. In this remarkable account can be seen a very primitive theory of evolution, so evolution was not something that sprang up suddenly in the 19th century as traditional thought would have us believe; it was very old indeed. Diogenes of Crete, who lived in the first half of the 5th century BC, carried Anaximander's theory further. He believed that living beings were formed out of inorganic matter by the effects of solar heat; the embryo was developed by the heat of the mother irradiating the semen of the father.

In the late 6th century BC, Xenophanes noticed fossilised marine animals high up in the mountains, and declared this constituted proof that the mountains were once under water. These ideas were too bizarre and revolutionary for the time and were discarded; it was not till the Renaissance that they came to be accepted. Empedocles, who is believed to have lived in the 5th century BC, considered it impossible that matter could arise from or return to nothing. Cosmic matter must always remain the same, he insisted, its component parts merely mixed in different combinations. Thus living creatures arise from the earth, plants first, then animals. Leucippus and his disciple Democritus (c. 470–370 BC) probably first formulated the 'atom' theory, a theory which sounds surprisingly like modern science. The latter claimed that change is merely a separation or aggregation of existing parts, everything has cause and necessity, and nothing exists except 'atoms' and space: everything else is an impression of the senses. It seems likely that much of the writing of Aristotle derived from Democritus. Certainly the distinction of red-blooded (vertebrates) and bloodless animals, the main principle of Aristotle's classification, derives from Democritus. However, the latter correctly regarded the brain as an organ of thought, in contrast to Aristotle, who considered it merely as an organ for cooling the blood.

It was under Hippocrates the Great (460–377 BC) that human anatomy was first seriously studied, and Plato (429–347 BC) believed that the world, having been created by an external and perfect god, was therefore constructed in the most perfect shape, a sphere. None of this, of course, is directly relevant to ornithology as a study in its own right, but demonstrates that much intelligent biological thought had begun in early times, in contrast to the long period of stagnation in the Dark Ages.

Scientific study of birds really began, as most disciplines did, with Aristotle. This remarkable person was born in Stagira in Thrace (Macedonia) in 384 BC and died in 322 BC, aged 63. His father, Nicomachus, died when he was still young, and at the age of 18 he was sent to Plato's academy in Athens where he remained for 20 years, at the end of which time he apparently found himself in disagreement with his master's views, to the latter's observed displeasure. After Plato's death, Aristotle failed to be elected his successor and left to go to Asia Minor, where he settled at the court of the Persian prince, Hermias. After a few years the prince was deposed in a revolution, and Aristotle fled back to his native country, where he was appointed by Philip of Macedon as tutor to his son (later Alexander the Great), a task which occupied three years (338–335 BC).

Aristotle: Greek Scholar, Philosopher and Naturalist

Aristotle was not the handsomest of men. He was described as of short stature, with thin legs and remarkably small eyes. He had a shrill voice and spoke with a stammer. But he was considered a man of fashion, dressed nattily, wore magnificent and valuable jewellery, and shaved his face in contrast to the practice of the time. He was twice married. For years he gave regular lectures, morning and evening, in Athens, and his high level of activity indicates tremendous energy and powers of organisation. After Alexander's death, he came under the criticism of the authorities, and fled to the island of Euboea, where he died a few years later. He has been revered as the father of biological thought, but his writings on other topics have suffered more criticism, many having seen deficiencies in his system of thought. Aristotle's system was based on Plato, who argued that eternity is an actual reality and that things of the earth are an imperfect shadow of the eternal form. Form is a reality, matter a potentiality which the form turns into a reality. Aristotle proposed that every lower stage or form of plant or animal life form has the potential to achieve a higher stage of development; thus the egg or embryo embodies the "potentiality" of the adult creature. Lower forms of life have a potentiality to become higher forms. Like Anaximander, Aristotle had arrived at a crude concept of evolution. This great thinker is said to have written 400 books, of which only about 48 survive. Aristotle wrote on an extraordinarily wide range of topics: logic, metaphysics, art, politics, psychology and biology. He believed that all things are eternal; no particle of matter will ever perish, but merely change place; from the remains of one thing another is made and thus the mass of the world always remains entire. All of the globe now covered by water was once dry land, and the present land will one day be water, as river erosion and silt deposition cause the mountains to be worn away and the shores gradually to advance.

Of Aristotle's known biological works, the following survive: ten books on *The History of Animals* (of which, however, three are considered to be spurious); four books *On the Parts of Animals*; five books *On the Reproduction of Animals*; and three books *On the Soul*. He recognised only two classification categories, the "genos" and the "eidos", the latter corresponding roughly to a species, as dog, cat, pig, etc., while the "genos" applied to all groupings of higher degree. In these works he most certainly borrowed heavily from the writings of previous authors, but it is impossible now to say to what extent, as most of their works have not survived, and he never quoted them directly except to criticise. His description of the crocodile, for example, is in the exact words of Herodotus, but without acknowledgement. This, and other instances, demonstrates an uncritical exploitation of sources. But it is the way he used and reworked his sources that give his work its value. His studies of anatomy and morphology were comprehensive and detailed, but those relating to human anatomy have, unfortunately, been lost. In physiology, as for example in his discussions of the digestive system, his views, perhaps inevitably, are vague and rather primitive. The food, he declared, is "cooked" in the intestine.

He recognised some 140 species of birds, and gave adequate descriptions of many of them, but about others he knew so little that his attempts at classification were unsuccessful. Nevertheless, his influence was so great that for nearly 1,500 years his book was the principal reference work on birds. In 1931, the American ornithologist Margaret Nice was distressed to realise that the incubation periods given for certain birds of prey in standard American ornithologies as late as Bergtold's *A Study of Incubation periods of Birds* (1917) were ridiculously short. Nice set herself the task of discovering the source of these errors, and found that they could be traced from book to book back through the centuries to Aristotle. Aristotle knew that the domestic fowl usually hatches her eggs in about three weeks, and the domestic goose in 30 days. He inferred that the length of incubation varied with the size of the bird, and concluded that medium-sized birds matched the hen and large ones the goose. His conclusions were followed unquestioningly by Pliny in the 1st century AD, Albertus Magnus in the 13th century, Gesner in the 16th and Aldrovandus in the 17th. They all quoted his 20- and 30-day incubation periods for small and large raptors respectively, and they made no attempt to supply incubation periods for birds not specifically mentioned by Aristotle. The first author to attempt to supply such periods for all species of birds discussed, was Bechstein in his *Gemeinnützige Naturgeschicte Deutschlands nach allen drey Reichen* [General Natural History of Germany] (Bechstein 1789–95), and to most of them he assigned incubation periods on Aristotelian principles. Margaret Nice showed that the first person to address the problem scientifically had been the Scottish naturalist William Evans in two papers published in the journal *Ibis* (Evans 1891, 1892). He had hatched the eggs of 79 species in an incubator, or under hens, pigeons or canaries. Margaret Nice published her findings in two short papers (Nice 1953, 1954) and commented that, at the time of writing, young American ornithologists were still copying errors from Bergtold, in other words, from Aristotle. Aristotle had made an inspired guess. He was wrong, but his error still influenced ornithological thought 2,000 years later.

However, Aristotle's influence was not merely negative. His ability to ask questions of significance in respect of his subjects of study made him one of the first scientific thinkers. Although he believed that swallows hibernated (a belief that was held into the 18th and 19th centuries), he was aware of the existence of migration. He considered, however, that it was a matter of choice rather than of instinct. According to Aristotle, everything had been created to fulfil a purpose, and the world was a hierarchy of purposive arrangements leading upward from the lowest to the highest. But the most significant thing about Aristotle is that he was not only the first serious thinker in zoology, but for many centuries the *only* one. William MacGillivray, after pointing out that Aristotle was far from free of errors, summed him up as follows:

> The observations of Aristotle, considering the period at which he lived, and the proneness of the human intellect to wander from the true path, are remarkable for the great proportion of truth which they present to us. Whatever may be their actual merits, they are certainly superior to those of any other naturalist whose works have come down to us from the remote ages of classical antiquity (MacGillivray 1834).

A search for ornithological material in *The History of Animals* and *On The Parts of Animals* reveals a convincing impression of Aristotle's attempt to explore the whole field of bird life. Although many of his species remain unidentified as a result of inadequate descriptions, it is as the first person to attempt to classify birds that he remains important. He divided the class into five groups:

1. Birds of prey
2. Swimming birds
3. Pigeons and doves
4. Swifts, martins and swallows
5. All others, i.e. all passerines except swallows, and various other groups as well

The closeness of his observation was remarkable. He noticed that quadrupeds that are viviparous (i.e. give birth to live young) have hair, those that lay eggs have scales. No single-hoofed animal has two horns, and no animal has both tusks and horns. He was well aware of the difference between whales and fish, and in this was very far in advance of his time. He had observed the legs and feet of birds closely, and knew that the wings corresponded to the forelimbs of mammals, though he did not pursue the matter deeply enough to understand the details of the similarity. (He also mistook the tarsus or shank for the shin, not realising that birds stand on their toes and that the 'leg' is a fused tarso-metatarsus). He knew that feathers are the equivalent of reptiles' scales, and that birds change their colours by means of seasonal moults, warning readers not to be deceived by these changes into thinking that the different plumages represented different species. He is known to have dissected parts of 110 species of animals, including dove, duck, goose, owl, pigeon, partridge, quail and swan. In these dissections he paid attention to the alimentary canal, knew about the existence of the crop, examined the liver, heart, lungs, kidneys, spleen, pancreas and gall bladder, though he did not have a clear understanding of their functions. However, he did understand the sex organs, for he spoke of determining the sex of doves, whose outward appearance is similar, by an examination of their "interiors". He was the first to put on record any accurate examination of the developing chick in the egg. He made comparative examinations of the development of the oesophagus, the stomach and the caecum; he studied lung capacity, and described the development of the chicken as exactly as was possible without the aid of optical instruments. He observed the breaking of the shell, but did not know whether the chick or the adult was responsible. It was Albertus Magnus, 1,500 years later, who first described how the young bird breaks the shell with its bill.

After Aristotle fled from Athens, his school was left in the hands of his disciple Theophrastus, primarily a botanist, who presided over it for 30 years, and subsequently by Strato, all of whose writings are lost. A later Athenian, Epicurus (342–271 BC), taught his pupils that the universe had a natural origin and was not divinely created. If an object or pursuit gave pleasure, then it was by definition good. This later became distorted into an invitation to unbridled worship of pleasure. As far as zoology is concerned, Aristotle's school withered, and no-one truly followed in his footsteps.

The next ornithological writer of note was Gaius Plinius Secundus, Pliny the Elder. Usually referred to simply as Pliny (AD 20 or 23–79), he was noted for his *Historia naturalis*, in 37 books. This is the only treatise of his that has survived, but he also wrote on subjects as diverse as military science, military history, rhetoric and linguistics. He was the most eminent of Rome's natural philosophers, and next to Aristotle, by far the most influential biologist of the ancient world. Unlike Aristotle, however, he has been harshly criticised by present-day biologists, having been described as nothing more than a sedulous compiler. Erwin Stresemann remarked scathingly of his book:

> For nearly fifteen hundred years this encyclopaedia, almost unusable as a source of zoological knowledge, was held in the highest esteem, even by noted naturalists (Stresemann 1975).

This is unfair. Pliny was a remarkable man in his own way, and suffered criticism because he was more honest than Aristotle and always quoted his sources. He has been ridiculed for quoting, as fact, accounts of fabulous creatures the existence of which no-one of his day questioned, possibly because of a 20th century desire to see Greek civilisation as greater than Roman. Nordenskiöld, who had a much higher opinion of Pliny than Stresemann, said:

> For fifteen hundred years his work was the main source of man's knowledge of natural history, and when during the Renaissance a Gesner or an Aldrovandi [Aldrovandus] revived the pursuit of zoological research, they at once began where Pliny had left off and carried on the work after his method. In this way present-day zoology, as regards the study of fauna and classification, takes Pliny as its starting point, just as in the matter of comparative anatomy and morphology it is based on Aristotle, and therefore the services of the one should in all fairness be recognised as much as those of the other, even if they refer to entirely different fields of study (Nordenskiöld 1928).

Pliny was born in Comum, now Como, in northern Italy, and, although he appears to have travelled over a great part of Europe in the service of his country, to have visited Africa, and perhaps Egypt and Palestine, no record of these travels has been preserved. Had it not been for occasional comments that occur in his writings, and especially information regarding his private habits and literary labours contained in the writings of his nephew, Pliny the Younger, posterity would have known nothing of his life. At an early age, the elder Pliny was sent to Rome where he attended the lectures of Appion. While still young, he was employed in the Roman armies in Germany, and served there under Lucius Pomponius, who became a lifelong friend, entrusting him with command of part of the cavalry. At about the age of 30, Pliny returned from Germany, and may have then spent some time in Spain; on his return to Rome he was courtier to the Emperor Vespasian, with whom he had been on intimate terms in Germany. He was in the habit of calling on the Emperor in his apartment before sunrise, a privilege reserved only for the latter's very close friends. Pliny appears to have been as amiable and affectionate as he was learned and studious. Throughout his work he expressed love for justice, respect for virtue and detestation of cruelty and baseness. Pliny the Younger, in a letter to his friend Macer, commented thus on his uncle's habits:

> You will wonder how a man so engaged as he was, could find time to compose such a number of books, and some of them, too, upon abstruse subjects. But your surprise will rise still higher, when you hear that for some time he engaged in the profession of an advocate; that he died at the age of fifty-six; that from the time of his quitting the bar to his death, he was employed partly in the execution of the highest posts, and partly in a personal attendance of those emperors who honoured him with their friendship. But he had a quick apprehension, joined to unwearied application. In summer he always began his studies as soon as it was night; in winter generally at one in the morning; but never later than two, and often at midnight. No man ever spent less time in bed; insomuch that he would sometimes, without retiring from his books, take a short sleep and then pursue his studies. Before daybreak he used to wait upon Vespasian, who likewise chose that season to transact business. When he had finished the affairs which the emperor committed to his charge, he returned home again to his books. After a short and light repast at noon, he would frequently in the summer, if he was disengaged from business, repose himself in the sun, during which time some author was read to him, from whom he made extracts and observations; as indeed this was his constant method, whatever book he read, for it was a maxim of his, "that no book was so bad, but something might be learned from it".

> When this was over, he generally went into a cold bath, and as soon as he came out of it, just took a slight refreshment, and then reposed himself for a little while. Then, as if it had been a new day, he immediately resumed his studies till supper time, when a book was again read to him, on which he would make some hasty remarks ... In summer he always rose from supper with daylight, and in winter as soon as it was dark; and this rule he observed as strictly as if it had been a law of state.

It is one of the tragedies of history, that a man of such erudition should have lived at a time when inevitably a great proportion of his writings would be lost to posterity. But the study of books cannot make an all-round naturalist, and therefore the writings of Pliny contain not so much a description of objects as a compilation of all that had been previously recorded. As such, however, they are of considerable value. The *Natural History* of Pliny is one of the richest stores of information, being composed of extracts from more than 2,000 books, written by authors of all kinds – travellers, historians, geographers, philosophers and physicians – authors of whose works only about 40 survive, and of several of which we have mere fragments, or works different from those Pliny used; and there are many whose names escaped oblivion only through the quotations Pliny made from them. Seventeen books are devoted to botany, and these form the most extensive portion of Pliny's writings; but the plants are discussed so vaguely that it is often impossible to identify the species. The last five books are devoted to describing metals, mining, the earth and stones; the descriptions of some of the precious stones in the last book (of amber and beryl for example) are said to be as good as those of many 19th century mineralogists. The work's most obvious defect is the lack of anything like system or classification; it is impossible to conjecture on what principle species are arranged. Pliny did not always select what was important, but seems to have had a predilection for bizarre or marvellous things. On many occasions he substituted for the Greek word used by Aristotle a Latin word referring to a quite different animal. Admittedly one of the greatest difficulties experienced by ancient naturalists was that of fixing nomenclature, and this shows itself in Pliny more than in any other. The tenth book deals with "Foules and flying creatures". Here Pliny is meagre and confused, but the description of the common cock would have done credit to Buffon, that most meticulous of 18th century writers; the account of the nightingale is also entertaining. The various birds and animals are treated without any sequence, but, generally speaking, the largest and most interesting are treated first. The author described their habits, utility, the damage they do, and the date of their first being exhibited in Rome.

Two passages demonstrate his style:

> Of all the birds which we know, the Aegles carry the price both for honour and strength. Six kinds there be of them. The first, named of the Greeks *Melaenaetos* (The Saker as some thinke. Translator's note), and in Latin, *Valeria;* the least it is of all others, and strongest withall, black also of colour: In all the whole race of the Aegles, she alone nourisheth her young birds: for the rest (as we shall hereafter declare) do beat them away: she only crieth not, nor keepeth a grumbling and huzzing as others do: and evermore converseth upon the mountains. Of the second sort is *Pygargus* (A kind of falcon. Translator's note). It keepeth about towns and plains, and hath a whitish tail. The third is the *Morphnos*, which Homer called also *Percnos*: some name it *Plancus* and *Anataria:* and she is for bigness and strength, of a second degree: loving to live about lakes and meers ... Of the fourth kind is *Percnopterus*, the same that *Oripelargus*, fashioned like to a Geire or Vulture: it hath least wings, a body bigger than the rest: but a very coward, fearfull and of a bastard and craven kind, for a raven will beat her. Besides, she hath a greedy and hungry worm always in her gorge and craw, and never is content, but whining and grumbling. Of all Aegles she only carieth away with her the dead prey, & feedeth thereupon in the air: whereas others have no sooner killed, but they prey over them in the place. This bastard buzzard kind maketh that the fifth, (which is the royal Aegle) & is called in Greek *Gnesios*, as one would say, true and kindly, as descended from the gentle and right airie of Aegles.

> This Aegle royal, is of a middle highness, and of a reddish colour, a rare bird to be seen. There remaineth now the sixth and last sort, and that it *Haliartos*. This Aegle hath the quickest and clearest eye of all others, soaring & mounting on high: when she spieth a fish in the sea, down she comes with a power, plungeth into the waters, and breaking the force thereof with her breast, quickly catheth up the fish, and is gone ... Now, as touching the *Haliartos* or the Osprey, she only before that her little ones be feathered, will beat and strike them with her wings, and thereby force them to look full against the Sunne beames. Now, if she see any one of them to wink, or their eyes to water at the rays of the sun, she turns it with the head forward out of the nest, as a bastard and not right, nor none of hers; but bringeth up and cherisheth that, whose eyes will abide the light of the sun as she look directly upon him. Moreover, these Orfraies or Ospreies are not thought to be a several kind of Aegles by themselves, but to be mongrels, and engendered of divers sorts. And their young Ospraies be counted a kind of Ossifragi: from them come the lesser Geires, they again breed the greater, which engender not at all.

Of the cuckoo:

> As touching the Cuckow, it seemeth that he cometh of some hawke changed into his shape at one certain time of the year: for then those other hawkes are not to be seen, unless some very few days. He sheweth himself also but for a small season in summer time, and afterwards appeareth no more. It is the only hawk that hath no talons hooked downward, neither is he headed as other hawks, nor like unto them, but in colour: and for bill, he resembleth rather the dove. Nay more than that, the hawk will prey upon him and devour him, if haply they be seen both together: and it is the only bird of all other that is killed by those of the own kind. He altereth his voice also. In the spring, he commeth abroad, and by the beginning of the dog-days [from 3 July onwards], hideth himself. These lay always in other birds nests, and most of all in the Stock-doves, commonly one egg and no more (which no other bird doth besides) and seldom twaine. The reason why they would have other birds to sit upon their eggs and hatch them, is because they know how all birds hate them: for even the very little birds are ready to war with them: for fear therefore that the whole race of them should be utterly destroyed by the fury of others of the same kind, they make no nest of their own (being otherwise timorous and fearfull naturally of themselves) and so are forced by this crafty shift to avoid the danger.

The first two centuries AD were a time of universal peace unique in the history of the Western world; but, paradoxically, during this period, science, and indeed the whole culture of the ancient world, declined. Galen (AD 131–c. 210) is described as the last great biologist of antiquity. He was born in Pergamum in Asia Minor, of Greek parents, and is believed to have died there, but this has not been ascertained. He studied medicine and philosophy at the feet of all the best teachers of the time and spent a great part of his life in Rome lecturing on these subjects. He wrote 256 books, 131 of which were on medicine. Of these, 83 are extant. His other works included philosophy, mathematics, grammar and law, but most of them are now lost. Galen rejected the evolutionary theories of Aristotle and his followers. To him, everything constituted one single hymn of praise to the wisdom of the Creator. The veneration in which Galen was held in the Middle Ages may go some way to explaining the resistance of the Church to evolutionary theories. When a character in Shakespeare's *Merry Wives of Windsor* says scornfully of another that he has no knowledge of Hippocrates and Galen, it demonstrates how deeply engrained these writers must have been in the culture of the day. Yet Galen's extrapolation of human anatomy from the study of that of apes implies an acceptance of a relationship, and therefore of evolutionary principles, quite alien to ecclesiastical thought. Galen rejected the claim of Epicurus that body organs develop with use and weaken with disuse (which we now know to be true) claiming that in that case energetic people would develop four arms and four legs, and lazy people only one of each. When asked why man has not long ears like a donkey, which would give better hearing, he replied that the Creator had given man small ears so that he could wear a hat! Yet Galen must not be seen as a figure of ridicule. As Nordenskiöld points out, there is something truly noble and biblical in words such as:

> In my opinion true piety consists not in sacrificing hundreds of beasts or offering quantities of spices and incense, but in oneself knowing and learning about the wisdom, power, and love of the Creator (Nordenskiöld 1928).

As an anatomist, Galen was far in advance of anything that had ever been written before. His ideas, however, were not held in high esteem in his day, and during this period interest in natural history declined more and more. A generation after Galen, Claudius Aelianus (Aelian) wrote *On the characteristics of Animals*, which liberally mingled myth and fantasy with misinterpretation of fact. Aelian lived in Rome in the first half of the 3rd century, and is believed to have died in AD 260. His book is a series of anecdotes, the sort of 'old wives tales' frequently believed by simple folk even in the 20th century. In these anecdotes the lowliest creatures are possessed by personal reverence for the Creator, so the ecclesiastical writers of the Middle Ages found them a collection of edifying sermons. Thus a cock with a broken leg crowed so piteously before a statue of a god or saint, that the latter showed his mercy by miraculously healing the leg, and the bird went gratefully on its way. According to Aelian, birds which lure enemies away from their nests by pretending to be crippled are displaying exceptional love for their young, although it evidently never occurred to him that animals act according to blind instinct.

During the 5th century AD, the Roman Empire finally collapsed, and the emerging nations had too much trouble establishing themselves to have any time for philosophical or cultural matters. The last remnants of classical culture were established in the then barbarous but peaceful country of Ireland, and, during the 6th and 7th centuries, this became the starting point for the renaissance of civilisation. In the Middle East, from the seventh century onwards, a number of Arabian philosophers and writers contributed to the disciplines of mathematics, astronomy, and, particularly, chemistry, but little towards zoology. However, Abdallatif (AD 1162–1231) wrote an account of the animals of Egypt, based partly on personal observation. *Animal Life*, by Muhammed el Damiri, written at the end of the 14th century, described nearly 900 animals, some from observation, but many from pure imagination. Sakarja ben Muhammed el Kasvini, who lived in the 13th century, wrote *The Wonders of Nature*, in which he proposed, among other things, a curious theory for the origin of fossils. He believed that they were turned to stone by steam rising out of the ground where they stood. Perhaps this is no stranger than the Norse myth of Trolls, petrified if exposed to the rays of the sun.

Chapter 2

The Renaissance of Ornithology

The natural history books written in the succeeding centuries were either further compilations, collections of zoological anecdotes and fables, or treatises purporting to interpret the natural world as objects of religious significance. For instance, Isidore (c. AD 560–636), Bishop of Seville, compiled an encyclopaedia called the 'Etymologies' in which he devoted a chapter to birds. It contains little of importance. The swan's long neck was said to be the reason for its sweet song, and the cuckoo's saliva ('cuckoo-spit') produced grasshoppers. No new thinker of significance emerged until the 12th century AD. The Emperor Frederick II of Hohenstaufen (1194–1250), Holy Roman Emperor, King of Sicily and King of Jerusalem, has long been acclaimed as a visionary who embraced the religious divisions within his empire and challenged the oppression of the medieval church. But unsurprisingly, in his time he was denounced as the Antichrist, a heretic and a heathen, and he died condemned by the papacy to eternal damnation. But whatever his status as statesman or philosopher, his expertise as a falconer and ornithologist has never been questioned. In between affairs of state and military expeditions, he sought relaxation in hunting in various parts of his empire. He satisfied his passion for zoology by building parks for animals, and creating a great artificial marsh fed by a regulated water supply from aqueducts, where he kept many species of water birds. Unfortunately as a committed opponent of the authority of the Church, after his death he did not receive the credit that was his due. His writings could have had a profound effect on ornithology. To the ecclesiastic naturalists of the Middle Ages he was a heretic and his writings were prohibited.

Frederick was born heir to two crowns, a descendant both of King Roger II of Sicily and of Barbarossa, Emperor of Germany. He acquired the Kingdom of Jerusalem by marriage. Part of the trouble seems to have begun with his accession to two crowns. Since the creation of a fully independent Kingdom of Sicily had been the work of Roger II, and it was in the interests of the papacy to keep the kingdom independent of the Empire, Frederick's very existence was a threat to this stability. His succession to both the ancestral crowns was resisted by the nobility of the time; he survived several assassination attempts, and was often ruthless, even brutal, in punishing enemies. When Otto IV, Emperor of Germany, so infuriated Pope Innocent III by his attempt to invade Sicily that he was excommunicated, Innocent turned his favour towards Frederick's cause in gaining both crowns. Under the next Pope, Honorius III, Frederick's relations with the papacy were almost cordial. At this time, Frederick had sworn an oath to go on a crusade to the Holy Land, but internal troubles required his continuing presence in the Empire long past the date he was supposed to leave. This was seen by Honorius's successor, Gregory IX, as vacillation; and it was with this pope that Frederick's problems really began. Gregory does not emerge as an endearing character; on several occasions he apparently lied about Frederick and encouraged his subjects to rebel against him. In Gregory's eyes, Frederick could do nothing right. When the latter finally departed on his crusade he was struck down by a serious illness soon after leaving port, and was forced to pause on the way. This was interpreted by Gregory as breaking a holy vow, and Frederick was immediately excommunicated. Eventually, after the crusade, Frederick rushed home to Sicily to defeat a rebellion stirred up by Gregory. The latter lost face, and was reluctantly persuaded to lift the excommunication.

Emperor Frederick II of Hohenstaufen

The Kings of Sicily were noted for their religious tolerance, and attempts to respect and integrate the Muslim, Jewish and Christian communities may have angered the bigoted Gregory. In the years 1236–39, Frederick's armies defeated the Lombard League, an association of north Italian cities pledged to resist the Emperor. Gregory had tried, and failed, to arrive at a peaceful settlement. Frederick was excommunicated again by Gregory in 1239 for reasons which can only be interpreted as vindictive. His replies to Gregory's fulmination against him strike us today as eminently reasonable and moderate responses to a vicious campaign of lies and insinuations. Evidently the 13th century viewed it differently. Gregory died in the heat of the summer of 1241, in a Rome besieged by Frederick; his successor, Celestine IV, survived only three weeks. Owing to pressure from Frederick to persuade the cardinals to elect a candidate amenable to his own way of thinking, it was two years before a successor to Celestine was found. Frederick had made a serious error of judgement in trying to force the cardinals' hands, for the new Pope, Innocent IV, was a man very much in the mould of Gregory. Behind him stood his adviser, Cardinal Rainier, a bitter enemy of Frederick. The Emperor proposed a face-to-face meeting with Innocent, to try to sort out differences, but the latter's cowardly response was to flee Italy to the city of Lyons. From there he continued to defy Frederick until the latter's death in 1250, still unabsolved.

Science and learning entered Frederick's life early. In Sicily there had been a large zoo. Such things were not unique: the Norman Kings of England had kept menageries, but that in Sicily (founded by a scion of the Norman dynasty) was more magnificent, and included such things as giraffes and camels. Frederick kept this zoo stocked with rare animals as evidence of the endless wonders of the natural world. He was also a great reader; the fact that he seems to have had no formal, or indoctrinating, education may actually have worked to his, and history's, benefit. He read, among other things, Aristotle - no doubt in a Latin translation for he apparently did not learn Greek till later in life. In addition, Frederick is credited with being the founder of Italian lyric poetry. It was a courtier of his, Giacomo da Lentini, who is believed to have invented the 14-line poem known as the sonnet. The surviving poetry that emanated from Frederick's court was not, however, of exceptional quality, and his achievements in this field in no way compete with those in ornithology. The first part of his great unfinished work, the *De Arte Venandi cum Avibus*, serves as an introduction to the whole field of ornithology. He classified birds according to ecology and choice of food, their behaviour through the day, their feeding, stages of migration and anatomy. All these aspects, and also flight, plumage and moult are described with painstaking exactness. What he records is probably a small part of his vast experience, for it was only in his later years that he was persuaded, after great insistence by others, to put pen to paper.

Page from 'The Art of Falconry'

Frederick was not afraid to criticise Aristotle who, for instance, had claimed that birds migrating in a V-formation did so as a result of hierarchical order which was maintained. Frederick had observed that the leading bird in a flight of cranes was relieved by another during passage, the original shifting back into the formation. He also corrected many of Aristotle's notions of anatomy, describing in detail the avian kidney, even though Aristotle had denied that birds possessed such things, presumably based on the fact that birds do not urinate as a separate function. The book contains a long chapter on flight, and although this subject has been much written about, it was apparently not till 1933 that Frederick's account was surpassed, in variety of experience and acuteness of interpretation, by Konrad Lorenz. Owing to the opposition of the Church, Frederick's work was not published until 1596, over 300 years after the author's death, and was not 'discovered' by ornithologists until 1788, nearly another 200 years later, by J. G. Schneider and Blasius Merrem. By this time, of course, a great deal had already happened in ornithology.

Albert von Bollstädt (Albertus Magnus), who lived from about 1206 to 1280, a contemporary of Frederick II, was a teacher at Cologne. He was a Dominican friar, highly renowned in his lifetime for learning and piety. Most of his writings deal with theology and philosophy, but he was also a chemist and was the first to make arsenic in a free form. In later life he turned to natural history and edited the works of Aristotle in a Latin translation, adding observations of his own. His writing is characterised by direct observation and caution, but in faithfully reproducing Aristotle in the 26 books of his *De Animalibus*, printed in 1478, he incorporated many of the latter's errors which had been corrected by other medieval authors. His disciple, Thomas de Cantimpré (fl. 1233–1247), in *De Naturis Rerum*, included accounts of 144 birds (including the bat), a remarkable number for that period.

❧

Giraldus Cambrensis (1146–1223), in *Topography of Ireland* (written in 1187 but not published till 1587), devoted ten chapters to the birds of that country, but many of his observations, even though they were made at first hand, are unreliable. Passing mention of birds is also to be found in the writings of William of Worcester (1480) and John Leland (1552). Elsewhere in Europe, information was as bad. Those few writers who commented on natural history were content to derive their information from the Greek and Latin authors, with little or no attempt to improve this information. Their books were written primarily for a medical purpose, and were intended to indicate the curative properties of various parts of the birds and other animals in question, rather than to describe the animals themselves. These writers included Vicentius Belovacensis, whose *Speculum Naturae* was published at Strasbourg about 1264, and Bartholomew de Glanville (fl. 1230–1255), whose work *De Proprietatibus Rerum* (c. 1470) gives a good idea of natural history in the Middle Ages. *Hortus Sanitatis*, ascribed to Johannes de Cuba, was published at Mainz in 1475. Although primarily a herbal, the third tract of it deals with 'Birds and flying things' and was the first printed book to contain illustrations of birds. The study of anatomy had had as its first notable exponent Leonardo da Vinci (1452–1519), but it was Girolamo Fabrizio, or Fabricius (1537–1619), who became the first person since Aristotle to address the study of embryology, illustrating embryonic development in a number of vertebrates, including birds. He described in detail the development of the egg, the shape and appearance of the placenta and embryonic tissue. Marc Aurelio Severino (1580–1656) is remembered for his *Zootomia Democritea* which contained a section on birds, the *Ornithographia*. This gave some anatomical details, particularly of their feet.

William Turner was born about 1500 at Morpeth in Northumberland. He is said to have been the son of a tanner, but of his early life there is no information. He studied in Cambridge and became a close friend of Nicholas Ridley, later Bishop of London (who instructed him in Greek), and also of Hugh Latimer, both of whom were burned at the stake by Mary Tudor. As a supporter of Protestantism, Turner fled to Ferrara and later travelled in Europe. He returned to England after the death of King Henry VIII, but was forced to flee again on the accession of Queen Mary, finally returning only when Queen Elizabeth I was on the throne. He died in 1568. In 1544 he had published a little book in Cologne entitled *A Short and succinct account of the principal birds mentioned by Pliny and Aristotle*. Turner was a prolific writer,

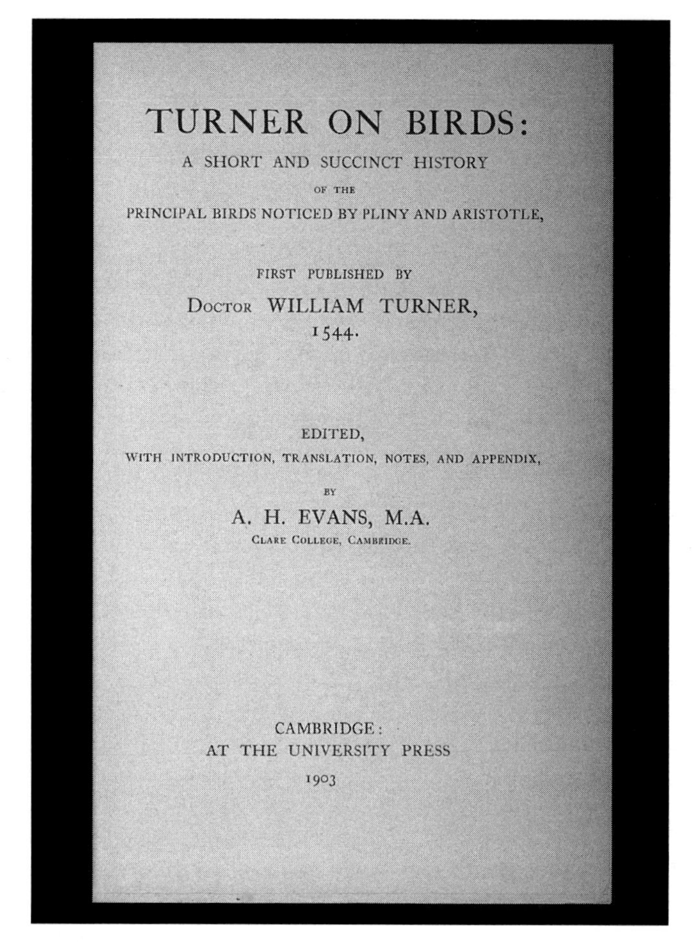
TURNER ON BIRDS:

A SHORT AND SUCCINCT HISTORY

OF THE

PRINCIPAL BIRDS NOTICED BY PLINY AND ARISTOTLE,

FIRST PUBLISHED BY

DOCTOR WILLIAM TURNER,

1544.

EDITED,

WITH INTRODUCTION, TRANSLATION, NOTES, AND APPENDIX,

BY

A. H. EVANS, M.A.

CLARE COLLEGE, CAMBRIDGE.

CAMBRIDGE:

AT THE UNIVERSITY PRESS

1903

Title page of Turner on Birds

mainly on ecclesiastical matters, but this is his only treatise on ornithology. The book was, of course, in Latin, but an English translation of most of it was published in 1903 with an introduction and copious notes by A. H. Evans, which makes it more accessible (Evans 1903). Evans translated all the important sections *specifically* on birds, but not all the introductions and general discourses. In his introduction he sums up the pros and cons of the book as follows:

> Doubtless Turner's work is not free from errors, as in the case of the very old story of the Barnacle Goose (which, however, he was most loth to credit even when assured of its truth by an Irish Divine), in his confounding of the *Onocrotalus* [Pelican] with the *Ardea stellaris* [Bittern] and the Cornish with the Alpine Chough; yet these are but small blots on a very excellent treatise, which compares most favourably with other writings of his time. It is quite evident from various passages that Turner was acquainted with Aristotle's works in the original Greek, and especially with his *History of Animals*; but he preferred quoting that author from the Latin translation of Theodorus Gaza of Thessalonica, the most celebrated Scholar of his day ... Exact transcription of a text was considered by no means necessary in those days: consequently we find many observations and explanations inserted in the text of Aristotle and Pliny, which had no place in the original ... Turner mentions other learned men by name and occasionally quotes from their works; while his pages also inform us of many places that he visited.

Commenting on kites, Turner remarked that the rufous variety is abundant in England, and was wont to snatch food out of children's hands, whereas the smaller, blacker form rarely haunted cities. He did not recall seeing this form in England. These observations are interesting in view of the oft-made claim that kites of medieval English cities were more likely to have been Black than Red, since Black Kites are recognised for their scavenging habits. Turner also commented on the godwit (*Limosa*), which was regarded as a delicacy in Medieval England, a fact confirmed by Ben Jonson in his play *The Alchemist*.

In 1546, the French King, Francis the First, sent an ambassador to Constantinople (now Istanbul), and the latter was accompanied by a naturalist named Pierre Belon. Belon had been born at Souletière, in Oisé, Le Maine, about 1517 but nothing is known of his parents. They were probably poor, and his education was financed by various bishops and cardinals, who enabled him to study at Wittenberg and Padua. The trip to the east took several years; from Constantinople, Belon explored a number of neighbouring lands, and, on his return to Paris, spent several more years writing an account of his travels, under the patronage of the powerful Cardinal de Tournon, minister to the King. These travelogues contain little on birds, as he saved this information for a later treatise, his *L'Histoire de la nature des Oyseaux*, published in Paris in 1555. In 1564 he was stabbed to death in mysterious circumstances in the Bois de Boulogne.

Belon is the first author whose work can be regarded in the modern sense as a classification. He did not arrange species into genera, but roughly grouped together those which appeared to have most affinity, and some of these roughly approximate to what would now be termed orders. The book consists of seven parts, and was intended to be a complete compendium on ornithology. The first part comprised a general survey, which leans heavily on Aristotle and other writers. The other six parts consist of detailed discussions on the six groups into which he divided birds. These were: raptors; waterfowl with webbed feet; marsh birds (in which he included kingfishers and bee-eaters); terrestrial birds (i.e. ostriches, gallinaceous birds, larks etc.); large arboreal birds; and small arboreal birds (including swallows). These groupings seem strange to us today, but they were based on Aristotelian principles, and were eminently sensible for an era which required nothing further than an arrangement of convenience. Belon had made many dissections of birds, in order to compare their anatomy with that of man. He made pertinent comments regarding analogies of structure between birds and mammals. However, in common with most writers of the time, he included bats with birds. Belon's book was much praised in later centuries, but attracted comparatively little attention at the time, for it had the singular misfortune to appear in the same year that marked the publication of the third volume (the one devoted to birds), of the *Historia Animalium* by the natural history giant of the 16th century, Conrad Gesner. Part of the *Historia Animalium* was translated into English by Edward Topsell, but unfortunately the volume on birds was not published and was long assumed not to have been written. However, a manuscript entitled *The Fowles of Heaven*, found at the Huntingdon Library at San Marino, California in 1933, proved to be the beginning of a treatise on birds by Topsell, and was, in all probability, based largely on Gesner's work.

Conrad Gesner was born in Zurich on 26 March 1516, the son of a poor tanner and furrier, Ursus Gesner. The latter was killed in the battle of Zug, and the family was separated, Gesner being sent to live with an uncle who awakened in him a deep love of literature and botany. At the age of 18 he went to Paris, where, to quote William MacGillivray (1834): "He indulged to excess his literary appetite, and devoured indiscriminately all kinds of knowledge." He was assisted financially by his tutor Oswald Myconius (1488–1552), a noted classical scholar and divine, who enabled him to study in Zurich, and through the patronage of Heinrich Bullinger (1504–1575), a famous Protestant writer, he studied at the Universities of Strasbourg and Bruges. He taught Greek for three years at Lausanne and spent a year at Montpellier where he met the French ichthyologist Rondelet. Soon after 1540, he had established a museum with 15 windows, on the glass of which he painted pictures of fish. He took his medical degree in 1541 at Basle, and then returned to Zurich as a physician, but his practice was never a large one. After

Conrad Gesner

his graduation there were few incidents in his life, but he worked hard and seldom a year passed when he did not publish something on literature or natural history. Among his publications is included a treatise on milk (1543). Most of his botanical works were published posthumously, but in 1545 he produced a *Bibliotheca universalis*, a great encyclopaedia of all contemporary knowledge. The bulk of his zoological writing is in the *Historia Animalium*. In this, the animals are arranged according to the principles of Aristotle. The first part includes quadrupeds, the second, birds, the third, fishes, and the fourth, reptiles and insects. Within these groups each species is described encyclopaedically on the lines of Pliny, but at far greater length, because of course Gesner had more centuries of accumulated literature to handle. Like Pliny, the book is purely descriptive, and lacks any comparison worked out between different forms of life. But Gesner had studied animals in the field, so his information did not come only from books. Gesner's greatest contribution to science, however, was the introduction of pictures. The use of woodcuts to provide illustrations for books was now coming into use, and Gesner utilised it to the full. (Belon had also introduced woodcuts, which, unlike Gesner's, were hand-coloured.)

The volume on birds was published in 1555, and contains 217 accounts of birds, illustrated with a woodcut for each species. The text includes external appearance, anatomy, distribution, habits, mythology and literature. Gesner included fabulous birds, but was not naive in this respect, and made it clear exactly what they were. He deserves the credit for having seen the need for a popular book on birds "for the people", and was the subject of many laudatory accounts and biographies. His admirers included Erwin Stresemann, but perhaps the latter's praise of Gesner should be treated with a little caution. A hundred and fifty years ago, Sir William Jardine considered Gesner's book to be little more than a literary curiosity. However, there is no doubt that this monumental work had a great influence on writers in the coming centuries. Gesner tried to include every known animal, and these are arranged alphabetically. He is believed to have been the first person to make a collection of natural history objects for the purposes of study. He was, however, primarily a botanist, and poor health and weak eyesight prevented him from acquiring a large working knowledge of birds. Nevertheless, he corresponded regularly with men of letters in many countries, including William Turner. Being of a delicate constitution, Gesner believed that he had not long to live, and throughout his life was compelled to seek relief by taking spa waters. He died on 13 December 1565 at the age of 50. After his death, his book went through repeated reprinting and editions, both in Latin and German. For nearly 200 years it was the German household book on natural history.

Owl picture from Gesner

Ulisse Aldrovandus or Aldrovandi (1522 or 1527–4 May 1605) of Bologna was of noble family and his inherited wealth enabled him to travel extensively to collect information for his books, and to employ artists in painting and suitable engraving. It is said that he was so liberal in expending his wealth in pursuit of his studies that he died in the workhouse in Bologna, but this is unlikely. Aldrovandus resembled Gesner in his capacity for work, and as he lived longer and worked under more comfortable conditions, he achieved more. His work is not always, however, an improvement on Gesner. He was less critical and wrote with less style, but his illustrations are better and his classification more advanced. Aldrovandus spent most of his long life in his native city. He lost his father while still a child, so at the age of six his mother sent him as page to a wealthy bishop. At 12 he was apprenticed to a merchant in Brescia, but disliked this work, and, after a period of travel in Spain, studied at the Universities of Bologna and Padua. He took his degree in 1553, and the following year was appointed professor of philosophy at the University of Bologna. In 1560, he was awarded the chair of botany, but the extent of his learning caused the University to award him the title of professor of natural history, in which post he was considered to be the most distinguished of those who had ever held it. Having made his name, he conceived the plan to write a natural history embracing all aspects of nature. In summer and autumn he went 'into the field' (and woods) to observe and study live creatures. Draughtsmen accompanied him to make drawings and write down his descriptions, and collectors prepared specimens which were taken back to the University.

These studies were to a great extent financed from his own pocket. By 1568, interest in his teaching had grown so great that the city established a botanic garden with Aldrovandus as its first director. He established a series of public lectures, by which he became extremely popular with the citizens. However, in his position as inspector of drugs, he earned the enmity of the apothecaries when he sought to introduce regulations in the drug business. He received the aid of Pope Gregory XIII (1572–1585), and, as a result of his experiences, Aldrovandus wrote a treatise on drugs (1574).

Aldrovandus

He began his great work on *Ornithology* early in his career, but so great was the amount of labour involved, that he was 77 before it was published, the three volumes duly appearing in 1599, 1600 and 1603. He gave the names of his birds in Greek, Hebrew, Arabic and Italian, and described them in detail, their nature; habits; food; how to capture and preserve them; their culinary properties; their use in medicine; and their employment as emblems, symbols and images, in sacred and profane mythology, on coins, in proverbs and hieroglyphics. The anatomy and the incubation of the egg of the different species was discussed; in sum the work contains far more information on birds than had ever appeared in print before. Birds were allocated to different groups, beginning with hawks and owls, then birds of a mixed nature (bats and ostriches), then fabulous birds, and finally parrots and crows. There are many anatomical studies in his work, such as the controlling muscles of the moveable upper mandible in parrots. His illustrations, though better than Gesner's, have been criticised. Although he engaged distinguished artists, they had clearly little idea of portraying living birds, and many of the pictures look grotesque by today's standards. His groupings are strange to us, but it is difficult to imagine what may have seemed reasonable to scholars of his day. He divided birds into those with hard, powerful beaks (raptors, parrots, ravens, woodpeckers, treecreepers, bee-eaters and crossbills); those that bathe only in dust or in dust and water (pigeons and buntings); songbirds (finches, larks and canaries); waterfowl; and shorebirds. Later writers have been scathing about these arrangements; but many laymen today make similar mistakes.

The three volumes of Aldrovandus's work on *Ornithology* are divided into the following books. The first volume contains 12 books, as follows: 1. Of eagles in general; 2. Of eagles in particular, where different kinds are described; 3. Of vultures in general; 4. Of accipitres in general; 5. Of accipitres in particular (including the cuckoo); 6. Of falcons in general; 7. Of falcons in particular; 8. Of nocturnal birds of prey (i.e owls); 9. "Birds of a middle nature, between birds properly so called and quadrupeds", such as the ostrich and bat; 10. Fabulous birds: gryphons, harpies, etc.; 11. Of parrots; 12. Of ravens in general, and other birds with powerful beaks, such as other corvids, birds of paradise, toucans, etc. The second volume contains six books: 13. Of wild gallinaceous birds, pheasants, partridges, etc.; 14. Of tame gallinaceous birds (domestic fowls); 15. Of birds similar to the preceding, but which yet seek water, such as pigeons and some passerines which live near streams; 16. Thrushes, blackbirds, etc.; 17. Insectivorous birds, such as wrens, swallows, etc.; 18. Singing birds, such as the nightingale. The third volume has only two chapters: 19. Palmipedes (the swan, etc.); and 20. Birds frequenting banks and shores, cranes, herons, flamingo, woodcocks, etc. Aldrovandus's entire works fill 13 folio volumes, but only four were published in his lifetime, the three on birds and one on insects. One later critic (Cuvier) pronounced his work to be "an enormous compilation without taste or genius" and considered that if the useless parts were removed, the writings would be reduced to one tenth of their bulk. After his death, his manuscripts, specimens and paintings were left to the city of Bologna, where they formed the embryo of the Bologna Museum. The Senate of the city arranged for the publication of a number of his still unpublished works.

Aldrovandus's pupil, Volcher Coiter (1534–1576), was the first to make a series of daily examinations of incubated eggs, describing and drawing each phase of development. His results were incorporated into a major paper entitled *Tables of the Principal External and Internal Parts of the Human Body and Various Anatomical Exercises and Observations Illustrated with New, Diverse and Very Ingenious Figures, Extremely Useful Especially to Those Devoted to Anatomical Study*, which was published at Nuremberg in 1572. He published a number of papers on birds, including a very rare work, the *De Differentiis Avium*. This contains extremely accurate drawings of the skeletons of crane, starling, cormorant and parrot, and the skull of a woodpecker. Coiter's remarks indicate that he was a good observer of bird's habits and adaptations. He pointed out, for example, that birds which live in marshes have elongated toes, but only slightly webbed feet, so as to enable them to walk on slippery soft ground. He also discussed birds' bills, pointing out the different types which were suited to different types of food, and enumerating (1) a straight bill for seizing, (2) a hooked bill for tearing flesh, (3) a broad bill for birds of placid habits which foraged in grass, (4) bills with saw-like edges in lieu of teeth, as in ducks, (5) bills which are neat and delicate for collecting small items like insects and seeds. He observed bird's necks and decided that it was impossible for birds with short legs and a long neck (or vice versa) to feed off the ground, and that no bird with hooked claws had a long neck. Although he made a few errors, for a writer of his day his writings are remarkably accurate, and he knew a surprising amount about birds' structure and habits. He was born in the Netherlands in 1534 in Gröningen, the son of a lawyer. Nothing is known of his childhood, but he attended college at St Martin's School at Gröningen where he showed an aptitude in medicine. He left his home town at the age of 21 and spent the rest of his life travelling the continent in order to study under leading professors in his field. In 1559 he reached Bologna, where he studied under Aldrovandus. He was gifted in the art of teaching, but his success earned him many enemies, who in jealousy plotted against him and hampered his work. He was fearless in the fight against quackery, but unfortunately his conversion to Protestantism barred him from any position in Italy. After a year's imprisonment for his beliefs (1566–67), he was able to go to Bavaria, where he obtained a post as physician to Duke Ludwig. It was in his last years that most of his writings were published.

The last major writer from this period is John Jonston (1603–1675) whose *Natural History of Birds* was published in 1650–53. It was translated from Latin into German, English, Dutch and French, and was last reprinted in 1773. The anonymous English translation of 1657 was much esteemed, but, according

Page of illustrations from Jonston

to one critic, exhibited more learning than judgement. To MacGillivray, it was a mere compilation of the writings of Gesner, Aldrovandus and others. But R. O. Cunningham pointed out that Jonston included original information on the gannet – the fact that its flesh is hard and dry – based on experience as a result of a visit to Scotland in 1623 (Cunningham 1866). MacGillivray also claimed that most of the plates were copied from the works of these authors, only a few being original. But they are not without merit, having been engraved by the well-known artist Matthew Merian. Stresemann dismissed Jonston's book on natural history as a hack work, adding bitterly that it has always been the worst books which sell the most copies (Stresemann 1975). (Perhaps this could be interpreted as meaning that the books which are naively written appeal to an undiscriminating public.) However, even he had to admit that the most attractive feature of Jonston's book was the illustrations. As in the case of many books today, it was clearly the pictures which sold it. Jonston was the son of Simon Jonston, who had migrated to Poland at the beginning of the 17th century. After attending schools in Prussia in 1622, Jonston proceeded to the University of St Andrews. He spent the next few years abroad, but returned to Britain in 1629 when he studied in Cambridge and in 1630 in London. He took an MD at Leyden in 1632 where he practised medicine for several years. In 1640, he was offered the chair of medicine at the University of Leyden, and two years later he received a similar offer from the Elector of Brandenburg. However, he preferred to study privately. In 1655, he returned to an estate in Silesia where he died on 8 June 1675.

Information also came back as a result of foreign travel and this was incorporated in the writings of Gesner, Aldrovandus and others. When Christopher Columbus, back from America, entered Barcelona in April 1493, he brought exotic birds (probably Amazon parrots) to astonish and delight the crowds. The turkey was introduced between 1530 and 1541 and in little more than a decade had been bred so successfully that Archbishop Cranmer had to forbid his clergy to serve more than one turkey at a banquet. Christopher Columbus (1481–1506), like all master mariners of the time, carefully observed birds while at sea. It was believed that the sight of known terrestrial birds indicated that land was near. His journal of his first voyage to America notes almost daily sightings, although many of the birds he saw and supposed to be landbirds are now known to be pelagic. For instance, on 12 September 1492 he saw two tropicbirds, which he thought did not fly more than 25 leagues from land and therefore probably came from the Azores. In fact he was then well to the south of these islands. Later in the month he saw more tropicbirds, boobies and terns which followed the ship scavenging anything edible that was thrown overboard. On the 23 September a (presumably) migrating dove was observed; this proved a very useful sign, for the crew was beginning to become restless and desired to turn back. When he reached the longitude of Bermuda, multitudes of small birds were seen, though he did not actually see the island, which was well to the north.

In the early years of the 16th century, parrots began to be brought in large numbers from many parts of the world, and they became popular with the upper classes as pets, their novelty and bright colours becoming important status symbols. In 1522, the first bird skins were brought back by Elcano, the commander who took over the ship of discovery after the death of Magellan. Ferdinand Magellan (c. 1480–1521) had been born into a noble Portuguese family, and in 1504 volunteered to join an expedition to the East, serving until 1514 in voyages to India, Cochin, Malacca, Java and the Spice Islands. Following an unjust prosecution by his country (for which he was acquitted) he renounced his nationality and went to Spain. In 1519, he sailed from Seville to try to find a westward route to the Spice Islands via South America. On the voyage was the careful chronicler Francesco Antonio Pigafetta (c. 1490–c. 1534). He was one of only 31 survivors to return to Spain. The journal he kept on the voyage notes skuas persecuting other birds, though he assumed that the meals eventually disgorged were "ordure". Along the coast of southern South America, he met natives who decorated themselves with parrot feathers, and obtained ten parrots from one of the natives in exchange for a small mirror. He also saw Roseate Spoonbills and rheas. Near the Straits he found penguins which were loaded on board in numbers as food. The bird skins brought back by Magellan's ship were skins of the Lesser Bird of Paradise given him by the Sultan of Batjan. People were amazed not only by the birds' plumage, but also by the fact that they had no flesh, no bones, and even more incredibly, no feet. It was believed that these creatures from Heaven never settled on the earth, fed only on dew as it 'fell', and that the female laid her eggs in a depression in the male's back. The error of thinking that these birds were footless (hence the name *Paradisaea apoda* given to the Greater Bird of Paradise) was because all the skins known at that time had had the legs removed. Belon, however, had not been deceived, for in his 1555 publication, he described a method of skinning, primitive by our standards, but quite advanced for the time. This involved removing the entrails, and saturating the rest of the body with salt and vinegar to preserve it from worms. He used this technique on his voyage to the Levant in order to preserve specimens for later study. In the following 150 years, many hundreds of birds, dead and alive, were shipped back to Europe, but most of them disappeared into the collections of the wealthy and only a few came to the notice of specialists. None of the bird skins survived long, for the primitive methods of preservation could not have protected them from the attacks of moths and beetles. A few achieved immortality by being made the subjects for paintings, but one must sigh with frustration at the vast quantity of material which disappeared without trace, probably including species now extinct and which did not even survive long enough to be scientifically described.

Throughout the 16th century these great collections continued to be maintained, but the greatest was that of the Emperor Rudolph II (1552–1612), an eccentric recluse who neglected affairs of state to devote himself to the study of art and science. His collections would, however, have vanished without trace had he not commissioned two Dutch painters, Georg Hoefnagel (1542–1600) and his son Jacob (1575–1630), to paint his zoological treasures. These pictures include 90 paintings of birds, and are still in the Vienna Hofbibliothek. But it was not until 1868 that they were rediscovered by Georg von Frauenfeld and it was only then realised that the Emperor had had a Dodo in his collection, as well as many other treasures not known elsewhere for many years subsequently. This collection of pictures includes the only known likeness from life of the Red Rail *Aphanapteryx bonasia* which formerly inhabited Mauritius, but became extinct there about a century later (Appendix A).

Picture of the Red Rail from Rothschild

Towards the end of the 16th century, the Dutch began to trade with the East Indies on a considerable scale. Their first ship returned from Java in 1597 carrying a young cassowary which was exhibited in a number of places in Europe before finding its way into Emperor Rudolph's menagerie. Much of this information was gathered by Carolus Clusius, who, at the age of 80, set himself the task of making an illustrated book out of everything exotic that he could find. Based at Leyden, he was well placed to receive immediate news from returning Dutch ships of exploration and the result was published in 1605. His book, the *Exoticorum libri decem*, contains the first descriptions of the cassowary, the Magellan Penguin, the Greater, Lesser and King Birds of Paradise, the Hawk-headed Parrot, the Chattering Lory and the Scarlet Ibis. In 1604, Clusius had received from Henrik Højer a report on birds of the Faeroes, which he printed in full. Among 12 species was included, as a rare visitor to the islands, the Great Auk. This remarkable man, described as of eminent ability, extraordinary memory and universal education, was born at Arras on 19 February 1526, and died at Leyden on 4 April 1609, having studied at Wittenberg, Strasbourg and Montpellier. He was primarily a botanist; between 1573 and 1587 he travelled extensively in Spain, England and Germany, and is believed to have been the first person to introduce the tulip to the Netherlands from Turkey, via Vienna, in 1594.

But Clusius was unaware of a remarkable manuscript which lay in the Imperial library of Spain, withheld from public view. It was the work of Francisco Hernandez (1517–1578) who spent seven years in Mexico from 1570–1577, sent there by King Philip II to collect information on the resources of the kingdom, including plants, animals and minerals. Hernandez died without finishing the work, whereupon Philip gave it to an Italian, Nardi Antonio Recchi, to work up. Recchi made a Latin summary of it which he delivered to Philip, and also a duplicate which he took to Italy, but he too died without making use of the work. The several copies of the manuscript eventually resulted in two editions of Hernandez's work, published in 1635 and 1651 respectively, and they contain many first descriptions of Mexican birds. Hernandez described some 230 species, but unfortunately most of the coloured pictures which he prepared to accompany the work were lost in a fire, and, in many cases, identification of his birds has been difficult. Fortunately, he had supplied all their Aztec names, many of which remain in use in Mexico to this day. For 200 years he remained the only authority on the fauna of Mexico, partly because the Spanish tried to prevent exact knowledge of the resources of their colony from becoming public. Nothing further was known about the birds of Mexico until that country became independent in 1821; two years later, William Bullock brought back the first bird collections ever made there.

Hernandez's work had not long been published, when the sphere of interest shifted to Brazil. In 1637, Count Johann Moritz von Nassau-Siegen (1603–1679), the Dutch governor based at Olindo, near Pernambuco, had plans for a scientific exploration of the lands which stretched to the south. Following the advice of Dr Johannes de Laet (1593–1649) he appointed Willem Piso (1611–1678) and two assistants. One of the assistants died soon after arrival in South America, the other was Georg Marcgraf, born in 1611, who spent seven years exploring the interior of Brazil and making copious notes on the subjects of natural history, geography, ethnology and astronomy. Tragically he died of fever aged only 34. De Laet undertook to edit Marcgraf's notes, and decipher the code in which they were written; in 1648 the *Historia naturalis Brasiliae* appeared with Marcgraf's material on birds in volume 5. After de Laet's death

Spread of Birds from Hernandez

the following year, Piso rewrote the entire section on natural history, omitting Marcgraf's name and much of his information. This work, *De Indiae utriusque re naturali et medica*, appeared in 1658. While in the field, Marcgraf had been accompanied by a painter who made detailed coloured pictures of the animals and plants (presumably in lieu of the actual collection of specimens) and these were preserved by the Governor of Olindo who had instigated the exploration. However, the woodcuts made from them, and subsequently published in the books, were so crude that for long many of Marcgraf's species could not be identified with certainty. The coloured paintings, however, passed to the library of the Elector of Brandenburg. They were discovered by J. G. Schneider in 1786 (it was he who two years later discovered the great book of Emperor Frederick). The Marcgraf paintings were studied first by Illiger (see p. 86), from 1817–19 by Heinrich Lichtenstein, and finally in 1938 by Adolph Schneider. During First World War, the paintings were 'lost', but were recently rediscovered by the late Peter Whitehead (Whitehead *in litt*). Marcgraf's work remained for 150 years the only source of information on the Brazilian fauna. He described 133 species of birds, his descriptions being subsequently used by such authorities as Ray, Brisson, Buffon, Linnaeus and Gmelin. Soon after Marcgraf, the Dominican missionary J. B. du Tertre arrived in the West Indian islands, and his account of the fauna of that area contains a great deal of valuable information. His work has been undervalued because he was not a scientist, and his writings lack scientific precision, but he described a number of species which subsequently died out and were never officially accepted. The writings of the Dutch doctor Jacob Bontius, who died at Batavia (Djakarta) in 1631, have been similarly undervalued, and for much the same reasons; he, too, described birds which no-one has been able to identify, and which may well represent extinct species.

5

GEORGII MARCGRAVII,

DE REGIONIBVS,
& Indigenis, Braſiliæ, & Chili
ejuſdem Continentis,

LIBER.

CAPVT PRIMVM.

De Regionis nomine, magnitudine, ſitu, ejuſque diviſione.

HÆc regio primo à Luſitanis appellata fuit SANTA CRVZ, quod nomen poſtea mutarunt in TERRA DO BRASIL, ob ligni copiam, quod hic cæditur & in Europam infertur, vulgo notiſſimum, ita ut hodie communi & vulgari nomine ab omnibus Europæis BRASILIA dicatur.

Sita porro eſt Braſilia in zona torrida & quidem in medietate illius Auſtrali, extenditurque ultra Tropicum Capricorni in zonam temperatam meridionalem.

Non autem omnes conſentiunt in illius magnitudine ſeu latitudine definienda inter Septentrionem & Auſtrum. Nos ejus initium ſtatuimus à latitudine Septentrionali unius gradus & ſemis, ſeu à Flumine *Para*, & finem ad latitudinem auſtralem quatuor & viginti graduum & ſemis, ſeu uſque ad fluvium *Capibari*, duabus leucis ultra oppidum S. Vincentii; ita ut magnitudo illius ſit inter Boream & Auſtrum trium & viginti graduum. Inter ortum & occaſum in quantum extendatur, adhuc indefinitum eſt, cum pauci adhuc in interiora hujus terræ penetraverint & quædam tantum obiter luſtraverint.

Quantum judico, adſumendo longitudinem *Limæ* Metropolis Peruviæ, quam Hiſpani ponunt eſſe duobus & octuaginta gradibus Occidentaliorem *Toleto* (Toletum autem novem-decim gradibus & quinque ſcrupulis Occidentalius eſt *Vranoburgo*) ita ut longitudo Limæ ſit ducentorum & nonaginta quinque graduum & quadraginta ſcrupulorum. Sumendo itidem longitudinem *Mauriciæ* civitatis, in Orientaliſſima fere ora Braſiliæ, quam ipſe ſæpius obſervavi per Eclipſes & inveni trecentorum & quadraginta graduum & ſemis reſpectu *Vranoburgi*, & hinc deducendo differentiam quatuor & quadraginta graduum & quinquaginta ſcrupulorum. Sub parallelo autem 8, 40 reſpondent ex mea obſervatione uni gradui 28175 decempedæ Rhynlandicæ, quæ per 1500 diviſæ (tot enim perticæ milliarium horarium efficiunt) dabunt horaria ſeu Hollandica milliaria XVIII. $\frac{1117}{1500}$, id eſt, milliaria octodecim & tres quadrantes. Differentia ergo 44.50. dabitur 1263197 decempedarum & duorum pedum, quæ efficiunt milliaria 842, $\frac{128}{1500}$, detractis porro centum milliaribus, quæ *Cuſco* Orientalior eſt *Lima* & in confiniis Peruviæ & Braſiliæ ſita, magnitudo Braſiliæ inter ortum & occaſum eſſet circiter ſeptingentorum & quadraginta duorum milliarium; quod intervallum omnes Tabulæ Geographicæ circiter centum & octuaginta octo amplius faciunt. Ponunt enim Braſiliam multo Orientaliorem, quam revera eſt, ut ex propriis obſervationibus ſæpiſſime comperi.

In gratiam eorum, quibus Phænomena cæleſtia, & immobiles Planetarum vices noſcere ſtudium eſt, apponam Eclipſis Solaris faciem, pro temporis momentis aliam atque aliam.

Apparere Mauriciam urbem incolentibus cœpit ea in finem anni quadrageſimi ſupra milleſimum & ſexcenteſimum, Novembris die decima tertia, hora decima ante meridiem, & ad ſummum provecta fuit hora undecima, treſque Solis tertias obſcuravit, cum minutis viginti octo; ſic ut minus quarta ſui parte hic mundo luxerit. Verum quadraginta ſeptem ab hora duodecima minutis, labore ſuo tenebriſque evolutus Titan pleno refulſit lumine.

a 2 A B. Ecli-

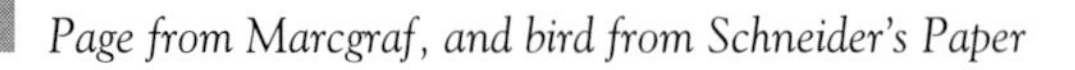

Page from Marcgraf, and bird from Schneider's Paper

Georg Eberhard Rumpf (1628–1702), who later Latinised his name to Rumphius, has been described as the Father of East Indian natural history. He was probably born at Münzenberg, but his father, a builder, moved to Hanau soon after his birth, so that his son could have a good education. The gesture was wasted, for the boy was seized with an urge to travel. His first trip, however, was disastrous. He was lured into enlisting in the Venetian army; the ship set sail for Brazil, but was captured and taken to Portugal. Three years elapsed before Rumphius could get home. Nothing daunted, in 1652, he set out for Batavia as an employee of the East Indies Company. Ending up in Amboina, where he was caught up in a Moluccan rebellion, he managed under very difficult circumstances to conduct scientific studies while carrying on the duties of a trader and administrative officer. Smitten with blindness, he continued with dogged persistence and with the aid of secretaries and his devoted son, to write a number of books on Amboina, but his zoology, the *Ambonsch Dierboek*, was incomplete at his death and remained unpublished. It is now lost, but was read by Rumphius's 'friend' Frans Valentijn (1666–1727) who used much of it in the third volume of his own encyclopaedia *The East Indies Old and New* (Valentijn 1726), but, perhaps predictably, without giving any credit to Rumphius. Valentijn had visited Amboina at least twice; in 1686–94 and again in 1707–12, and evidently studied the manuscript at great length, reproducing large sections from it. A number of these are so different in style from the rest of his work, that they were probably copied verbatim from Rumphius's text.

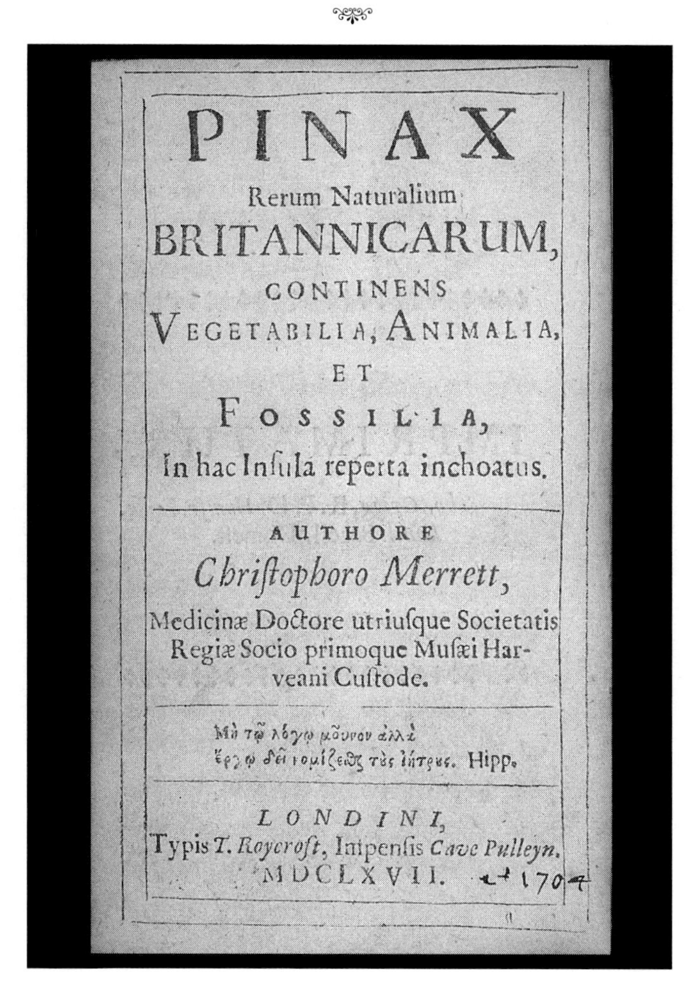
PINAX
Rerum Naturalium
BRITANNICARUM,
CONTINENS
VEGETABILIA, ANIMALIA,
ET
FOSSILIA,
In hac Insula reperta inchoatus.

AUTHORE
Christophoro Merrett,
Medicinæ Doctore utriusque Societatis Regiæ Socio primoque Musæi Harveani Custode.

Μὴ τῷ λόγῳ μοῦνον ἀλλὰ
ἔργῳ δεῖ νομίζεσθαι τοὺς ἰητρούς. Hipp.

LONDINI,
Typis *T. Roycroft*, Impensis *Cave Pulleyn*.
MDCLXVII.

Title page from Pinax

But in spite of the wealth of new material being gathered in previously unexplored parts of the world, classification was still in a primitive state. Christopher Merrett (1614–95) published in 1666 what is probably the first classic of British ornithology, the *Pinax Rerum Naturalium Britannicarum* (usually called the *Pinax* for short; the word meaning a list or index). Merrett, who was primarily a botanist, was born at Winchcombe, Gloucestershire, on 16 February 1614, graduated as BA from Oxford in 1635 and devoted himself to medicine. He settled in London, where he became a Fellow of the Royal College of Physicians in 1651 and subsequently Librarian of the College. However, the bulk of the library and Merrett's own house and books were destroyed in the Great Fire of London, and the rarity of the first edition of the *Pinax* is thought to be as a result of a great many copies having then perished. A second edition was printed the following year, 1667. Merrett died at his house in Hatton Garden on 19 August 1695 and was buried "12 feet deep" in the church of St Andrew's, Holborn. He was the author of a number of medical books, and papers in the *Philosophical Transactions*. In an article published in the journal *British Birds*, the historian W. H. Mullens reprinted in full the section on birds from the *Pinax*, together with a translation from the Latin, and explanatory notes (Mullens 1908). There was not a great deal of original observation in Merrett's book, which was based largely on Aldrovandus and Jonston, but as the first attempt to list the birds occurring in Britain it is of historic importance, and was highly thought of by his contemporaries.

Walter Charleton *Walter Charleton*

Walter Charleton (1619–1707), was also a physician, the son of the Rector of Shepton Mallet in Somerset. At the age of 22 he graduated as MD from Oxford, and became physician to King Charles I. After the Restoration he served Charles II in the same capacity. He published a large number of works, including, we are told, one called the *Chorea Gigantum* in which he attempted to prove that Stonehenge had been erected by the Danes and used as a Royal coronation place. He was the first English writer to append illustrations to a list of birds, which he did in the main work to concern us, the *Onomasticon*

Zoicon (1668). He included foreign as well as British birds, but relegated the former to an appendix. Considerable time was spent observing the birds in the Royal aviaries at St James's Park and the museum of the Royal Society; and he was well versed in previous ornithological literature, referring to Gesner, Belon, Aldrovandus, Turner, Jonston and Merrett. As a result, the book is described as "A List, with the English, Latin and Greek names of all known Animals, including an account of Charles II's menagerie in St James's Park, followed by certain anatomical descriptions and a general account of fossils". In classification, Charleton relied, as did all his predecessors, on the principles of Aristotle, and divided his birds into two principal divisions of Land and Water Birds (see Appendix 1). Land birds were divided into meat-eaters, seed-eaters, berry-eaters and insect-eaters. The first group included all the birds of prey, owls, shrikes, cuckoos (!), parrots (!), corvids, ostriches, cassowaries and bats. The second group contained the game birds, bustards, rails, thick-knees (ducks appear here as well as under water birds!), pigeons, sandgrouse and hard-billed seed-eaters. The berry-eaters included thrushes, starlings, the Hawfinch, the Nuthatch and the Cardinal. The insect-eaters included woodpeckers, warblers, wagtails, pipits, titmice, swallows and swifts, chats and the Hoopoe. Water birds were divided into Palmipedes (webbed-footed) and Fissipedes (non-webbed). The first contained pelicans, gulls, cormorants, gannets, geese, swans and ducks, divers, grebes, coots and gallinules, and petrels. The Fissipedes consisted of storks, ibises, herons, flamingos, cranes, waders, kingfishers and dippers. Among his various pertinent comments, he remarks that the Peregrine Falcon was called Peregrinus from its habit of moving from one district to another, and because its nest could never be found. A revised edition of his book, published in 1677, contains what must be the first colour code, on which W. H. Mullens comments as follows:

> This is not only a very remarkable composition for the date at which it was written, but contains much that is of importance and value even now; while its completeness may be shown by the fact that the author is able to classify no less than eleven shades of white (with their Latin and English names), eighteen of yellow, and sixteen of red. Charleton, in short, seems to have been a man of high ability and greatly in advance of his day.

However, unlike that of Merrett, Charleton's book did not receive the recognition it deserved in his day, largely because it was not considered to be of sufficient importance to translate from Latin into English, and because it had the misfortune to appear very soon before one of the landmarks of ornithological writing, the *Ornithologiae* of Francis Willughby, an English country gentleman of considerable means. It was with Willughby's great book, published in 1676, that the classical systematics were finally abandoned, in favour of a system based on form rather than function. Francis Willughby (1635–1672) never lived to see this book published, for he died after a long illness on 3 July 1672, at the age of only 37, leaving a mass of notes and papers which were edited by his faithful friend and tutor John Ray (1628–1705). It is probable that Ray was as much, if not more, responsible for the content of the book than Willughby, but he refused to take any credit. Willughby's book is important, too, for another reason. It was the last attempt to write everything that was known about birds in one work. After this time, the volume of information increased so enormously that no one person was capable of digesting it all (Stresemann 1975).

John Ray (originally spelled Wray) was born on 29 November 1628 at Black Notley, near Braintree, Essex, and educated at Braintree School and Cambridge University. His literary work was extensive and embraced many subjects: sermons, religious essays, discussions on the classics and folklore, as well as the natural sciences. Ray and Willughby had met at Cambridge, where Ray lectured on Greek and Mathematics. His mind and personality impressed Willughby, and the two became friends. In the summer of 1662 they visited the west coast of England where they made a study of breeding seabirds. In the spring of 1663 they departed for the Continent, with two other pupils, to extend their knowledge of natural history. First they visited the Netherlands, where they saw the great breeding colonies of herons, spoonbills and cormorants. They journeyed up the Rhine Valley to Zürich and thence to Venice, where they spent much of the winter. The following year they took ship from Leghorn to Naples. There they

Francis Willughby

John Ray

split up, Willughby and one of the others travelled to Spain and so back to England. Ray and his other pupil visited Sicily (where they climbed Etna) and then travelled back through France, returning to England in 1666. Soon after the continental tour, Ray was elected a Fellow of the Royal Society. His last contribution to ornithology was the *Synopsis Methodica Avium* [*et Piscium*], completed in 1694, but only published in 1713, long after his death in 1704.

In the *Ornithologiae*, birds were still divided firstly into Land and Water Birds, the latter being divided into those which frequent watery places, and those which swim in water, each subdivision being further broken up into sections and a key provided. Land Birds were subdivided into those with a crooked beak and talons, and those with a straighter bill and claws. Those with crooked beak and talons were divided into 1. carnivorous (i.e. raptors); and 2. frugivorous. The frugivorous were the parrots, divided into great (the macaws), medium (parrots and popinjays) and small (parakeets). The carnivorous were divided into diurnal and nocturnal, the latter consisting of owls, again subdivided into horned and unhorned species. The diurnal were divided into greater and lesser; the greater being further divided into the "more generous" (eagles) and the "more cowardly and sluggish" (vultures). The lesser were also divided into "generous" and "cowardly", each group here being further divided. The more generous (i.e. "that are wont to be reclaimed and manned for fowling") consisted of long-winged (falcons) and short-winged (goshawks and sparrowhawks); while the more cowardly and sluggish "or else indocile, and therefore by our falconers neglected and permitted to live at large" consisted of the greater (buzzards) and the lesser (shrikes and birds of paradise). The birds with straight bills were divided into great, medium and small. The greatest were those which were too big to fly (ostriches, cassowaries and the Dodo), the small were divided into soft-beaked, which fed on insects, and hard-beaked, which fed on seeds. The medium birds were also divided into those with large strong bills and those with smaller weaker ones. The large-billed were divided into omnivorous (crows), fish-eaters (kingfishers) and insect-eaters (woodpeckers); while the small-billed were divided into poultry, pigeons and thrushes. For a long time Willughby's system remained unsurpassed, and represented true relationships rather better than Linnaeus's so-called 'natural

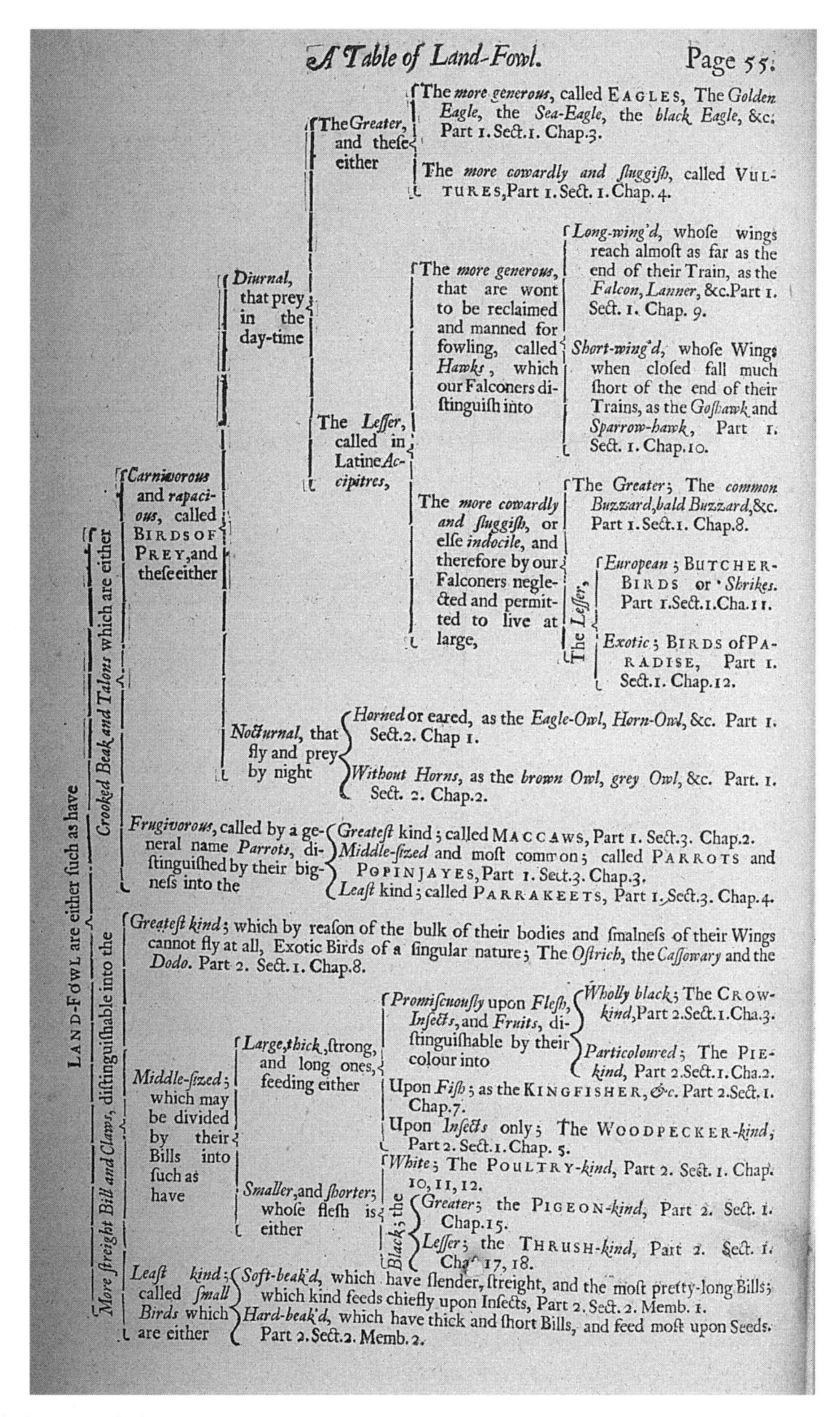

A Table of Land-Fowl. Page 55.

LAND-FOWL are either ſuch as have

- *Crooked Beak and Talons* which are either
  - *Carnivorous* and *rapacious*, called BIRDS OF PREY, and theſe either
    - *Diurnal*, that prey in the day-time
      - The *Greater*, and theſe either
        - The *more generous*, called EAGLES, The *Golden Eagle*, the *Sea-Eagle*, the *black Eagle*, &c. Part 1. Sect. 1. Chap. 3.
        - The *more cowardly and ſluggiſh*, called VULTURES, Part 1. Sect. 1. Chap. 4.
      - The *Leſſer*, called in Latine *Accipitres*,
        - The *more generous*, that are wont to be reclaimed and manned for fowling, called *Hawks*, which our Falconers diſtinguiſh into
          - *Long-wing'd*, whoſe wings reach almoſt as far as the end of their Train, as the *Falcon*, *Lanner*, &c. Part 1. Sect. 1. Chap. 9.
          - *Short-wing'd*, whoſe Wings when cloſed fall much ſhort of the end of their Trains, as the *Goſhawk* and *Sparrow-hawk*, Part 1. Sect. 1. Chap. 10.
        - The *more cowardly and ſluggiſh*, or elſe *indocile*, and therefore by our Falconers neglected and permitted to live at large,
          - The *Greater*; The *common Buzzard*, *bald Buzzard*, &c. Part 1. Sect. 1. Chap. 8.
          - *The Leſſer*,
            - *European*; BUTCHER-BIRDS or *Shrikes*. Part 1. Sect. 1. Cha. 11.
            - *Exotic*; BIRDS of PARADISE, Part 1. Sect. 1. Chap. 12.
    - *Nocturnal*, that fly and prey by night
      - *Horned* or eared, as the *Eagle-Owl*, *Horn-Owl*, &c. Part 1. Sect. 2. Chap 1.
      - *Without Horns*, as the *brown Owl*, *grey Owl*, &c. Part. 1. Sect. 2. Chap. 2.
  - *Frugivorous*, called by a general name *Parrots*, diſtinguiſhed by their bigneſs into the
    - *Greateſt* kind; called MACCAWS, Part 1. Sect. 3. Chap. 2.
    - *Middle-ſized* and moſt common; called PARROTS and POPINJAYES, Part 1. Sect. 3. Chap. 3.
    - *Leaſt* kind; called PARRAKEETS, Part 1. Sect. 3. Chap. 4.
- *More ſtreight Bill and Claws*, diſtinguiſhable into the
  - *Greateſt kind*; which by reaſon of the bulk of their bodies and ſmalneſs of their Wings cannot fly at all, Exotic Birds of a ſingular nature; The *Oſtrich*, the *Caſſowary* and the *Dodo*. Part 2. Sect. 1. Chap. 8.
  - *Middle-ſized*; which may be divided by their Bills into ſuch as have
    - *Large*, *thick*, ſtrong, and long ones, feeding either
      - *Promiſcuouſly* upon *Fleſh*, *Inſects*, and *Fruits*, diſtinguiſhable by their colour into
        - *Wholly black*; The CROW-*kind*, Part 2. Sect. 1. Cha. 3.
        - *Particoloured*; The PIE-*kind*, Part 2. Sect. 1. Cha. 2.
      - Upon *Fiſh*; as the KINGFISHER, *&c.* Part 2. Sect. 1. Chap. 7.
      - Upon *Inſects* only; The WOODPECKER-*kind*, Part 2. Sect. 1. Chap. 5.
    - *Smaller*, and *ſhorter*; whoſe fleſh is either
      - *White*; The POULTRY-*kind*, Part 2. Sect. 1. Chap. 10, 11, 12.
      - *Black*; the
        - *Greater*; the PIGEON-*kind*, Part 2. Sect. 1. Chap. 15.
        - *Leſſer*; the THRUSH-*kind*, Part 2. Sect. 1. Chap. 17, 18.
  - *Leaſt kind*; called *ſmall Birds* which are either
    - *Soft-beak'd*, which have ſlender, ſtreight, and the moſt pretty-long Bills; which kind feeds chiefly upon Inſects, Part 2. Sect. 2. Memb. 1.
    - *Hard-beak'd*, which have thick and ſhort Bills, and feed moſt upon Seeds. Part 2. Sect. 2. Memb. 2.

Table from Ornithologiae

system' of some while later. Aristotle had never actually worked out a system of classification – what passes for his was compiled by others from his writings. In the centuries that followed, there was no need for a more elaborate classification, the animals and plants known during the Middle Ages were few enough that a very simple one sufficed. Not until the discoveries of the 16th and 17th centuries were a sufficiently large number of creatures known for a more sophisticated classification to become necessary. During the 16th century natural philosophers were mainly university teachers, such as Vesalius and Galileo. In the 17th, and more so in the 18th, universities ceased to be centres of scientific progress and became seats of conservatism, while the pioneers were private scholars, like Willughby. During this period, plant classification made considerable progress, while zoological classification made virtually none. It was Ray who made the first attempt to deal with both plants and animals. In those days, it was quite possible to learn much without going into 'the field'. Unfamiliar birds could be seen in markets and on fowlers' stalls. Such birds as Hazel Grouse, Black Stork, Great Bustard and Grey Crane, were found by Ray and Willughby quite easily in fresh condition. These were dissected and detailed descriptions made of their anatomy.

Illustration from Ornithologiae

Important as it was, Willughby's *Ornithologiae* by no means included all the information that was in existence at the time. It did not include the observations of Dr Robert Hooke (1635–1703), born in the Isle of Wight, and primarily an experimental physicist. He had worked for Robert Boyle, and used his skill in the construction of his air pump. For many years he was curator of experiments at the Royal Society, but his interests were said to be too wide. He initiated much, but perfected little. In his *Micrographia* (1665) he had examined bird feathers under the microscope and drawn detailed illustrations of the way in which the distal and proximal barbules of a goose's feather work together and how the iridescent barbules of the peacock feather are constructed. He remarked:

> [The] whole form and shape cannot easily be determined, since they change continually, and seem very different in different positions to the Light: nay, only interposing one's hand between them and the light, or even putting up or pulling down a sash very much changes their appearance.

This examination of feather structure seems to have been Hooke's only contribution to ornithology.

Ole Wormius

At this time the preservation of birds in the form of skins for study was not common. To make a drawing, or paint a picture of the creature was a much easier, as well as cleaner, practice. The need to retain bodies for study soon became apparent, as did the need for more careful methods of preservation; collections such as those shown in the frontispiece of Ole Wormius's *Museum Wormianum*, placed on shelves or hung from the ceiling, cannot have lasted long, for they would soon have fallen prey to moths and beetles. Wormius's collection was famous throughout Europe, and the catalogue of the museum was published at Leyden the year after his death in 1655. Twenty-two pages were devoted to birds, indicating that there were a fairly large number in his collection. Among the birds depicted were the now extinct Variegated Raven of the Faeroes (most probably a colour-morph rather than a species), the Emu, the Rhinoceros Hornbill, birds of paradise, the Roller, the Little Owl, peafowl, guineafowl, various hummingbirds, the Mute Swan, Merganser, Great Auk (the famous picture of a Garefowl with a white ring round its neck has been generally dismissed as a bird with an artificial collar, but it may well have been a different species from *Pinguinus impennis*), *Anas islandica* (? Barrow's Goldeneye), Black Stork, Bittern, White Heron, Curlew, Scarlet Ibis and Flamingo. Nehemiah Grew's *Museum Regalis* (1681)

Ole Wormius's Museum Wormianum

recognised, probably for the first time, that bats were anomalous creatures standing between mammals and birds "The Bat stands in the Rear of Mammals, and in the Front of Birds". Grew's treatise lists and comments on a wide range of items in the collection of the Royal Society held at Gresham College. It includes one of the first catalogues of an egg collection, including a swan's egg with another egg inside it. Grew's nomenclature is interesting, many Linnaean names were used by him long before they were used by Linnaeus, e.g. the Gannet he calls 'Anser bassanus'. His work reveals, interestingly, that the name Murre was apparently originally applied to the Razorbill, rather than to the Guillemot, the bird for which the name is now commonly used, at least on the American side of the North Atlantic. Grew was born in 1628, the son of a clergyman, and died in 1712. Most of his work concerned botanical anatomy.

Albert Seba, a famous Amsterdam collector, kept many of his birds in alcohol, while Frederik Ruysch (1638–1731, author of *Thesaurus Animalium primus ... etc.*, 1710) kept his collection 'dried', each bird in a wide-necked bottle, its cork sealed with a fish bladder and red velvet. Seba was born on 13 May 1665 in Etzel, Germany, and died on 2 May 1736 in Amsterdam. As a young man, he travelled extensively in Germany and the Netherlands, to learn his trade as an apothecary, and in 1696 he settled permanently in Amsterdam, where he became a prominent citizen of the town. In his trade as apothecary and druggist, he became rich, and as a result was able to indulge his interest in natural history. His collection of specimens on all natural history subjects became world-famous. Based in the busy port of Amsterdam, he was in a good position to obtain specimens arriving from ships from all corners of the world. He sold his first collection to Peter the Great of Russia in 1717, but immediately began to collect again. Seba's book, privately published at vast expense, and ostentatiously entitled *Accurate Description of the Most Richly*

Frontispiece from Seba's Thesaurus, (1710)

Endowed Treasury of Nature, and an Illustration with the most Skilful Pictures, for a Universal History of the Physical World, was a multi-volume work in which may be found plants, snakes, monsters, birds, butterflies, mammals and other insects in wild disorder, with superficial descriptions written in what was, even for that time, an archaic language. Among the birds are a number which have never been identified, and which may represent extinct species. Although Seba's text is very poor, the richly coloured engraved plates make his book an important one for its time, and many authors based their species on plates in Seba's book. After being lost for many years, all Seba's original copper engravings, from which his plates were printed, were found in 1826 in the Museum d'Histoire Naturelle de Paris. It is not certain how they got there, but it seems likely that they were incorporated in the collection of the Prince of Orange and carried off to Paris in 1795 with this collection by revolutionary French occupying troops, after the Prince had fled to England.

Seba's unidentified hummingbird

Levin Vincent's *Descriptions and Plates of cabinets of Curiosities and Marvels of Nature* (1719) contained a number of tropical birds. The engraving of the museum accompanying the book shows shelves crowded with bottles containing birds and animals. Unfortunately the descriptions are so vague that in most cases it would be impossible to identify the species with certainty.

At this time there developed for the first time a demand for large books, lavishly illustrated in colour, for the libraries of the rich. Interest in nature was 'fashionable'. Among the earliest of these sumptuous works were Mark Catesby's *Natural History of Carolina* (Catesby 1731–43), and Eleazar Albin's *Natural History of Birds* (Albin 1731–38). Even better than either was the work of George Edwards (1694–1773), who published four volumes of *Natural History of Birds* (Edwards 1743–51); three more were later added, the *Gleanings of Natural History* (Edwards 1758–64). These books lacked any sort of arrangement or classification, but this hardly mattered to the rococo society for which they were intended.

Mark Catesby was born on 3 April 1683 and died in December 1749. His father was a lawyer, landlord and gentleman farmer. Little is known of Catesby's early life but that he studied natural history in London before departing to stay with his sister in Virginia at the age of 29. He subsequently settled in South Carolina and visited other parts of eastern North America and the West Indies. He collected seeds and botanical specimens which were sent back to England, and birds which he either preserved in alcohol, or dried in an oven and then stuffed. It is said that he attempted to preserve the freshness of the colours of the plumage by covering the specimens with tobacco dust. Many of these specimens were sent to his mentor, Sir Hans Sloane. Catesby returned to England in 1726 and devoted the next 17 years to producing *The Natural History of Carolina*, the first published account of the fauna and flora of North America. He published the work himself with the help of an interest free loan from a Fellow of the Royal Society, the Quaker botanist Peter Collinson. The plates are often stiff and flat, but were highly praised at the time, and considered to be extremely lifelike. Many of the birds he described for the first time were incorporated by Linnaeus in the later editions of his *Systema Naturae*. No portrait of Catesby is known to survive, and the only description of his physical appearance that has come down to us was made towards the end of his life; he was described as having a:

> Tall, meagre, hard-favoured and sullen look and was extremely grave or sedate, and of a silent disposition; but when he contracted a friendship was communicative and affable (Allen 1951b).

Little is known about the life of Eleazar Albin, who died c. 1741–42. He was a professional water colourist who taught drawing and painting. Although there is one picture attributed to him in a gallery at Kassel, Germany, he is remembered only for his book illustrations. He painted a wide range of natural history subjects, including insects, spiders and fish. His first book, *A Natural History of English Insects*, began publication in 1714. The three volumes of *A Natural History of Birds* were published in 1731, 1734 and 1738. This work, illustrated with 306 copperplate engravings of indifferent quality, remains of importance because some of his pictures and descriptions are the first of those particular species, and therefore count as the 'type specimens'. (The 'type' is the original specimen on the basis of which a new species is described. In the absence of a real specimen, an accurate published picture may also fulfil this function). Some of the plates remain unidentified, and may represent unknown extinct species. Much of his information was taken from Willughby's *Ornithologiae*, but was often somewhat garbled in the process of transcription. William Swainson described Albin as "without much knowledge of natural history and a very indifferent artist", which is partly true, he knew little of bird's habits, and though he claimed to have drawn "from the life" many of his pictures are clearly of dead specimens. Evidently, "from the life" had a different meaning in Albin's time to that which it has today. Following the success of *A Natural History of Birds*, Albin embarked on a second ornithological work, a small book titled *A Natural History of English Song Birds* (Albin 1737). All but three of the plates also depict the eggs – the first time eggs had been illustrated. Some copies of this book had hand-coloured plates, in other copies the plates were left uncoloured.

George Edwards

George Edwards (3 April 1694 to 23 July 1773) was born at Stratford, east of London, and went to school at Leytonstone, after which he was apprenticed to John Dod, a tradesman in Fenchurch Street. However, he found this uncongenial, and soon left. At this time a Dr Nicholas, a relative of his master, died, and his library was removed to the apartment where Edwards lived; the latter was thus able to pass his leisure reading volumes on natural history, sculpture, painting, astronomy and antiquities. His parents were obviously well-to-do, and Edwards decided to travel. In 1716 he made a trip to the Netherlands, after which he remained idle in London for two years, until he met a man who offered him a trip to Norway and Sweden. In the latter country he was arrested by the Danes, who took him for a spy, but he talked his way out of this difficulty. He returned to England, and, in 1719, visited France, where he narrowly escaped being transported to the American colonies. On his return to Britain, Edwards began to study natural history, to which he had become much addicted, and devoted his time to drawing and painting animals. Edwards' biographer, A. Stuart Mason (Mason 1992), however, considered that Edwards' skill lay in accurate representation rather than in artistic impression, and found little merit in the collection of preliminary sketches made by Edwards in the preparation of his paintings, which are held by the Zoological Society of London. In December 1733, Edwards was chosen by Sir Hans Sloane as librarian to the College of Physicians (a post held by Christopher Merrett about 70 years previously), and given apartments in the College. Access to a library of rare and choice books on natural history enabled him to become one of the leading ornithologists of his day. To Edwards, with this job of librarian, or more correctly 'bedell', came a rent-free house of considerable size in the college complex. The salary was £12 a year, topped up with gratuities to £22. After five years as bedell/librarian, Edwards decided to publish a book on birds, probably goaded on by the fact that Albin's paintings were attractive but lifeless. Edwards was also critical of Albin's accuracy. In 1743, the first volume of Edwards' *History of Birds* appeared. He also contributed a number of papers to the *Philosophical Transactions* and other journals, and several of the essays which form prefaces and introductions to his various books were collected and published as an octavo volume. He also prepared a detailed catalogue of Sir Hans Sloane's entire collection, including 1,172 birds and eggs. He left an edition of Willughby's *Ornithologia* with manuscript notes. Edwards retired to a little house at Plaistow where he died of cancer. He was buried in West Ham, but his tombstone has long since disappeared.

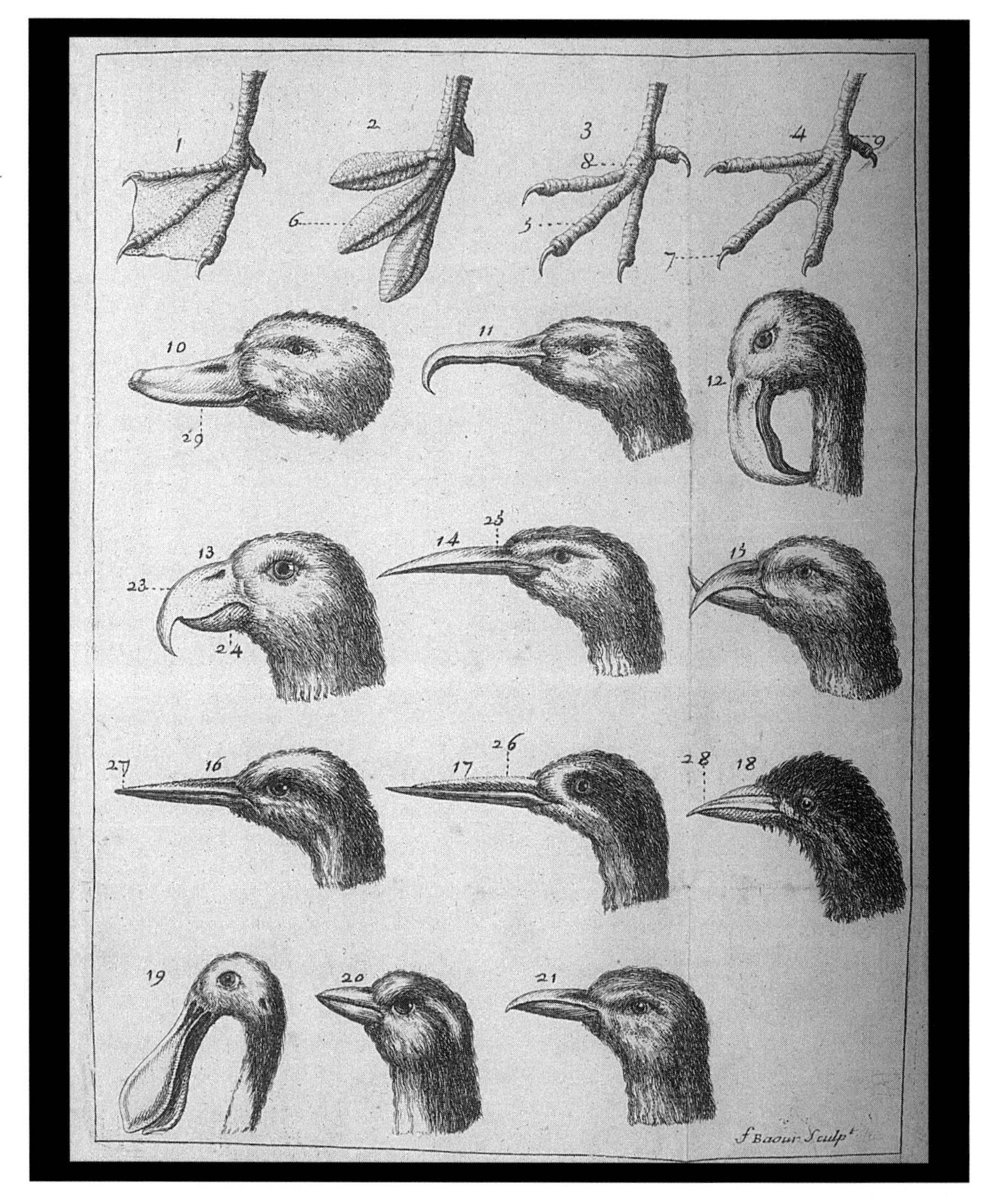

Plate of heads and feet from Barrère

Few words of praise have ever been lavished on those systematists who followed Willughby, and filled the 18th century with books which seem bizarre today. Pierre Barrère (1690–1755), whose book *Ornithologiae Specimen novum* appeared at Perpignan, France in 1745, has been ridiculed for a seemingly incomprehensible system. In the words of Stresemann:

> The bustard is placed between swallow and bunting, the latter between bustard and pelican, the bird of paradise between cassowary and ostrich, and so on, exactly as if the author wanted to put all the birds in one cupboard, and, in order to save space, filled the inevitable holes between the large birds by fitting in small ones.

This is unfair. In fact, Barrère's system was one based entirely on the shape of the feet. Thus he divided birds into four classes: Palmipedes (webbed), Semipalmipedes (half-webbed), Fissipedes (unwebbed) and Semifissipedes ("half-unwebbed" (!)). Within these four groups, the genera are listed more or less at random. Within these rules (if you can call them that), Barrère was fairly consistent, but the result was that instead of profiting from the work of Willughby and Ray, he produced a system which was totally artificial, as indeed, every system must be which relies on only a single anatomical feature. Barrère, described as a naturalist and doctor, was born at Perpignan about 1690 and died there on 1 November 1755. He began medical practice in his native town in 1717, but in 1722 left for Cayenne in French Guiana, where he spent five years, returning to France where, in 1727, he obtained the chair of botany in Perpignan as well as the position as doctor at the military hospital. He wrote a number of books on botany and medicine. The artificiality of his ornithological system may have been the result of his looking at birds through the eyes of a botanist.

Jacob Theodore Klein, a member of several learned academies, published a system (1750) which is as artificial and unsatisfactory as that of Barrère. It was based on the number of toes, which led him to class in one family birds different in all the rest of their organisation and mode of living. Klein divided the Class of birds into eight families, as follows:

1. Two toes, no hind toe (ostrich)
2. Three toes, no hind toe (rheas, cassowaries, bustards, some Limicolae)
3. Four toes, two forward, two back (zygodactylous birds)
4. Four toes, simple, hind toe present (raptors, passerines, game birds (!))
5. Four toes, webbed, hind toe present
    1. Plain billed: ducks, geese
    2. Cone billed: gulls (part), divers, etc.
    3. Anomalous billed (!): avocets, skimmers
6. Four toes, all connected (Steganopodes)
7. Three toes, webbed (auks, penguins, tube-noses, gulls (part))
8. Four toes, lobed (coots, grebes, etc.)

More elaborate was the *Avium Genera* (Möhring 1752) of P. H. G. Möhring (1710–1792), physician to the Prince of Anhalt. This, too, divided birds into four classes (see Appendix 2). The first (Hymenopodes) are divided into two orders, the *Piciae* which contains, as well as our Piciformes, a considerable number of passerines; and the *Passeres* which contains the remaining passerines. The second (Dermatopodes) is also divided into two, the *Accipitres* containing raptors, owls, parrots and nightjars, and the *Gallinae* containing gallinaceous birds and pigeons. The Brachypterae contains ratites, bustards and the Dodo. The Hydrophilae is divided into five orders, the *Odontorhynchae* (flamingos, ducks, darters), *Platyrhynchae* (penguins), *Stenorhynchae* (pelicans, cormorants, tropicbirds, auks, gulls, skuas, tube-nosed swimmers), *Urinatrices* (divers and coots), *Scolopaces* (cranes, ibis, herons, crakes, storks, guineafowl, the plains wanderer, hummingbirds, waders, plovers, dippers). In addition, there is a list of "Fragmenta Generum" (birds he could not classify). These included a number of unidentified species taken from

Seba and other older authors. This system, although inadequate by our standards, is certainly an improvement over previous arrangements, and shows some beginnings of the groupings we know today.

In 1774, J. C. Schaeffer published a "methodical distribution" of birds, in which he, too, used characters furnished by the feet. This was the *Elementa Ornithologica* (Schaeffer 1774, see Appendix 3). In 1789 he followed this with *Museum Ornithologicum* (Schaeffer 1789) describing the contents of his museum. Schaeffer was born in 1718, died in 1790, and wrote a considerable amount on various zoological subjects, but chiefly insects. He divided birds into two classes, Nudipedes (with naked legs) and Plumipedes (with feathered legs). These two groups virtually correspond to the Land and Water Birds of earlier authors, and by and large, Schaeffer's arrangement seems very sensible. The first class is divided into seven orders categorised by the nature and the number of toes, and, where further divisions are required, the form of the bill. Thus we have Fissipedes (divided), Pinnipedes (lobed) and Palmipedes (webbed). Under the first order, Fissipedes Didactyli (divided with two toes), there is only one genus, the ostrich. Under Fissipedes Tridactyli (three toes) come rhea, cassowary, bustards, stone curlews, oystercatchers and stilts. Order 3, Fissipedes Tetradactyli (four toes), includes plovers, jacanas, crowned cranes, pratincoles, spoonbills, turnstone, rails, swamphens, herons, storks, Hammerkop, boatbill, waders, cariama, screamers and Dodo. Order 4 consists of gallinules, coots, phalaropes and grebes. Here the rigidity of the treatment has caused the gallinules and coots to be separated from the swamphens and the rest of the rails. (This problem had previously arisen in Charleton's arrangement). Order 5 consists of auks and albatrosses; Order 6 of divers, penguins, shearwaters, gulls, terns, skuas, skimmers, avocets, mergansers, geese, ducks and flamingos. Order 7 comprises the present Pelecaniformes: gannets, anhingas, tropicbirds, cormorants, pelicans; there is no mention of the frigatebirds. In the second class (Plumipedes), Order 1 consists of woodpeckers, wrynecks, jacamars, barbets, cuckoos, toucans, trogons, anis and parrots. Order 2 consists of birds of prey and owls; Order 3 of game birds (turkeys, pheasant family, guineafowl and grouse). Order 4 comprises finches, buntings and allies and also colies. Order 5 consists of pigeons, birds of paradise, American orioles and blackbirds, rollers and the crow family. Order 6 is made up of oxpeckers, starlings, cotingas, thrushes and shrikes. Order 7 comprises larks, flycatchers, titmice, nightjars and swallows (with which are presumably included swifts). Order 8 consists only of nuthatches. Order 9 consists only of hummingbirds with straight bills. Order 10 comprises hoopoes, promerops, treecreepers and hummingbirds with curved bills. Order 11 consists of cocks-of-the-rock, manakins, todies, kingfishers, bee-eaters and motmots. Thus some forms crystallise into what we now recognise as natural groupings, while others do not.

One of the first writers with a sound biological education was Caspar Schwenckfeld (1563–1609) who was born at Greiffenburg in Silesia. As a child he was fascinated by animals, and this did not wane when he took up medicine as his profession. In 1603, he published *Theriotropheum Silesiae*, in which he described all the local birds of Silesia, but credulously took over statements from Gesner and Aldrovandus, many of which were incorrect. After Schwenckfeld, it was a century before another man of similar bent appeared in the field. This was Freiherr Johann Ferdinand Adam von Pernau, Lord of Rosenau, born on 7 November 1660 at Steinach in Lower Austria. When 16, he entered the University of Altdorf in Bavaria, where he graduated. He then travelled widely through Italy, France and Holland, but soon after 1690 he bought the Rosenau estate near Coburg. There he retired from the world and devoted himself to studying natural history until his death on 14 October 1731. He was particularly attracted to birds, which he kept in considerable numbers in his aviaries, and observed in the fields. Many of his birds he tamed and trained with great patience and skill to regard their cages as their territories, and to return to them, after flights into the neighbouring countryside. In this he has been compared to the 20th-century scientific writer Konrad Lorenz, whose methods were surprisingly similar. In 1702, he published (anonymously) the first record of his experiences, entitled *Lesson on what can be done, for pleasure and amusement, with those delightful creatures, birds, in addition to catching them, only by inquiring into their qualities and taming or otherwise training them*. It was a success, and enlarged editions followed in 1707 and 1716. Encouraged, he wrote another book, *Agreeable Country Pleasures*, in 1720. It was Pernau who first discovered 'territory' in a bird's world, and established that some bird song is not inherited, but must be learned by listening to the songs of others of the species. He was also the first to recognise that migration is not triggered by hunger or cold – birds tend to be well fed before setting off on their journeys – but by a hidden impulse that they must obey. Thus, Pernau may be described as the first student of causal analysis. Already in the early 18th century he was taking an interest in aspects of ornithology that were not to become a standard topic of study till the 20th. When he was writing, the only books available were those of such as Willughby (1676), which were solely concerned with describing external characteristics, or Gesner (1555) which treated life histories and habits in a very superficial way. Pernau was a true lover of nature in every modern sense. "It is not my intention", he wrote, "to describe how to capture the birds; I resolved, instead, to show how greatly Man can delight in these lovely creatures of God without killing them." Therefore, he may, perhaps, be described as the first conservationist.

The knowledge that birds migrate is very old. In the Book of Job, God asks "Doth the hawk fly by thy wisdom and stretch her wings toward the south?" (39:26), while an oft-quoted passage from the prophet Jeremiah says: "Yea, the stork in the heaven knoweth her appointed times; and the turtle [dove] and the crane and the swallow observe the time of their coming" (8:7). Other early writers mention bird movements in such a way as to imply that their readers would be fully conversant with these facts. Homer's *Iliad* described the Trojan army as "like the cranes which flee from the coming winter and sudden rain", while the poet Anacreon suggested that swallows wintered along the banks of the Nile. Aristotle was the first person to write with any clarity on the subject. In Book 8 of *The History of Animals* he explained that some creatures are sedentary during the winter, others leave after the autumn equinox to avoid the cold, and move north again in spring to avoid the southern heat. He was also aware that some species migrate over short distances, others, such as cranes, undertook journeys to "the ends of the world". He noted that the cranes flew from the steppes of Scythia to the marshlands of the Nile, south of Egypt. Aristotle also noticed the formations in which migrating pelicans flew, and deduced that the birds in the front waited for those behind, lest those in the rear should lose their way. This, of course, may explain how a flock stays together, but not how those in front find their way.

Willughby (1676) mentioned several migratory species. A century later Gilbert White, Thomas Pennant and George Edwards kept records of the dates of arrival and departure of breeding birds. Even so, at this time many superstitious beliefs prevailed. To the peasants the appearance and disappearance of various bird species was mysterious. In 1740, J. G. Gmelin was told by the Tatars of Krasnoyarsk that migrating cranes carried corncrakes on their backs. Others believed that migrating birds flew to the moon and back, and that they required no food for this journey as they travelled in rarefied ether. Another superstition was the supposed hibernation of some species. Aristotle believed swallows to hibernate, stating that they spent the winter in caves or hollow trees in a torpid state. Birds found, however, were almost certainly dead, not hibernating. Although one American bird, the Common Poorwill *Phalaenoptilus nuttallii*, is known to hibernate or at least achieve a torpid state, no species that any of the early writers would have known, does. Pliny believed that hibernation and waking were determined by the stars. The wheatear, he claimed, went into hibernation on the first day that Sirius was visible over the horizon, and woke on the day Sirius set. Olaus Magnus, Archbishop of Uppsala, in 1555, believed that swallows spent the winter under water, rolling themselves up into a ball. Magnus described how fishermen had often caught these lumps of swallows in their nets. He continued: "If that lump be drawn out by ignorant young (for old and expert fishermen put it back) and carried to a warm place, the swallows, loosened by the access of heat, begin to fly about, but live only a short time." Many distinguished writers such as Linnaeus, J. R. Forster, Cuvier and Geoffroy Saint-Hilaire, all believed in hibernation: over the centuries more than 200 papers have been written describing supposed cases. It was explained in great detail how swallows gathered in reedbeds in autumn, their combined weight causing the reeds to bend until the birds were submerged. This may be partly explained by the fact that swallows do before migration regularly roost at night in marshes, and during storms many may fall into the water and not revive. Daines Barrington in the 18th century was a believer in the 'immersion' theory, and his views influenced Gilbert White, who, while always acknowledging migration, admitted that hibernation may also occur. However, John Hunter (1728–1793) conducted experiments in which he captured swallows and confined them in greenhouses furnished with large tubs of water and reeds. When none of the birds even attempted to enter the water, Hunter doubted the truth of the theory. As late as 1824, Edward Jenner had to argue against hibernation.

Transmutation was also a theory used to explain bird disappearance. Aristotle suggested that the European Robin changed into the Redstart, and that the "Beccafico" (probably the Garden Warbler) into the Blackcap. This confusion probably arose from similarity of form and difference of plumage, and the seasonal changes in plumage which do occur in many species means that, though incorrect, it is by no means such a fanciful explanation as might appear. Even in our own time, letters in local newspapers ask what happens to Black-headed Gulls in winter, and what happens in summer to the similar-sized gulls with the white head flecked with sooty marks. One of the most bizarre examples of alleged transmutation was that of the Barnacle Goose, supposedly hatched from the goose barnacles, which grow on waterlogged timber. The barnacle is a crustacean, related to lobsters and crabs, which has evolved to lie on its back and kick food into its mouth. The modified feet look rather like feathers, and the shell, which grows on a long stalk, bears a remarkable resemblance to the head and neck of the Barnacle Goose. The myth was handy, in that it could be claimed that the goose was "fish" not "flesh", and could therefore be eaten during fasts.

Many explanations have been put forward to explain migration. One of these was a seasonal food supply. According to this theory, the approach of winter in high latitudes brings with it a reduction in food, and this causes birds to travel towards the tropics. In spring, the food supplies in the tropics become insufficient to support large numbers of young birds, and so the birds move out again to colder zones. Another explanation is that migration is a response to cold, from which many species retreat in autumn. But as we have seen, Pernau in the early eighteenth century had been the first to recognise that migration is not triggered by cold or hunger – birds are well fed when they set out on their journeys. Some have claimed that these birds originated in the north, were driven south by the advance of the Pleistocene ice age, and then returned to their original habitat with the arrival of milder conditions. Against this must be stressed that many species migrate early, before there is any climatic necessity. This has given rise to the suggestion that birds originated in the south and some species spread north as a result of a natural struggle for food and space.

The main trouble with these and other theories is that they attempt to explain the phenomenon by means of a single factor. However, it seems likely that over the millions of years birds have been evolving on earth, their instincts have been moulded by many differing factors. It is therefore probable that migration is due to a great many causes some of which may have affected one species and some another, but no one of which can be said to explain the entire phenomenon. As late as the mid-19th century, Heinrich Gätke, who lived on Heligoland for much of his life studying bird migration, and who published *Heligoland as an Ornithological Observatory* (Gätke German 1891, English 1895), was unable to furnish any clue as to what caused it. But a great mass of material had accumulated, reliable and unreliable, although unevaluated and uninterpreted.

Pernau was followed by the Protestant pastor Johann Heinrich Zorn (1698–1748) whose book with the self-explanatory title, *Ornithotheology, or an Attempt to Encourage Men, through closer Observation of Birds, to the Admiration, Love, and Reverence of their most powerful, wise, and good Creator*, was published in two volumes (Zorn 1742, 1743). In order to achieve his purpose, Zorn searched all fields of ornithology for evidence of the purpose of Creation. A careful observer, he noticed that species of bird which usually live and breed on the ground are earth-coloured, so that they will not be observed – Zorn was recognising the principle of camouflage. He was surprised to notice that such birds seemed to understand that their colour protected them, and crouched when danger threatened, not trying to fly away. The colour of eggs laid in the open was also protective, he decided, whereas those laid in holes are usually white, so that parents can easily see them. Although this last suggestion is unverified, no-one has yet improved on it. Zorn was clearly asking himself many questions about cause and effect, and answering some of them correctly. But many subsequent writers ceased to ask such questions, finding the mere recording of facts to be sufficient. For example, the three volumes of Johann Matthaeus Bechstein's *General natural history of Germany* (Bechstein 1789–1795) were nothing but a meticulous catalogue of information.

Chapter 3

Eighteenth Century Systems: Linnaeus, Brisson & Buffon, And Their Legacy

A revolution was on the way, something that would render all the systems of the past obsolete. A young man aged 27 left Uppsala, Sweden, in 1735, to settle in Holland. His name was Carl Linnaeus. He was born on 25 May 1707 in Smaland, the son of a poor pastor, and made little progress at school, being too devoted to natural history, especially botany, to want to bother with the more boring classroom subjects. This performance at school was so poor that at one stage apprenticing him to a shoemaker was considered. His habit of wasting his time collecting plants and gluing them on to sheets of paper greatly distressed his parents, who felt that he should be doing something more useful like learning to preach the Word of God. Linnaeus's ideas seem to have formulated slowly in the course of his investigations of the great amateur collections of Europe, including the chaotic assemblage of Albert Seba. It was over 20 years from his arrival in the Netherlands before the book which is now regarded as the definitive starting point for binary nomenclature, the tenth edition of the *Systema Naturae*, was to make its appearance. His early ideas on classification were still primitive. In the sixth edition (1748), for instance, he divided birds into six orders, as follows:

1. Accipitres - raptors, owls and parrots
2. Picae - woodpeckers, hornbills, cuckoos, crows, birds of paradise and hoopoes
3. Anseres - all birds that swim
4. Scolopaces
5. Gallinae - fowls, ostriches, bustards and coots
6. Passeres - pigeons, thrushes, larks, hummingbirds, crossbills, treecreepers, wagtails, tits and petrels

The superficiality of this system should be easily determinable, but it is symptomatic of Linnaeus that he never progressed beyond this basic arrangement. He apparently believed in some regularity of proportion in creation, the six orders of birds corresponded to the six orders of mammals.

By 1756, two private ornithological collections had outstripped all others. One of these was the collection of Sir Hans Sloane, which eventually became the foundation of the British Museum. Sloane (1660–1753) had been a disciple of John Ray, and as a young man spent the years 1687–89 in Jamaica, where he collected natural history specimens, and produced a folio *A Voyage to Jamaica* in two volumes, published in 1707 and 1725. He also financially assisted Mark Catesby in his journey to Carolina. His collection and library were vast by the standards of his day, and they were bequeathed to the nation,

Sir Hans Sloane

enabling the British Museum to open in 1759. His ornithological collection consisted of 1,172 items, including bird skins, skeletons, nests and eggs. Unfortunately, none of this ornithological material survives. (In the case of the skeletons this seems incomprehensible). The other collection referred to above, was the French cabinet of René Antoine Ferchauld de Réaumur, born at Rochelle in 1683. In 1708, aged only 24, he read papers on geometry to the Academy of Sciences in Paris, on the strength of which he was elected a Fellow. He read and researched widely into many branches of science, and wrote many papers. Between 1734 and 1742, he published a large six-volume work on insects, but this was never completed. MacGillivray, writing in 1834, said that for minute observation, especially in entomology, the writings of Réaumur had not been surpassed by those of any subsequent author. The latter carried on an enormous scientific correspondence, and received specimens from Europe and the colonies abroad. He was the first naturalist to form an extensive natural history collection in France, and the value of his collection was the care he took (unusual in those days) to obtain more than one specimen of each species, together with a sample nest and information on habitat and behaviour. Even so, at this time birds constituted a small part of natural history collections, shells were more fashionable – they were also, of course, easier to collect and preserve. Réaumur himself had pointed out the problems of preserving and studying birds:

> The study of birds, has remained as yet very imperfect, nor has it yet made them sufficiently known to us, because no considerable collections have hitherto been made of them; and those who had begun to make any soon became weary of going on, having had the mortification to see them every day destroyed by ravenous insects, in spite of all the care that had been taken to preserve them against their teeth (Réaumur 1748).

Réaumur never solved this problem, but he devised a method of drying birds in an oven, using the residual heat after the bread had been baked. In this way he built up the largest collection seen in Europe up to this time. It was worked on by a man who, though a far greater ornithologist than Linnaeus, is largely forgotten today. His name was Mathurin Jacques Brisson (1723–1806). His *Ornithologie* (Brisson 1760), was based largely on his work as curator of the Réaumur collection, and it was a more complete catalogue than any written previously, because he had access to a far greater collection than anyone before him. In addition to Réaumur's collection, he also examined a number of other private collections in and near Paris. The sources of his specimens were all carefully noted in his text, and he described the nest and eggs when these were present in the collection. But, being handicapped by a total lack of detailed field information, his work remained a museum catalogue. It consists of six volumes of detailed bird classification, and was the first of its kind, but set a pattern for the century to come. One of the last of the genre of catalogues of great bird collections was published between the years 1874 and 1895, the 27 volumes of the *Catalogue of the Birds in the British Museum*. The difference in the sophistication and volume of information between these two works, little more than a century apart, is truly staggering. The progression in ornithological knowledge between the beginning and end of the 19th century far surpassed that of any previous era, and in some ways exceeded the increase in knowledge which took place in the 20th century.

Mathurin Jacques Brisson

Brisson had realised the deficiencies of Linnaeus's system, and described his birds in as much detail as possible, whereas Linnaeus had been brief to the point of obscurity. Linnaeus had used large genera, Brisson, by defining groups more sharply, reached a total of 115 genera, which he grouped in 26 orders, not the mere six Linnaeus had used (see Appendix 4). Brisson's first order recognised the distinctness of pigeons, and his second consists of game birds. Raptors and owls constitute the third, and the fourth corvids, birds of paradise, rollers and icterids. The fifth order consists of shrikes, thrushes, cotingas and flycatchers, the sixth of starlings and oxpeckers. The seventh order consists of the hoopoes and the sugarbirds (*Promerops*). Though these groups are not related, the sugarbirds have always been problematical, and Brisson deserves considerable credit for recognising the distinctness of the hoopoes. Other authors

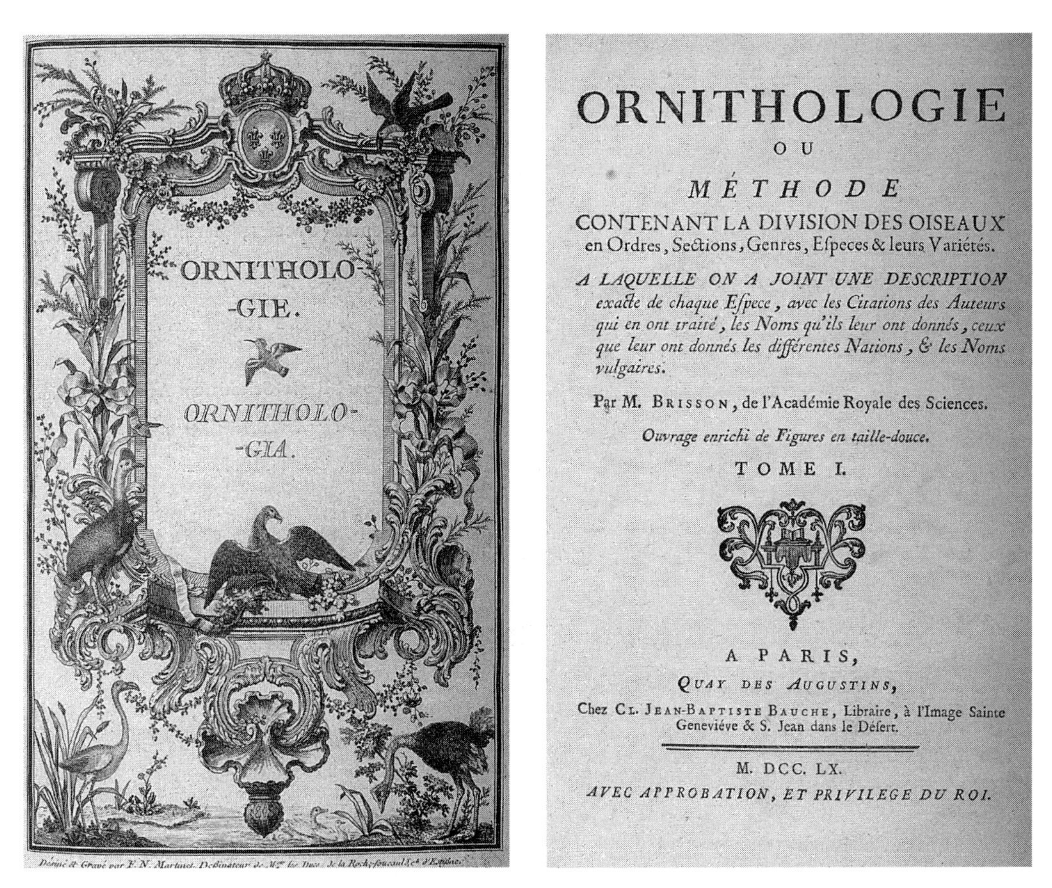

ORNITHOLO-
-GIE.

ORNITHOLO-
-GIA.

ORNITHOLOGIE
OU
MÉTHODE
CONTENANT LA DIVISION DES OISEAUX
en Ordres, Sections, Genres, Eſpeces & leurs Variétés.

A LAQUELLE ON A JOINT UNE DESCRIPTION
exacte de chaque Eſpece, avec les Citations des Auteurs
qui en ont traité, les Noms qu'ils leur ont donnés, ceux
que leur ont donnés les différentes Nations, & les Noms
vulgaires.

Par M. Brisson, de l'Académie Royale des Sciences.

Ouvrage enrichi de Figures en taille-douce.

TOME I.

A PARIS,
Quay des Augustins,
Chez Cl. Jean-Baptiste Bauche, Libraire, à l'Image Sainte
Geneviéve & S. Jean dans le Défert.

M. DCC. LX.
AVEC APPROBATION, ET PRIVILEGE DU ROI.

Title pages from Brisson's Ornithologie

of the time associated them with the treecreepers on account of the long, downcurved bill. Order 8 would appear to be a blunder, consisting of nightjars and swallows, but since the swifts are certainly included with the swallows, the blunder is not so great as would at first appear. The ninth order comprises finches, buntings and tanagers, the tenth larks, titmice and wagtails. The eleventh order consists of nuthatches only, the twelfth of treecreepers and hummingbirds. Brisson recognised the kinship of the two groups of hummingbirds, though Schaeffer later separated them. Order 13 consists of woodpeckers, jacamars, puffbirds, cuckoos, trogons and parrots, number 14 of cocks-of-the-rock, manakins, motmots, kingfishers, todies, hornbills and bee-eaters. The fifteenth order consists of the ratites (including the Dodo), the sixteenth of bustards, stilts and others. The seventeenth order was a large one, comprising waders, the stork and heron group and some rails, while the remaining orders consist of the seabirds, the duck tribe and other rails. Even more importantly, Brisson's generic and sub-generic divisions were far more correct than those of Linnaeus, and therefore capable of better diagnosis. He was also the first ornithologist, possibly the first zoologist, to introduce the concept of the 'type', although he did not actually use the term. Unfortunately, almost as soon as Brisson had established himself as an important writer on ornithology, he was forced to leave the discipline for ever. Réaumur died on 18 October 1757, aged 74, and a year later his collection was transferred to Paris and incorporated with the Cabinet du Roi, in direct defiance of his will: he had bequeathed it to the Académie des Sciences. In charge of the transfer was a man who hated Réaumur, and certainly had no intention of permitting his protégé Brisson to continue to work on the collection. Brisson was therefore obliged to abandon ornithology and turn to physics, where he is said to have had a successful though not brilliant career.

Illustration from Brisson's Ornithologie

The 'villain' responsible for Brisson's departure from the study of birds was Georges-Louis Leclerc, Comte de Buffon (1707–1788). He was the leading zoological historian of the latter part of the 18th century, and may well be the only ornithologist to have had a street named after him, the Rue de Buffon in Paris. He had won the confidence of King Louis XV, and diverted Réaumur's collection to his own custody at the Cabinet du Roi, by Royal command. Buffon has often been dismissed as a superficial writer. However, he was painstaking in the extreme in collecting data with which to embellish his multi-volume work, the *Histoire Naturelle*, and he aimed to produce, not merely a taxonomic treatise, but a compendium of everything he could find out about the subject. He was, however, fully aware of the enormity of the task; as he wrote by way of introduction:

> The first thing that one must undertake when one sets out to elucidate the history of an animal is a rigorous critique of its nomenclature: to specify clearly the different names which have been given to it in all languages and at different times, and to distinguish, as much as possible, the different species to which the same name has been applied. This is the only way that will allow us to make use of part of the knowledge from antiquity and to join it usefully to modern discoveries, consequently it is the sole manner to make true progress in natural history (translated in Farber 1982).

When the first volumes of Buffon's work appeared, the elegance of their style had a prejudicial effect on the popularity of Réaumur's writings, although many of the specimens had come from the latter's collection. Buffon had nothing but contempt for purely taxonomic studies, such as Brisson's, which he considered to be sterile exercises, and did his best to efface the memory of his rivals. But he combined this arrogance with the humility necessary for a truly enquiring mind. In describing the newly discovered Secretary Bird, he remarked:

> To what class can one relate a being in which are united characteristics so opposite? Here is another proof that nature, free in the midst of limits that we think prescribe it, is richer than our ideas and vaster than our systems.

Georges-Louis Leclerc, Comte de Buffon

He realised that, however thorough his work, it would be no more than a starting point, a foundation stone to the science of ornithology. No extant collection, however vast, could begin to supply all the necessary material for the task. He pointed out that to provide a minimally adequate description of a species, one would need a male, a female and two juveniles, a total of 8,000 specimens, which was ten times the actual size of the Cabinet du Roi in 1770. It is evident that Buffon read most of the earlier authors, and borrowed readily from them, but on the whole he did not admire their work, and relied heavily, as Réaumur had done, on a network of correspondents. He realised the inadequacy of language to describe the colours of birds in detail. For this he resorted to colour plates. The series of volumes produced as an accompaniment to his work was the *Planches enluminées*, produced under the supervision of Edmé Louis Daubenton, and this work appeared between 1765 and 1773 (Daubenton 1787). It comprised a series of hand-coloured engravings, of which a total of 1,939 represented birds. Although the plates are without artistic merit, they are sufficiently accurate for identification, and the book was a project without equal for the time. It was also of tremendous importance in an era when specimens in collections had a life expectancy that was uncertain to say the least. There are a number of extinct species, for which a plate in the *Planches enluminées* is the only surviving evidence for their former existence.

In 1783, an identification key to the *Planches enluminées*, the *Table des Planches enluminées*, assigning scientific names to Daubenton's plates, was provided by Pieter Boddaert (1730–1796). Little is known about him, and on this work alone rests his ornithological importance. Only 50 copies of the work were printed, and only one of these is now believed to survive, but in 1876 a reprint was issued by W. B. Tegetmeir (an authority on bees, fowls, pigeons, pheasants and drama). Mathews and Iredale assess this book scathingly, pointing out that typographical errors abound and suggesting that all the names in it should have been rejected:

Plate from Planches enluminées *(Blue Cotinga)*

> Throughout the book there is such a quaint disregard of genera, species and subspecies as regards the determination of the names of these plates, that it seems most absurd that our "priority-mad" predecessors should have accepted the nomenclature utilised in it as valid (Mathews and Iredale 1915).

Under current nomenclatural regulations, the first scientific name proposed for a species that can be confidently identified with it, is the name that must be used. Many older books proposed names which were too vague for identification. Mathews and Iredale felt that Boddaert's work was too sloppy, and his nomenclature too lax for his names to be accepted. Nevertheless, they have been accepted and a considerable number of current scientific names are credited to Boddaert.

By comparison with Brisson and Buffon, the ornithological achievements of Linnaeus, who was primarily a botanist, seem almost paltry. Linnaeus (1707–1778) began to study at the University of Lund but, after only a year, transferred to the University of Uppsala and in 1730 was engaged by Professor Olof Rudbeck Jr, as resident tutor for his younger sons. In this way, Linnaeus had access to Rudbeck's excellent library and the botanic gardens, which had been laid out by the elder Rudbeck. Rudbeck Jr was a scholar of wide training, primarily in medicine, but also in botany, and had a lifelong interest in birds, giving a number of lectures on ornithology while Linnaeus was in the city. Rudbeck Jr was the first Swedish ornithologist, and was born at Uppsala on 15 March 1660 where he died on 23 March 1740. As a result of an expedition to Lapland in 1695, he discovered a number of new species of birds and planned a full report on the expedition, to include the descriptions of these. But only the first part was published, the rest of the manuscript was destroyed in a fire in 1702 which flattened much of Uppsala. However, he made coloured pictures of many of the birds seen on this expedition, 215 in all, and these seem to have survived, though a good many of them were painted after the fire. These were used by Linnaeus, a number of whose bird names were based on Rudbeck's paintings. The texts of the latter's various lectures also survive, but none of his writings on birds were ever published, and so he never really received the credit that was his due. In 1732, Linnaeus went on a collecting expedition to Lapland, and returned six months later with stores of botanical, zoological and mineralogical material. Although he had been lecturing extensively on botany, and was regarded as a genius in Uppsala, he still had no medical degree after seven years as a student, and thus no security of tenure. So in 1735 he made the break and went to Holland, to take his degree at the University of Harderwijk, and, if possible, to get his writings published. Here he had the good fortune to meet the botanist Jan Frederick Gronovius (1690–1760) and showed him his *Systema Naturae*. So impressed was Gronovius that he had it published at his own expense. Linnaeus was then sponsored by the wealthy George Clifford, President of the Dutch East India Company, as a result of which he was able to begin publishing several of his early works.

It was in 1758 that Linnaeus first proposed binomial nomenclature, which was to make his name for all time, and many immediately realised the advantages offered by this clear and unambiguous system. He retained his six orders (listing 930 species), but moved some of the groups. Parrots, hummingbirds and treecreepers were now placed with the picarian birds; bustards, ostriches and coots under the gallinaceous birds, and petrels with the geese and ducks (see Appendix 5). In the 1758 arrangement, note that the missing numbers in the list of genera refer to genera not introduced till the later edition of 1766). In fact, Linnaeus's treatment of birds (and indeed, also of mammals) was based on very imperfect knowledge of the subject. Furthermore, as a result of isolation, he was prevented from examining collections of mammals and birds in the cities of Europe. Thus both his sixth and tenth editions are inadequate in many ways. It is somewhat paradoxical, therefore, that the *Systema Naturae* became the foundation of zoological classification and nomenclature, while the work of Brisson, who was an expert in both groups, had access to a fine library and possessed one of the finest collections then in existence, was bypassed. Linnaeus's sixth edition was effectively a synopsis of the fauna of Sweden filled out by exotic species taken almost entirely from descriptions of other authors. His vagueness as regards classification and appellation is demonstrated by the fact that even in the twelfth edition (1766), one of the two species of penguin known to him is classed as a tropicbird, and the other as an albatross. Alfred Newton (1885) was to remark acidly that, "He for the most part followed Ray, and where he departed from his model he seldom improved upon it."

On the other hand, Brisson was a most painstaking author, and the excellence of his work was so obvious to the 1842 Committee on Zoological Nomenclature that, although they adopted the twelfth edition of Linnaeus as the starting point for nomenclature, they made a special recommendation for the retention of Brisson's genera. Under present rules, no special dispensation is necessary, for, with the subsequent adoption of the tenth edition as the starting point, Brisson's generic names (though not his specific names which are not consistently binomial), fall within the requirements. Although Brisson's work was published in 1760, much of it had evidently been printed earlier, as he cites only Linnaeus's sixth edition through the first four volumes, and only the tenth in the last two. This explains his failure to utilise the binomial system, for which he has often been criticised. Yet his work is far superior. By using 26 orders in place of Linnaeus's six, and 115 genera in place of the 51 of Linnaeus's sixth edition or 63 of his tenth, Brisson was able to give a much better classification than had ever been given before. His work owed little to Linnaeus. On the other hand, the indebtedness of Linnaeus's twelfth edition to Brisson was very great. Of the 386 new species in this edition, no fewer than 240 are taken from Brisson, and of the 15 new genera, all but one were taken from him, but, presumably to make this pilfering less obvious, he renamed most of them. Linnaeus and Buffon were now leaders of two disparate schools of thought. The latter's disciple, Cuvier, was later to contrast the two as follows:

> Linnaeus and Buffon, in fact, seem to have possessed, each in his own way, those qualities which it was impossible for the same man to combine, and all of which were necessary to give a rapid impulse to the study of nature. Both passionately fond of this science, both thirsting for fame, both indefatigable in their studies, both gifted with sensibility, lively imaginations, and elevated minds ... Linnaeus seized on the distinguishing characters of beings, with the most remarkable tact; Buffon, at one glance, embraced the most distant affinities: Linnaeus, exact and precise, created a language on purpose to express his ideas clearly, and at the same time concisely; Buffon, abundant and fertile in expression, used his words to develop the extent of his conceptions. No-one ever exceeded Linnaeus in impressing every one with the beauties of detail with which the creator has profusely enriched everything to which he has given life. No-one better than Buffon ever painted the majesty of creation, and the imposing grandeur of the laws to which she is subjected (Lee 1833).

The very idea of Linnaean classification was repugnant to Buffon. It seemed to him to be an arbitrary dissection of the greatness of nature into little bits. Linnaeus died in 1778. In 1829, his papers were purchased by the executors of James Edward Smith (1759–1828), the founder, in the year of Linnaeus's death, of the Linnean Society of London.

The way was now open for many others to carry on the work that had been begun. The first of these new authors was the Bishop of Bergen, Erik Pontoppidan, who had written a *Natural History of Norway* (Pontoppidan 1755), to be followed by the first volume of his *Dansk Atlas* (Pontoppidan 1763), which included the first description of the Herring Gull *Larus argentatus*. Pontoppidan had been born at Aarhus on 24 August 1698, and studied divinity at the University of Copenhagen. For a time, he was a travelling tutor to several young noblemen, and later chaplain to the king. He became professor of theology at Copenhagen in 1738, and Bishop of Bergen in 1745. He died on 20 December 1764. Most of his considerable output of writings was theological. Pontoppidan's friend, Morten Thrane Brünnich (1737–1827) is generally regarded as the founder of Danish zoology. He was born in Copenhagen on 30 September 1737, the son of a portrait painter. He took his BA at the age of 20, and then studied theology and oriental languages. He was, however, influenced by the writings of Linnaeus and so turned to the study of natural history. Soon, he was making original observations on insects, and contributed the entomological section to Pontoppidan's *Dansk Atlas*. In 1762, he became curator of the large natural history collections of Counsellor Thott and Judge Christian Fleischer and became more absorbed in ornithology. The collections contained many species from the Scandinavian countries, and the study of these provided the basis for Brünnich's *Ornithologia Borealis* (Brünnich 1764), published when he was 26, which separated for the first time the two species of Atlantic guillemots, although both had been named before. The second one continues to bear his name as Brünnich's Guillemot, though in America it is known as the Thick-billed Murre. *Ornithologia Borealis* reads as a catalogue of a collection, since, for several species different individuals are listed, other species are placed in footnotes implying that no specimens were present. A few of Brünnich's species have not been identified, but generally speaking it is a very accurate list of the avifauna of northern Europe (see Appendix 6). In 1763, Brünnich had published a *History of the Eider Duck* (Brünnich 1763), and in the same year as the *Ornithologia Borealis*, his *Entomologia* also appeared. Through the influence of Pontoppidan and other naturalists, he was appointed lecturer in natural history and economy at Copenhagen University. Before taking up this post he made a long tour of Europe, visiting European mines and quarries in Cornwall, Hungary and other places. He founded the museum of Copenhagen University, building up a fine collection in a short space of time. Much of his

BRÜNNICH. *DANISH NATURALIST WHO WROTE THE* ORNITHOLOGIA BOREALIS. *THIS PORTRAIT WAS PAINTED WHEN HE WAS SIXTY-TWO YEARS OLD.*

Morten Thrane Brünnich

later life, however, was devoted to mineralogy. He died on 19 September 1827. Other writers included P. L .S. Müller, the publication of whose German edition of the twelfth edition of the *Systema Naturae* (1773), added a number of new species gleaned from other authors, to which binomials were assigned for the first time. This found general acceptance for Linnaeus's system in that country. Müller (a prolific writer, though few of his writings related to ornithology), was considered an uncritical hack by Stresemann, though Cassin (1864) spoke more highly of him.

Giovanni Antonio Scopoli (1723–1788), the son of a lawyer, was born at Cavalese in the Val di Fiemme on 13 June 1723 (Violani mss.). At the age of 20, he obtained a degree in medicine at Innsbruck University. For some years he practised medicine in Cavalese and later in Venice, but spent much of his time botanising in the Alps and making large collections of insects. He was then recommended as private secretary to the Count of Seckan, with whom he stayed for two years, and was able to make use of the Count's library in his spare time. This helped him to prepare for an important examination in Vienna, which he passed successfully in 1754, thus entitling him to practise medicine anywhere in the Austrian Empire. He was highly praised by Van Swieten, the Head Commissioner to the Medical Faculty and personal physician to the Empress Maria Theresa, who promised to recommend Scopoli for a high medical position at Linz. However, he was not appointed, but was instead sent to Idria, a small village in Carmiola, as physician to the mines. He spent 16 years there on very little pay, constantly clashing with the Director of Mines, who felt that he was spending too much time studying plants and insects. From a natural history point of view the period was productive, Scopoli published major works on the flora and entomology of the area. As a result of these publications, he entered into correspondence with Linnaeus, who thought highly of the Italian's work. Scopoli's most important ornithological work, however, was the *Anni Historico-Naturales*, issued in five parts (Scopoli 1769–1772). This contains descriptions of new birds in his own collection, in the museum of the Count Francesco Annibale Della Torne and in the Imperial menagerie.

These included *Strix alba*, *Strix noctua*, *Sturnus collaris*, *Hirundo rupestris*, *Ardea ralloides*, *Emberiza melanocephala*, *Anser albifrons*, *Procellaria diomedea*, *Anas leucocephala* and *Psittacus krameri*. As a result of his publications he became well known in Europe. In 1769, after turning down the appointment of the chair of mineralogy at the Academy of St Petersburg (owing to his patriotism to the Austrian Empire, which seems to have done little to deserve Scopoli's loyalty), he was offered a senior lectureship in Scheunitz, then in Hungary. In 1777, he transferred to Pavia, Italy, where he spent the rest of his life, but this was not an entirely happy period, owing to differences of opinion with another member of staff, Professor Lazaro Spallanzani (1729–1799). Scopoli's last work, and one of ornithological importance, was the *Deliciae Flora et Fauna Insulicae*, published in three parts, in which he described a great many objects of natural history. It is of particular importance in that it gave scientific names to many of the birds and mammals described by Sonnerat (see p. 82) in the accounts of his various voyages. Scopoli died in 1788 after a heart attack, said to have been brought on by a series of rows with Spallanzani. The influence of Scopoli on Italian ornithology was great. He published a number of regional works on birds, which started a tradition carried on by other writers. His adherence to Linnaeus's rules consolidated the use of the binomial system in Italian ornithology.

In other countries such as Britain, the system was not so quickly adopted. One of the most remarkable of semi-popular, non-systematic writers at this period was Thomas Pennant. Pennant (1726–1797) came of a very old Welsh family, based at Downing in Flintshire. It was a beautiful retreat, surrounded by many square miles of woods, fields and glades. As a young man, Pennant had been much influenced by Willughby's *Ornithologiae*, a copy of which he had received as a gift when only 12 years old. At the age of 20 he began a series of tours which made him a famous and popular writer, succeeding to the estate in 1763 on the death of his father. Like other wealthy young men of the time he did the 'Grand Tour of Europe', an account of which he wrote in 1765. In this he made detailed notes of birds seen in the field. In Holland he met Pallas (see next chapter) with whom he discussed natural history at length, and by whom he was also inspired to write his *Synopsis of Quadrupeds* (Pennant 1771). He was elected a Fellow of the Royal Society in 1767. Although he wrote many books, such as the *British Zoology* (Pennant 1761–66), *Arctic Zoology* (Pennant 1784–85) and *Indian Zoology* (Pennant 1769), as well as various books of travels, his greatest importance may lie in his vast network of correspondents and associates, which included Linnaeus and Gilbert White. He was a friend of such important figures as Buffon, Voltaire and Albrecht Haller. The four parts of *British Zoology* appeared between 1761 and 1766, illustrating in colour the vertebrate animals of Britain. *Arctic Zoology* (two vols. 1784–85) is Pennant's most important work, for in it many species are described for the first time. The second of the two volumes, concerned solely with birds, contains a number of species which have not been identified. While preparing this work, he corresponded with such authorities as Brünnich, Otto Müller and Fabricius, and it remained the standard work on the area till Richardson's *Fauna Boreali-Americana* of 1829–37. The period between the death of John Ray in 1705 and the publication of Pennant's *British Zoology* was a lean period in the history of British ornithology, but this publication seems to have given a boost to other works, and among those which appeared in quick succession were John Berkenhout's *Outlines of the Natural History of Great Britain* (Berkenhout 1769), William Hayes's *Natural History of British Birds* (Hayes 1775), John Walcott's *Synopsis of British Birds* (Walcott 1789), William Lewin's *Birds of Great Britain* (Lewin 1789), Thomas Lord's *Entire New System of Ornithology; or Æcumenical History of British Birds* (Lord 1791) and Edward Donovan's *Natural History of British Birds* (Donovan 1794). These were all little more than compilations. Pennant was an indefatigable writer on many subjects, and was a strange mixture of humility and conceit. On the other hand, Jardine (Jardine 1833) said that Pennant's works, "contained the greater part of the knowledge of their times". Pennant was in his time, a law student, traveller and antiquary, businessman, man of letters and naturalist. His interests were probably too diverse to bring him to the top rank as an ornithologist.

Marmaduke Tunstall

Marmaduke Tunstall's *Ornithologia Britannica* (Tunstall 1771) followed Linnaeus's general grouping but retained Willughby's classical division into land and water birds. This work used binomials, and was probably the first British work to do so. It was also the first work to use names such as the Dipper for the bird formerly known as the Water Ouzel. Following the book's publication, Tunstall (1743–1790) was elected a Fellow of the Royal Society, four years after Pennant. His museum was at Welbeck Street in London, and was much used by Pennant as a source of specimens for illustration. In 1776, Tunstall moved to his estate at Wycliffe in Yorkshire, and four years later the museum was also moved there. Tunstall died on 11 October 1790, and his collection was bought by George Allen of the Grange, near Darlington, for £77. Much of this collection is now lost, though some specimens survive in the Hancock Museum in Newcastle-upon-Tyne.

Between 1781 and 1785 the three volumes of Latham's *General Synopsis of Birds* (Latham 1781–83) had appeared. John Latham (1740–1837) was born at Eltham, Kent on 27 June 1740, the son of a surgeon. He attended Merchant Taylor's School, became a general practitioner, and was a regular correspondent of Pennant, Ashton Lever, Joseph Banks and others. He was elected to the Royal Society in 1775, the Royal Society of Stockholm, the Natural History Society of Berlin and helped to found the Linnean Society of London. *General Synopsis of Birds* was his first ornithological work, and it contained 106 plates, all executed by him. It described many new species, based on specimens he found in the various private collections which he studied. Like the work of Buffon, Latham's made no attempt to supply scientific names. Almost at once, however, Latham realised that this would mean that posterity would not accord him the necessary acknowledgement in having his name attached to those of the new species he had described. He hastened to rectify this and, in 1790, produced the *Index Ornithologicus* in which he assigned Latin binomials to all his species. But he was too late. In 1788–93, Johann Friedrich Gmelin had produced his own edited edition of Linnaeus's *Systema Naturae* (Gmelin 1788–93; the volumes concerning birds were published in 1788–89), in which he added a great many new species, plundered from the accounts of Latham among other writers. He was not in fact a zoologist at all, but belonged to a dynasty of famous German chemists among whom were a number of naturalists. Gmelin was born in

1748 at Tubingen and died in 1804 at Göttingen. He obtained a medical education at Tubingen and was awarded the MD in 1769, at the age of 21. He then undertook a 'scientific tour' of Holland, England and Austria, returning the following year to become professor of medicine at Tubingen. The next year, still aged only 25, he became professor of philosophy and medicine at Göttingen, and in 1778 became professor of chemistry, botany and mineralogy. His literary output on chemistry was important, but the *Systema Naturae* was his only contribution to ornithology. It was not unimportant, however, for a great many of his names are still in use today. Latham's arrangement of birds was based largely on Willughby, and was thus rather old-fashioned but it was serviceable enough for the time (see Appendix 7). It will be seen that, while following Willughby's general order, Latham utilised the Linnaean groupings, but with the addition of two orders for the pigeons and the ostrich respectively. He borrowed a third order from Schaeffer, the Pinnatipedes, or lobed feet. Latham has been criticised in that he did not possess the inborn faculty for picking out diagnostic characters, and that he made many errors, often describing the same species more than once, under different names. This last charge, however, is unfair, for it merely reflected the state of knowledge of the time. Virtually every systematic author of the period made similar errors.

Latham's third important work is the ten-volume *General History of Birds* (Latham 1821–28), when he was already in his eighties. In this he added many more species which he had seen in collections. Sadly, many of these specimens no longer exist, and Latham's descriptions have not always been identified. Some of these are undoubtedly extinct species of which we know nothing. The *General History* quoted existing binomials but made no attempt to supply these for birds described for the first time – showing that Latham had never been completely won over to the Linnaean system. However, he was the first ornithologist since Willughby to achieve anything like completeness in his listing of the known birds of the world of his day. A high proportion of his new species were Australian, as a result of the intensive colonisation of the continent being undertaken at that time. Indeed, Latham is often regarded as the father of Australian ornithology.

John Latham

Johann Friedrich Gmelin

The very first book on any aspect of Australian ornithology was George Shaw's *Zoology of New Holland* (Shaw 1784), but only 12 plates and 33 pages of text were issued before the work was discontinued. Only five of the plates represented birds, so that its ornithological import was slight. (For a bibliographical note on this very rare book, see Mathews 1912). Following Shaw's book were the *Journal of a Voyage to New South Wales* (White 1790), and *The Voyage of Governor Phillip to Botany Bay* (Phillip 1789), in which several birds were represented. However, Latham was the first to describe a large number of Australian birds. He had described many species from the area in 1802 (*General Synopsis of Birds: Supplement 2*) on the basis of a collection of paintings formed by a Mr A. B. Lambert, but they were rather crude in style, and this resulted in many errors in Latham's account. His work was followed by a major paper by Vigors and Horsfield based on material in the Linnean Society collection, and there was a small book by Lewin, *Birds of New South Wales* (Lewin 1813). A few more species were described by Quoy and Gaimard in the *Voyage de l'Uranie* (1824–26) and the *Voyage de l'Astrolabe* (1830–35), and by Lesson in *Voyage autour du monde sur La Coquille* (1826–30) and *Journal de la Navigation autour du Globe de la frègate Thetis* (1837). This was all that had been written on Australian birds by the time John Gould arrived and spent two years there, culminating in his large illustrated work *Birds of Australia* (Gould 1840–48). In addition to ornithology, Latham had many interests, including antiquarian study, and he wrote a treatise on the history and architecture of Romney Abbey; though the manuscript of this survives, it was never published. At the time of his death on 4 February 1837, at the age of 97, he was at work on a revised edition of Pennant's *Indian Ornithology*.

Meanwhile, writers such as Gilbert White, Thomas Bewick and George Montagu were enriching the English regional faunas. The Rev. Gilbert White was born in Selborne on 18 July 1720. He was educated privately at Basingstoke and at Oriel College, Oxford, where he took his MA in 1746 and was ordained the following year. For the rest of his life he was curate of various parishes, travelling a great deal, but for a large part of his life was able to live at his grandfather's house, 'The Wakes' in Selborne, now the Gilbert White Museum. *The Natural History and Antiquities of Selborne* (White 1789) consists of a series of letters addressed to Thomas Pennant and Daines Barrington. It is for this that he is largely remembered. The book has gone through many editions over the years, and has hardly, if ever, been out of print. It is said to have sold more copies than any other book in English except for The Bible and the complete works of Shakespeare. Although his literary style is charming, and his observations meticulous, the book made little impact on the progress of scientific ornithology except in its documentation of the former distribution of species whose occurrence in Britain is now very different. However, it inspired generations of amateur ornithologists who have greatly enriched British ornithology. White died on 26 June 1793. Although White's Thrush (*Zoothera dauma*) was named after him, he had no direct connection with the bird.

The influence of Thomas Bewick (1753–1828) on the scientific study of ornithology was slight, his chief contribution being to popularise the subject via his skills as artist and wood engraver. He illustrated many books on a wide variety of subjects; from the ornithologist's point of view his most important work being the *History of British Birds* in two volumes (1797 and 1804). It passed through many editions, delighting thousands of readers, and, in the opinion of Alfred Newton, did more to promote the study and interest in ornithology in Britain than any book other than White's *Natural History of Selborne*. For most of his life Bewick lived near Newcastle-upon-Tyne, and the transfer of Tunstall's Museum to the latter's seat at Wycliffe in Yorkshire, though a great loss to London authors, was a gain to Bewick.

Gilbert White

Thomas Bewick

George Montagu

George Montagu (1751–1815) was of a noble family, and began his career in the army, where he rose to the rank of Lieutenant-Colonel, but his career was not particularly distinguished. When he left his wife and took to living with another (already married) woman, he was court-martialled and expelled from the army; the rest of his life was devoted to natural history. He also amassed a considerable collection of specimens, which were eventually sold to the British Museum. His main contribution to ornithology was the *Ornithological Dictionary* (1802) which "set a new standard for others to follow" (Mearns and Mearns 1988, p. 268). He was the first to separate Montagu's Harrier (*Circus pygargus*) from the Hen Harrier (*Circus cyaneus*) and was involved with the first British records of the Cirl Bunting (*Emberiza cirlus*), Cattle Egret (*Bubulcus ibis*), Little Crake (*Porzana parva*) and Gull-billed Tern (*Gelochelidon nilotica*). He also described for the first time the Roseate Tern *Sterna dougallii* from a specimen sent him by Dr Peter McDougall of Glasgow; and the American Bittern (*Botaurus lentiginosus*), from a vagrant found at Piddletown in Dorset. (In the last mentioned case, however, another description of Montagu's specimen, by Thomas Rackett, appeared only a month before that of Montagu (Macdonald and Grant 1951).)

In Italy, similar writers included Bonelli and Savi, after whom an eagle and two warblers were to be named. In Sardinia, the priest Francesco Cetti produced his work *Natural History of Sardinia* (Cetti 1776), for many years the classic work on the island. Cetti (1726–1778), after whom Cetti's Warbler (*Cettia cetti*) is named, was born in Mannheim in Germany, but his parents were Italian. He spent most of his life as a Jesuit missionary on Sardinia. Franco Andrea Bonelli (1784–1830) was born at Cuneo in Piedmont, but his family soon moved to Turin, where, in the valley of the River Po, young Bonelli learned to hunt birds and mammals, to prepare specimens, and to keep copious ornithological notebooks. He later became engrossed in entomology, and at the age of 25, published one of his first works, on subalpine fauna, followed by a monograph on the ground beetles. In the autumn of 1811 after a period of study in Paris, Bonelli became professor of zoology at Turin University, and that year published a *Catalogue des Oiseaux du Piedmont* (Bonelli 1811). In 1815 he discovered a leaf warbler, which was named *Phylloscopus bonelli* in his honour, and the eagle *Hieraaetus fasciatus* is known as Bonelli's Eagle in English, French and Italian. Bonelli was one of the most respected naturalists of his day, and raised the status of Turin University to equal that of Pavia. Paulo Savi (1798–1871) was the son of a lecturer in botany at the University of Pisa, and graduated as doctor of physics and natural science at the age of 19. He became particularly interested in zoology and mineralogy, and spent much of his time preparing specimens. At the age of 23, as well as a teaching post, he was appointed curator of Pisa Museum and began to expand and improve the then rather mediocre collections. He spent a lot of time studying birds in the coastal marshes and olive groves, where he discovered an unknown warbler which he described in 1824 as *Sylvia* (now *Locustella*) *luscinioides*. It is known as Savi's Warbler. In 1823, he published a catalogue of the birds of Pisa province. He worked hard at improving the museum, and by 1826 it had extended into buildings in the adjoining street. He found less and less time to study birds. Although on his retirement he returned to the study, he never managed to finish his magnum opus, the *Ornitologia Italiana*, which was edited and published by his son (Savi 1873–76).

Chapter 4

Pallas and the New Awakening

Peter Simon Pallas

At the end of the 18th century many ornithologists felt that they had to choose between Buffon and Linnaeus, because the contradictions seemed impossible to resolve. The man who succeeded in synthesising the two schools was Peter Simon Pallas (1741–1811). He was born in Berlin, but as a young man took up a post at the Academy of Natural Sciences at St Petersburg, filling the gap left 20 years earlier by the death of Georg Wilhelm Steller, who had taken part in Bering's ill-fated expedition to the sea that now bears his name. Before reaching St Petersburg, however, Pallas had travelled extensively in Europe, and had made the acquaintance of Thomas Pennant. The two men corresponded for years, and much of the material for Pennant's *Arctic Zoology* was supplied by Pallas. To place the career of Pallas in context, however, it is first necessary to discuss the series of Siberian expeditions undertaken by Steller and others which opened up that country and marked the beginning of the study of natural history in the Russian Empire.

Steller (1709–1746) was born at Windsheim, near Nuremberg, in Germany, the son of a cantor, and was interested in natural history from an early age. He did well at school and took theology at Wittenberg University, fully expecting to return as a preacher. However, at university he began to attend lectures in anatomy and botany, eventually changing his course from theology to medicine, tutoring privately to pay for his fees: his botany lectures attracted a lot of students. Failing to obtain a professorship in botany at Halle, he travelled to St Petersburg under his own volition by a difficult overland route. In that city, he was befriended by an archbishop, whose physician he became, and it was this archbishop who recommended Steller to join the second Pacific expedition of Vitus Bering. The first had been under the auspices of Peter the Great, when Bering sailed through the strait, later named the Bering, thus proving that Siberia and America were not joined. The second expedition, planned by the Empress Anna, was to chart the Siberian coast, the Kuril Islands, Japan and other parts of eastern Asia.

Steller's predecessor at the Academy of Natural Sciences had been J. G. Gmelin, who had participated in a Siberian expedition from 1733 to 1743. Johann Georg Gmelin (1709–1755) was born in Tubingen, a precocious boy who went to university at the age of 14 and studied medicine and science (particularly botany). He graduated in 1727 at the age of 18. After he was awarded his MD in 1728, he was recommended by two of his tutors to go to St Petersburg, where Peter the Great had founded an academy in 1725. He was given a teaching post in 1730, and the following year, aged only 22, became professor of chemistry and science. After publishing several papers on geology, chemistry, etc., he was sent on an expedition to Kamchatka. This departed in July 1733 across Siberia, returning to St Petersburg in February 1743, when Gmelin returned to his former duties. The expedition resulted in his monumental *Flora Sibirica*. In 1747, he returned to Tubingen where he died.

Johann Georg Gmelin

Steller set out in January 1738 across Siberia to join Bering in Kamchatka, exploring, and collecting specimens which were sent back to St Petersburg. On the way he met Gmelin who was returning. The latter, travelling in a huge caravan with lots of servants, was amazed at Steller's simple needs and method of living. It took two years for Steller to reach Okhotsk, where Bering was greatly impressed by the young man's learning and enthusiasm. The coasts of southern Kamchatka, parts of Alaska and the Aleutian Islands were explored. Steller, as a result of eating gentian and other plants when he went ashore for collecting purposes, was almost the only man on the expedition to escape scurvy. They were stranded for the winter on what is now Bering Island, where Steller found the huge, flightless, and now extinct Spectacled Cormorant *Phalacrocorax perspicillatus* (he was the only naturalist ever to see it alive), and other creatures such as a white "sea raven" (possibly a now-extinct gannet) which nobody has subsequently identified. While on the island, he wrote much of his most famous work, the *De Bestiis Marinis*, posthumously published (Steller 1751). After escaping from the island, Steller set off back across Siberia on foot, but died of a fever in 1746 aged only 37. His companions buried him on a high bluff overlooking the River

Tura, but his grave has long since been eroded away by the river. Among the birds named after him are Steller's Eider *Polysticta stelleri*, Steller's Sea Eagle *Haliaeetus pelagicus* and Steller's Jay *Cyanocitta stelleri*.

The expedition of 1768, organised by the Empress Catherine the Great, involved a number of zoologists, each team exploring different parts of her empire. Anton Güldenstädt travelled to the Caucasus, and his accounts remain, after over 200 years, the most authoritative on the birds of that area. Ivan Lepechin moved further east. They were followed by Pallas and by Samuel Gottlieb Gmelin (1744–74, nephew of J. G. Gmelin and cousin of the compiler J. F. Gmelin), who joined Güldenstädt in the Caspian provinces, where he discovered and described several new species of birds. Like his uncle, S. G. Gmelin began medical studies in Tubingen when very young, and received a doctor's diploma at the age of 19. In 1768, when only 24, he was appointed professor of natural history at the Imperial Academy of St Petersburg. On his return journey he was captured by the Khan of the Chaitakens and held to ransom. The negotiations in regard to this were protracted for several months and he died of dysentery before it could be paid. Johann Anton Güldenstädt (1745–1781) was born in Riga, and at the age of 18 began a medical career in Berlin, later studying at Frankfurt and obtaining his doctorate in 1767. His father had been secretary to the Imperial Cabinet, and it was most likely as a result of his influence that his son was invited to take part in the expedition. Among his discoveries in or near the Caucasus Mountains were the Terek Sandpiper (*Xenus cinereus*), the Great Rosefinch (*Carpodacus rubicilla*) and Güldenstädt's Redstart (*Phoenicurus erythrogaster*). On his return, he published several papers on zoological and botanical subjects, but the bulk of his researches (including his ornithology) were not published until after his death, when they were edited and issued in two volumes in 1787 and 1791, by Pallas. On 23 March 1781, at the age of 37, Güldenstädt died during an epidemic, while carrying out his duties as a doctor in St Petersburg. Ivan Lepechin (1740–1802) was primarily a botanist. Russian by birth, he had first studied at the Imperial Academy of Sciences, St Petersburg, followed by a course in medicine at Strasbourg where he took his doctorate before becoming an associate of the Imperial Academy. On 8 June 1768 he left for the lower Volga and Caspian basins. He spent August of that year in the province of Kasan studying the course of the River Tscheremschan, which separates the department of Kasan from the province of Stavropol. The following year, he crossed the Urals, returning in the winter of 1773, via the White Sea. He apparently travelled in Siberia again in 1774–75. He was the first permanent secretary of the St Petersburg Academy of Sciences, and one of the first native Russians to become a great naturalist.

Pallas had probably the most extensive itinerary of all. Part of his route seems to have crossed that of Lepechin, a certain rivalry developing between them though they were only slightly acquainted. Some of the same species were described and named almost simultaneously by both men, and the resulting confusion of nomenclature was not resolved until the 1970s. The most famous case is that of the Caspian Tern, which was named *tschegrava* by Lepechin and *caspia* by Pallas. Peter Simon Pallas (1741-1811) was born in Berlin, where his father, Simon Pallas, was professor of surgery and chief surgeon of the public hospital in the city. The son learned Latin, French, English and German under private tutors, and had been interested in natural history since before he was 15. After studying at Göttingen from the age of 17, he took his doctorate at Leyden University two years later at the age of 19. He then travelled through the Netherlands, visiting many important museums and hospitals. In July 1761, he went to London for ten months to study medicine, but devoted a great deal of his time to zoology. He subsequently persuaded his father to allow him to settle in The Hague where he continued to study. There he met Thomas Pennant who was making the 'Grand Tour' of Europe. So great was Pallas's reputation by now that in 1764, when he was only 23, he was elected a Fellow of the Royal Society of London, and, the following year, a member of the Académie des Curieux de la Nature.

Plans for a voyage to South Africa and the East Indies under the auspices of the Prince of Orange were prevented by his tyrannical father who insisted on Pallas's return to Berlin. However, such was his fame that he was invited to Russia by Catherine the Great, and Pallas had the presence of mind to defy his father's opposition and accept this offer. He arrived in St Petersburg in August 1767. On learning of the Empress's plans for the expedition, Pallas immediately volunteered to accompany it, and was eagerly accepted. He set off in June 1768, crossing the plains of European Russia, the northern shores of the Caspian Sea, travelling through the Urals, to Tobolsk, the Altai, east to Lake Baikal, and south to northern China. It was six years before he returned home. Pallas's accounts of the expedition covered a very wide range of subjects – little seems to have escaped his attention, and he reported on geology, mineralogy, ethnology, zoology and ornithology. After the expedition, Pallas settled in St Petersburg, the favourite of the Empress Catherine, their friendship possibly helped by their mutual German origins. She ordered that all herbaria collected by the other naturalists should be made accessible to Pallas, who was engaged to write the *Flora Rossica* (Pallas 1784–88). While at St Petersburg, Pallas commenced his great work, *Zoographia Rosso-Asiatica* (Pallas 1811) which was to occupy him for the rest of his life. In it he described for the first time Pallas's Grasshopper Warbler (*Locustella certhiola*) and Pallas's Warbler (*Phylloscopus proregulus*), among others which he had obtained many years earlier.

Pallas published many books, but his most important contribution to ornithology, the *Zoographia Rosso-Asiatica*, on which he laboured for 15 years, remained unpublished at the time of his death on 8 September 1811. Pallas had been collecting material for this book from 1767 to 1795. In the latter year he retired to the Crimea to arrange and complete the work. It was not till 1803 that it was half-finished, and he arranged with his friend Geissler of Leipzig for the engraving of the plates. The first half of the manuscript (the mammals and birds) was delivered to the Academy in 1806 and the printing was finished by 1807, but the plates were not completed. Geissler got into financial difficulties, and the plates were taken to the pawnbrokers. By the time of Pallas's death, the problem was still unresolved, and the work was not officially released till 1826, without the plates. It was only in 1831 that the complete work was issued, with plates. Two copies of the 1807 printing (dated 1811) had been released and used by ornithologists. However, it was not until the 20th century that 1811 was accepted as the date of first publication of Pallas's work, and the consequent acceptance of the names proposed in that work, for a number of species which had been re-described by others between 1811 and 1826. In this work, Pallas had formed a skilful synthesis of Linnaeus and Buffon. From the former he took over general systematics and binomial nomenclature, while from the latter he took the dynamic outlook and the idea that species are mutable, particularly under climatic influence - a concept of which the birds of the steppes of Central Asia apparently afforded striking proof. It was Pallas who first called attention to geographic variation, and thus paved the way for the theory of Charles Darwin. For a complete list of the species described by Pallas, their arrangement and (where known) their modern identification, see Appendix 8. Pallas's ideas were quickly adopted by Constantin Lambert Gloger in a popular little book, *Schlesiens Wirbelthier-Fauna* (Gloger 1833). It was Gloger who first recognised the structural difference between swallows and swifts. His arrangement for European birds only may be found in Appendix 9.

Another zoologist who was deeply influenced by the work of Pallas was Blasius Merrem (1761–1824), a pupil of Johann Friedrich Blumenbach (1752–1840). The latter was an outstanding zoologist, but his influence on ornithology was mainly as a tutor. Blumenbach was born at Gotha on 11 May 1752, studied medicine at Jena and graduated at Göttingen in 1775. He died at Göttingen on 22 January 1840. He was interested in nature from an early age, and as a child, delighted in assembling skeletons from bones he had collected. Professor of anatomy at Göttingen for nearly 60 years, his lively lectures are said to have been interspersed with witty remarks, the same jokes being repeated year after year to the delight of succeeding generations of students. His writings are rather heavy, but spiced with dry humour, and he is said to have been very largely responsible for stimulating his countrymen's interest in nature. He is best known as an anthropologist, of which study he has been described as the founder, and first divided mankind into five races: Caucasian (white), Mongolian (yellow), Malayan (brown), Negro (black) and American (red). Another of Blumenbach's pupils, Samuel Thomas Sömmerring (1755–1830) was born in Thorn in Poland of a German family. He was for a time a Rosicrucian, dabbling in alchemy and spiritualism. His main work was on the anatomy of the human brain, but in the year of his death, the Copper Pheasant (*Syrmaticus soemmerringi*) of Japan was named after him. Merrem was born in Bremen on 4 February 1761, the son of a merchant. He originally planned to be a theologian, and studied Arabic as well as Greek and Latin. In 1778, he entered the University of Göttingen, where, under the influence of

Blumenbach, he took up zoology, particularly ornithology. He planned a vast monograph, *An attempt at an outline of the general history and natural classification of birds*, but sadly only fragments of it were ever published. In order to earn a living, Merrem was obliged to turn to law. In 1785, he was given a professorship at Duisburg, for a miserable salary, and which obliged him to teach mathematics and physics, leaving little time for zoology. Nevertheless, his publications are impressive, and Stresemann was firmly of the opinion that:

> Zoology would have made great strides if it had received his undivided services, but difficult circumstances and the need of money forced him to scatter his considerable gifts (Stresemann 1975).

Merrem died, worn out by illness and financial worry, on 23 February 1824, but his influence on the subsequent history of ornithology was considerable. In the opinion of Alfred Newton, his classification, published in the *Abhandlungen* of the Academy of Sciences of Berlin (Merrem 1813), must be regarded as the starting point for consideration of all more recent classifications. The arrangement was superior to anything that had gone before, and better than several that were to follow (Appendix 10). The most important innovation of this arrangement is the division of birds into Ratites (running birds) and Carinates (flying birds), based primarily on the presence or absence of a keel to the sternum. But Merrem was careful to point out other differences between the two groups, mainly concerning feather structure and the bones of the wings, which in the Ratites are degenerate and not adapted for flight. His perception was less good in relation to the smaller groups, where there are a number of anomalies, such as placing the todies and rollers with the passerines. He also failed (as did most of his contemporaries) to distinguish between swallows and swifts. Unfortunately Merrem's arrangement attracted little notice at the time, and it was not until many years later that its merits were appreciated.

In 1815, Henri Marie Ducrotay de Blainville (1777–1850) read a memoir on classification to the Academy of Sciences in Paris, and it was published five years later in the *Journal de Physique* (de Blainville 1821). It was not a success, since it was supposedly based on the condition of the sternum, but because this bone is buried in the body, his interpretation of it was based on external characters. It became obvious that the features of the sternum on which de Blainville relied (although he claimed the contrary) were entirely based on the posterior margin, a feature of very slight value, since the notches which occur there are, in some groups of birds, extremely variable. However, he deserved credit for regarding the parrots and pigeons as full-ranking orders; he realised that the rollers were not passerines, but belonged near the bee-eaters, and he was probably the first to recognise that lyrebirds are not gallinaceous birds (Appendix 11).

In 1827, Ferdinand Joseph L'Herminier (1802–1866), a native of Guadeloupe in the Leeward Islands, and a pupil of de Blainville, contributed a revision of his tutor's work, also based on the same feature, the sternum. The result was much superior, and, though there is no evidence that he was aware of Merrem's work, he, too, divided birds into two main divisions, the oiseaux "normaux" and "anomaux" corresponding to the Carinates and Ratites. L'Herminier was the son of Felix Louis L'Herminier (1779–1833) who was born in Paris but settled in Guadeloupe, where father and son made a special study of the birds of the island. After his period of study in Paris, the younger L'Herminier returned to Guadeloupe, where he remained for the rest of his life, becoming chief medical officer at the island hospital. However, the museum specimens and all his notes were destroyed in the disastrous earthquake on the island in 1843. (This demonstrates the unsuitability of housing important collections and archives in areas vulnerable to natural disasters.) Access to the collections in the Paris Museum was of great help to L'Herminier, as he was able to see many obscure forms not accessible to Merrem, and was able to recognise 34 families of birds, most of which accord well with those recognised today, though the actual sequence of the families, is, of course, very different. Some of the groups are composite and therefore need further splitting, but there are virtually no examples of actual incorrect grouping. Of particular note are L'Herminier's placing of the rails ("coots") close to the cranes, of dissociating the cranes from the herons, and his union of the snipe, sandpipers and plovers into one group. Unfortunately, the excellence of this work was ignored or scorned at the time, and had little effect on the development of ornithology (Appendix 12).

In 1820, Christian Ludwig Nitzsch published a treatise on the nasal glands of birds, on which he based his classification. It covered very few species not found in Europe, and although unsatisfactory it did contain some interesting features. He included some definite groupings such as the Passerines and the picarian birds which subsequent research showed to be natural groups. But his work underlined the vulnerability of basing any classification on a single feature or aspect of anatomy. In 1829, after he had been studying the vascular system, Nitzsch published the *Observationes de Avium arteria carotide communi* (Nitzsch 1829), and the result of his labours was surprisingly similar to that of L'Herminier, although the two had not collaborated. While the actual order of Nitzsch's families is rather different, many of the groupings are similar (Appendix 13). By contrast, the arrangement produced by Johann Georg Wagler in 1830 seems very strange (Appendix 14). Wagler (1800–1832) was director of the zoological museum at the Munich University. He died at the age of 32, after accidentally shooting himself when cleaning his gun. Scarcely better, indeed much worse, was the arrangement of Berthold, who devoted a long chapter in his *Beiträge zur Anatomie* (Berthold 1831) to classification, once again based on the sternum, and he came to the astonishing conclusion that this feature cannot be used in classification as it would result in such anomalies and 'obvious' errors as the separation of the swallows from the swifts, and the alliance of the latter with the hummingbirds. He apparently assumed that the arrangement of Linnaeus should not be challenged. Although he was an excellent anatomist and gave many accurate drawings of the avian sternum, he failed to grasp the significance of the structure in relation to flight.

In 1779, Blumenbach had attacked the followers of Linnaeus in his handbook of natural history, and had called for an animal's entire anatomy to be considered in classification, not merely one character. In 1788, Blasius Merrem thought much the same, but Lacépède (1799) used a much elaborated version of the Linnaean system. Whereas Linnaeus had used six orders and 81 genera, Lacépède employed considerably more, but used the same somewhat superficial divisions of beak and foot – thus hoopoes, treecreepers and hummingbirds were grouped together. Other unusual features to emerge from this treatment were the separation of *Orthorhynchus* from the rest of the hummingbirds, and the placing of the Secretary Bird with the waders. This arrangement of birds, in the *Tableaux Méthodiques des Mammìfères et des Oiseaux* (Lacépède 1799) was based on bill structure (see Appendix 15). Bernard-Germain-Etienne Médarde de la Ville sur Illon, Comte de Lacépède, was born at Agen on 26 December 1756. An aristocrat, he did not value his ancestry; but conducted his life on the principles of good manners and honourable conduct. His politeness was proverbial. Many of his family were musicians, and he grew up with a great love of music. He was quite an accomplished composer, though now forgotten as all his music is lost. Gluck is said to have admired his compositions. He devoted part of his early years to music, composing an opera, and publishing in 1785 a work entitled *The Poetry of Music*, soon followed by a treatise on electricity. However, at an early age he had read Buffon's *Histoire Naturelle*, and Buffon offered him a position at the Cabinet du Roi in succession to Daubenton, so that he might continue the *Histoire Naturelle*. Only a few months before Buffon's death, Lacépède published the first volume of his *History of Reptiles* (Lacépède 1788). The French Revolution almost proved fatal to him, but it is a measure of the estimation in which he was held, that when he naively applied for permission from Robespierre to return to Paris, the tyrant replied, "He's in the country? Tell him to stay there." He did not return to the capital until the terror was over (he was at this time aged 40), and was then appointed professor at the Jardin des Plantes, where he published several works on fishes and cetaceans. When the new government of France was established, Lacépède was appointed senator in 1799, president of the senate in 1801, chancellor of the legion of honour in 1803 and minister of state in 1804. His administrative duties occupied a great deal of his time and he conducted them with a facility and rapidity which astonished all his acquaintances. At first, he refused any salary for his services, and was both liberal and self-effacing in his monetary aid to those in need – a sharp contrast to most politicians. His requirements were simple. He never possessed more than one suit at a time, never drank wine, ate sparingly of the simplest food, worked hard and slept little. He died on 6 October 1825, and great numbers of the poor people of Paris attended his funeral.

Chapter 5

Explorations of the Eighteenth Century

Pallas had got about as far as was possible in describing the Eurasian avifauna. In the meantime, the Pacific Ocean was being opened up by the British and the French. The most significant expeditions of the latter part of the 18th century were those of Captain Cook. James Cook was born in October 1728 at Marton, Cleveland, in northeast England, the son of a farm labourer of Scottish origin. He went to sea while still young, and in the course of an eventful career, served at Quebec with Wolfe. Joseph (later Sir Joseph) Banks, who accompanied Cook on the first Pacific expedition, was born in 1744, the son of a wealthy landowner. He attended Eton where he spent little time on his studies. However, a chance walk along a flowery lane suddenly aroused in him a passionate interest in botany, on which his knowledge became encyclopaedic, and the large herbarium he built up in the course of his life is now in the Natural History Museum. Banks inherited the family estates at the age of 21, and in 1766 set off on a voyage of exploration to Labrador and Newfoundland; in 1767, he collected in Wales. He was elected a Fellow of the Royal Society at the age of 23.

James Cook

Joseph Banks

Cook's first Pacific voyage, from 1768–1771, visited Madeira, Rio de Janeiro, Tierra del Fuego, the Society Islands, New Zealand and eastern Australia. Although Banks was primarily a botanist, while at sea no botanising could be done, and so full attention was turned to zoology. Most of the birds collected and described were therefore seabirds, which are of less interest as far as the Pacific Ocean is concerned, as they are wide ranging and there are comparatively few species. Land birds were neglected, and these are precisely what one would have wished to know about, as most of them were endemic to the islands on which they occurred, and are largely now extinct. None of the bird specimens collected on the first voyage is known to survive. Banks published little, but, after his return, his house in London was open to all who wished to consult his library and specimens. Every Thursday a 'breakfast' was held, at which some subject of scientific importance would be discussed. A light meal of rolls and tea or coffee was served, while a more elaborate meal would take place in an adjacent room for a smaller, more select, group of guests. Banks's correspondence was enormous, but it is now widely scattered, which hinders critical assessment of his work. Following a disagreement with the Admiralty, Banks was bypassed for Cook's second expedition and replaced by J. R. Forster (1728–1798) and his son Georg (1754–1794). Johann Reinhold Forster was descended from a Yorkshire family which had settled in Danzig in the mid-17th century, and was of mixed Yorkshire, Scottish, German and Polish blood. He proved to be brilliant at school, studying avidly such widely diverse subjects as classics, philology, biblical studies, chronology, ethnology, cosmology, natural theology and ancient geography. Pressed first into an ecclesiastical career, the years from 1753 to 1765 were spent in a cramped country parsonage on low pay studying history, geography, Egyptian languages and antiquities. Sermons were said to have been put off till Saturday night when black coffee kept him awake long enough to finish writing them, but not always long enough to deliver them in the pulpit. He also corresponded widely with men of letters, and spent all he could afford, and much that he could not, on books. He had a library said to number thousands of volumes.

Johann Reinhold Forster

The death of the Empress Elizabeth of Russia in 1761, and the usurpation of the throne from her successor Peter III by his wife Catherine, formerly a minor German princess, but soon to become the Empress Catherine the Great (who had very definite ideas about the government of Russia), may have been one of the things that prompted Forster to break away from his restraints. Another factor was probably his association with the botanist Gottfried Reyger (1704–1788) whom Forster and his sons assisted with his work on the flora of the Danzig area. This brought Forster into contact with the methodology of Linnaeus. Among Catherine's policies was the colonisation of remoter parts, such as the Volga delta, by German settlers. Under Elizabeth's reign (1741–61) much in Russia had declined, including the sciences, and Catherine was determined to revive Peter the Great's expansionist policies. Forster (accompanied by his son Georg) left for St Petersburg, and, on 15 May 1765, was appointed by Catherine commissioner to report on the new colonies, where the settlers were erroneously supposed by the government to be living in blissful happiness. Forster here exhibited one of the traits that was to dog him throughout his life. He was tactlessly honest, and his report was not what the authorities wanted to hear. Disaffected, Forster turned his back on Russia, and sailed to England.

Here he threw himself into literary, social and intellectual circles, and his learning, particularly in natural history, became phenomenal. He came into contact with intellectuals such as Thomas Pennant, Daines Barrington and many others. Contact with Joseph Banks seems to have put into Forster's mind the possibility of travelling on Cook's second voyage. It was in the summer of 1771, when Forster was busy overseeing a number of publications and translations of other travellers' books, that the *Endeavour* returned to London, and Banks and his assistant, the botanist Solander, were much feted. It was common knowledge in intellectual circles that Cook intended to return to the South Pacific, and Banks was inundated by requests to sail with him. The preparations for the second voyage were protracted; and Banks, being dissatisfied with the arrangements on board the ship refused to sail until these had been rectified. The Admiralty called his bluff and dropped him in favour of Forster and his son. On 26 June 1772 (Forster in the meantime had been accorded the honour of having been elected a Fellow of the Royal Society), they left London by post-chaise for Plymouth. Cook's expedition left Plymouth in July, and three months later reached Cape Town, where an additional member, Anders Sparrman, was recruited to assist with the zoological studies.

Sparrman had studied medicine at Uppsala, where he had attracted the attention of Linnaeus. In 1765, aged only 17, he made a journey to Canton in China as a physician, and on his return described the plants and animals collected on the voyage. He desired to continue travelling, but his poverty would have prevented this had not a friend obtained for him the office of tutor to the children of a person residing at Cape Town, where he arrived in 1772. When Cook called at Cape Town, the two Forsters visited Sparrman and persuaded him to join the voyage with all expenses paid. In July 1775, after the voyage in the *Endeavour*, he returned to Cape Town, where he practised medicine and earned enough to undertake an expedition into the interior. Later in 1775 he returned to Sweden, where he found that he had been awarded an honorary doctorate in his absence. He was elected a member of the Royal Society of Stockholm. In 1787 he participated in an expedition to West Africa, but this was not a success, and he returned to Sweden where he remained till his death in July 1820. He published a Swedish ornithology (Sparrman 1806); before that a catalogue of the Museum Carlsonianum (Sparrman 1786–89), but his best-known work was *A Voyage to the Cape of Good Hope, towards the Antarctic Polar Circle, and round the World with Captain Cook: but chiefly into the country of the Hottentots and Caffres, from 1772 to 1776* (Sparrman 1786).

Another pupil of Linnaeus, Carl Peter Thunberg, whom Sparrman had met soon after his arrival in Cape Town, was born in Sweden in 1743 and died in 1828. Although the two had been acquainted in Sweden, they travelled to South Africa separately, and it was quite by chance that they met there a few days after arrival. In 1770, Thunberg travelled in Europe, and was engaged by the Dutch East India Company to go to Japan, Java and Ceylon in a medical capacity, spending some time at Cape Town on the way. His travels occupied nine years, and were more extensive than those of Sparrman; he was the first visitor with a scientific training to travel far into the interior and publish a book of his travels, though this appeared later than that of Sparrman. Back in Sweden he was appointed director of the Uppsala botanic garden on the death of Linnaeus. He was a prolific writer; the printed catalogue of the Natural History Museum library lists 112 titles by him, though a number of these were probably papers rather than books. Thunberg was exceptionally well educated for his day, and combined this learning with enthusiasm, but is perhaps best remembered today because in 1828 the Grey-headed Wagtail, *Motacilla flava thunbergi* (a subspecies of the Yellow Wagtail), was named after him.

The itinerary of Cook's second expedition is well known. Cook travelled south into the ice of the southern Indian Ocean, past Kerguelen Island to New Zealand, Tahiti, and the Society Islands, back to New Zealand, then to the Marquesas, Easter Island and the Tuamotus. A number of birds were collected which have never been seen since. The earliest contributions to New Zealand ornithology were made by Forster on this voyage, and although a few additional species were described by the voyages of the *Coquille* and the *Astrolabe*, nothing much more was achieved till 1842, when Dr Dieffenbach sent a large collection to G. R. Gray, and the results were published in *Travels in New Zealand* (Dieffenbach 1843). On his return to London, Forster found to his dismay that his meticulous work did not find favour with the authorities, and he was maliciously deprived of all reward, and £1,000 that were his by contract. As a result he fell deeply into debt. His collections were scattered, his son Georg's watercolour paintings went to the creditors, who sold them to Sir Joseph Banks. He retained his manuscript *Descriptiones animalium*, later deposited in the Berlin library, where it was consulted and used, usually without acknowledgement, by a number of ornithologists. It was not published until 1844, long after Forster's death, and then by Lichtenstein, who made a mess of the editing (Forster 1844). Martin Heinrich Carl Lichtenstein (1780–1857) was an all-round naturalist, though in his youth insects had been one of his main preoccupations. He was born in Hamburg and after graduating spent some years surveying South Africa. In 1810, he was awarded the chair of zoology at Berlin University, even though he had only a superficial knowledge of some of its branches. In 1815 he was appointed director of the Berlin Zoological Museum. He was amiable, a fine musician and spoke English very well. But his refusal to accept the rule of priority in nomenclature, and his poor editing of Forster's *Descriptiones animalium* have been much criticised. Worse was to follow. Many of Forster's specimens found their way into the collection of Sir Joseph Banks, but many more into the collection of Sir Ashton Lever, a man with a passion for collecting – objects of almost every kind, both scientific and artistic. Over a number of years this curious character built up an enormous private collection, the Museum Leverianum, which flourished in London from 1774 until its final dispersal in 1806. By 1784, swelled by the fruits of Cook's third voyage, it contained 28,000 items and was world-famous.

Sir John Ashton Lever, the eldest son of Sir James Darcy Lever and Dorothy (née Ashton), was born at Alkrington, near Manchester, on 5 March 1729, and at the outset lived the life of a country gentleman of those times. Why he first began collecting is not known, but it seems to have commenced in 1760, when he purchased several hogsheads of foreign shells, and to these added fossils, stuffed birds, native costumes, weapons and other curios. The first mention of the museum is by a person who visited it at Alkrington Hall, Lever's family seat, in 1773, and described it as consisting of: "1,300 glass cases, containing curious subjects, placed in three rooms besides four sides of rooms shelved from top to bottom with glass doors before them." In 1774 he removed the material to London, where he started to charge for admission, in order to reimburse himself, and to enable him to purchase further specimens. At first, it was a huge success, but the receipts in no way covered the expenditure in which Lever indulged in order to add to his collection. His eccentricity was commented on by one visitor to the museum who described him as prancing about his garden, dressed in the green costume of a forester, with a bundle of arrows under his arm. Unsurprisingly, Lever got into financial trouble and in 1781 launched an appeal for support. In 1783, he petitioned the Government to purchase the entire collection at a fraction of its value and combine it with the British Museum collection. With the stupidity of which only governments are capable, this generous offer was refused, and Lever auctioned the entire collection by lottery, offering 36,000 tickets at a guinea each. In the end, only 8,000 tickets were sold, and this took two years. Eventually the draw took place, and one James Parkinson acquired the collection, but he too failed to make it pay.

By this time, Lever was dead. Devastated by the loss of his collection, and having drowned his sorrows rather too freely, he died on 24 January 1788 at the Bull's Head Inn in Manchester. Parkinson offered the collection to the Government again, but the latter referred the matter to Sir Joseph Banks, who hated Sir Ashton Lever, and therefore spurned his collection. A public auction took place in 1806 and the entire contents of the museum were dispersed. Some items went to another private collector, William Bullock; others to the Derby Collection at Knowsley Hall; still others were purchased by the Vienna Museum, including the Mascarene Parrot, *Mascarinus mascarinus*, and the White Gallinule, *Porphyrio alba*, of Lord Howe Island (both now extinct). But a high proportion of them vanished into smaller private collections, never to be seen again. Both Thomas Pennant and John Latham knew the Leverian collection well, and based many of their descriptions on specimens therein. A considerable number of described species were known only from specimens in the Leverian Museum, and in the absence of the types cannot now be satisfactorily identified; some of them probably represented extinct species. This tragedy was entirely caused by the arrogance and stubbornness of Banks.

Frontispiece of Sonnerat's Voyage to New Guinea *which includes the only known portrait of him*

At this period, the French were by no means idle, but the most widely circulated book of the time, Pierre Sonnerat's *Voyage to New Guinea* (Sonnerat 1776), has been so maligned as to obscure the true achievements of its author. In 1781, J. R. Forster assigned scientific names to some of the species described by Sonnerat, including a penguin *Aptenodytes torquata*, allegedly from New Guinea. This penguin has been ignored, for it is based only on the brief description of Sonnerat, notorious for his apparent inaccuracy and carelessness. Because a few of his descriptions cannot be identified with any known extant species, it was a natural assumption that they must have been erroneous. However, it is also possible that they may be based on extinct species. Furthermore, Sonnerat's carelessness has been greatly exaggerated. The locality of New Guinea for a penguin has always correctly been regarded as ridiculous. However, part of the story was unravelled by Averil Lysaght (1952) and this throws the onus for the original description away from Sonnerat and on to a much more reliable observer. Sonnerat included in his book three plates of penguins, all by implication from New Guinea, two of which are perfectly identifiable antarctic species, the King and the Gentoo Penguins, *Aptenodytes patagonicus* and *Pygoscelis papua* respectively. The odd name of the latter is based on the erroneous assumption that it came from New Guinea. All three species were copied from drawings, still in Paris, made by Jossigny of specimens collected by Philibert Commerson, the naturalist on Louis de Bougainville's world voyage, 1766–69.

Louis de Bougainville

The Seven Years' War had exhausted France economically and ended her domination of Europe. The French still felt the need to continue to explore uncharted territory in order to forestall the colonial ambitions of the British. But the French Government was in no position to sponsor expeditions, so it was left to individuals to seek what glory they could reap for themselves and for their country. In many cases the ships were ill equipped, and conditions on them unhealthy and unpleasant. Bougainville set off with two, the frigate *La Boudeuse*, and a storeship, *L'Etoile*. One of his first tasks was to hand over the Falkland Islands to the Spanish in recognition of their claim, and to forestall any British attempt to seize them. Having visited the Falkland Islands and conducted this ceremony, Bougainville, in *La Boudeuse*, retraced his tracks to Rio de Janeiro, where he was joined by *L'Etoile*, with Commerson aboard. Problems with the Portuguese authorities delayed the ships and it was not till November 1767, nearly a year since

Philibert Commerson

they had left France, that Bougainville was able to set out on his voyage round the world. He travelled slowly down the South American coast, following in the path of Magellan. Commerson, an avid collector, probably collected specimens of penguins on this journey and possibly specimens were also collected in the Falklands.

Philibert Commerson was born in Châtillon on 18 November 1727, and at the age of 13 was sent to the College of Bourg-en-Bresse. He was a delicate, sickly child, but interested in natural history, particularly botany and ichthyology, from a very early age. Originally intended for a career in law, he managed, not without difficulty, to persuade his father to let him study medicine and natural science. He studied medicine at the University of Montpellier, entering there in November 1748. Botany became his chief interest. His enthusiasm was such that he used to raid gardens to obtain specimens, but when he tried this in the Botanic Gardens, he was summarily forbidden by the University's professor of botany ever to set foot there again. It seemed as though his entire career in botany was destroyed. He married and settled down as a country doctor, but his wife died after only two years. Eventually he reached Paris, where he must have gained a reputation sufficient to recommend him to the position of naturalist on Bougainville's voyage. He was described as cranky, but bubbling with enthusiasm, and intolerant of the lack of interest shown by naval men in the natural sciences. This was not calculated to make him popular. When the expedition was on its way back to France, Commerson, and his assistants, Jossigny and Jean Baré, remained in Mauritius. Pierre Poivre (1719–1786), the Intendant of the island and a conspicuous figure in French colonial history, was a talented naturalist and an exceptional administrator. Born at Lyons on 23 August 1719, he had been a missionary in Canton and later in Cochin China, and had sent many specimens from Annam, Manila and Madagascar back to France for Réaumur's museum. On the return journey, his ship was attacked by the British, and he lost an arm, but this in no way dampened his zeal. Poivre persuaded Commerson to undertake a biological survey of Mauritius, and the former's nephew, Sonnerat, a civilian administrator to the Navy, became Commerson's assistant.

In 1771–72, Poivre organised an expedition to the spice islands to collect and transplant plants in order to break the Dutch spice monopoly. New Guinea was announced as the destination: but it was never visited; the expedition terminated at the island of Gebe, near Waigiou, off New Guinea's north-west corner. The expedition was a great success; Poivre returned to France heaped with honours, but Commerson, worn out with the fatigue of travel, died at Mauritius in 1773. Sonnerat returned to France the following year, taking with him a rich cargo of skins obtained in the Moluccas, which were illustrated by Daubenton in the *Planches enluminées*. The new Intendant of Mauritius, Maillard du Merle, jealous of Commerson, suspended official recognition and sent the manuscripts and specimens back to France, though some consignments never arrived. Commerson's collections were used in part by others, including his executor, Buffon, but never properly assessed or published. Fate and posterity treated Commerson particularly harshly. During his stay on the coast of South America, he had made large collections of plants and fish, which he sent home. His plants never arrived. The fish were unearthed from the attic of Buffon's house many years after Commerson's death by Lacépède, and used by him in his *History of Fishes*. Lacépède's rare integrity ensured that in this department at least, Commerson received full credit for his work. From the letters written to his friends, it is clear that Commerson was grossly undervalued by the French Government and considered to have been overpaid. Most of his funds went towards the acquisition of specimens, for which he never received adequate credit. Governments, as opposed to monarchs, have always undervalued pure science and scientists. Commerson is just one in a long line of dedicated but scorned enthusiasts. The parallels between him and J. R. Forster are too uncomfortable to be ignored. And like Forster, he was lively in conversation, but caustic and utterly fearless of consequences, which inevitably made him many enemies.

A total of 1,500 drawings by Commerson are said to survive in the Paris Museum, of sufficient quality to have made him, had they been published at the time, one of the most celebrated naturalists of his day. Sonnerat had been working with Commerson on a publication, and apparently, after the latter's death, took his material and passed it off as his own. Commerson's documents in the Paris Museum include some coloured plates apparently painted by Sonnerat with critical comments written by Commerson on the back. It is probable that Sonnerat redrew Jossigny's penguin plates under Commerson's supervision, and, after the latter's death, utilised them himself. He could not claim that the birds came from the southern oceans as he had never been there and so he was obliged to imply that they had been seen on his voyage to New Guinea. Zoological knowledge was not then sufficiently advanced for him to have understood that this was zoogeographic nonsense. Of 22 birds listed by Sonnerat apparently from New Guinea, only two, including the penguin, are not readily identifiable as known species. Of 61 species listed from the Philippines, 32 actually occur there, 8 are from South Africa, 2 from South America, and several others from tropical Asia. It seems clear that the species not found in the Philippines were based on figures and descriptions taken from Commerson. But Sonnerat could not possibly give the true locations since the birds came from places he had never visited. He was unscrupulous and dishonest, but there is no evidence that he was inaccurate or careless.

Bougainville's voyage, although important in that he was the one 18th-century French voyager in the Pacific to come safely home, yielded little for science, owing to the bungling of the authorities. Others in the 18th century fared even less well. De Surville (March 1769–August 1773) achieved almost nothing; Marion du Fresne (October 1771–March 1773) discovered the Crozet Islands (named after his second in command), but was murdered in New Zealand. Kerguelen (May 1771–July 1772 and March 1773–September 1774) was too cowardly and mendacious even to reach the Pacific, and only succeeded in finding the island that now bears his name, though he thought it was part of the southern continent. He was followed by the tragic voyage of La Perouse (July 1785–February 1788), whose ships vanished somewhere in the Pacific after sailing out of Botany Bay in Australia. The expedition of D'Entrecasteaux was briefed to find them, but without success. He, too, failed to come home.

Another Frenchman who subsequently had rather bad press was François Levaillant (1753-1824), whose notoriety surrounds his travels in southern Africa. Before discussing him, however, it may be appropriate to look briefly at ornithological exploration in Africa up to his time. The earliest traveller to devote any attention to African birds was Claude Jannequin Sieur de Rochefort, who was born at Châlons and lived for a time in England. In 1637, he travelled as a soldier to West Africa, and on his return, wrote an account of his travels, *Voyage de Lybie au Royaume de Senegra, le long du Niger* (1643), which contains a chapter on birds. The next traveller was W. Bosman, who was employed by the Dutch East India Company in Guinea. He spent about 14 years in the Gold Coast, and on returning wrote an account (1704, with an English translation the following year) in which a number of birds are figured rather crudely. The most important of the early travellers was Michel Adanson (1726–1806). He was mainly a botanist, and is commemorated in the name of the Baobab tree *Adansonia*, but also in the name of the Blue Quail *Excalfactoria adansonii*. He embarked for Senegal in 1739, returning to France in 1754, was elected a member of the Academy of Sciences in 1759, but lost all his money during the French Revolution. His account, *Histoire Naturelle de Senegal* (Adanson 1757, English translation 1759), contains interesting observations of the birds he encountered. The first account of South African birds was by Peter Kolbe (1675–1726). He was born in Bavaria and was commissioned to make astronomical and natural history observations in the Cape. He arrived in 1705 and remained for seven years. On his return he published a volume of travels, *Caput Bonae Spei Hodiernum* (Kolbe 1719), in which the twelfth chapter is devoted to birds. The book was such a success that it was translated into English (1731), Dutch and French. He incorporated much hearsay and material derived from other writers, but there is also much of truth in his writing. He was formerly commemorated in a bird name, *Gyps kolbii*, a now discarded name for the Cape Vulture, *Gyps coprotheres*. Abbé de Lacaille (1713–1762) was a learned French Jesuit and astronomer, who was sent by the French Government to the Cape in 1760 to measure an arc of the meridian in the southern hemisphere. He also visited Mauritius and Réunion. His *Journal historique du voyage fait au Cap de Bonne-Espérance*, was published posthumously in 1763. In this he made few references to birds, but collected a considerable number, and Brisson based many of his descriptions of South African birds on his specimens. The first naturalist to visit northeastern Africa was the botanist Pehr Forskal (1736–1763), Swedish by birth, who accompanied a Danish scientific expedition to Turkey, Egypt, the Red Sea and Yemen, and made a few observations regarding birds met with on the way. James Bruce (1730–1794) a geographer and traveller, discovered the source of the Blue Nile and recorded his adventures in five quarto volumes, *Travels to discover the source of the Nile in 1768–73*, published in 1790. He devoted a few pages of his last volume to birds. The specimens were sent to the King of France and described by Buffon in his *Natural History*.

François Levaillant (1753–1824) was born in Paramaribo, the capital of what was then Dutch Guiana. His parents were keen naturalists, and used to take their son into the jungle to collect. Levaillant soon began to shoot and stuff birds, and commenced his own 'cabinet'. After the family's return to Europe in 1763, he became friendly with Jean-Baptiste Bécoeur (1718–1777) the owner of one of the largest collections of birds at that time, and who is said to have been the inventor of arsenical soap which was until recently used in the preservation of bird skins. In 1777, Levaillant was in Paris, where he spent three years studying birds in the great private collections of his day, particularly the large collection of Mauduyt de la Varenne. In 1780 he went to Holland where he met Jacob Temminck, treasurer of the Dutch East India Company, and an enthusiastic aviculturist. The latter was so impressed with the young man that he provided him with money and sent him to South Africa to collect. Levaillant left Holland in December 1780, and remained in South Africa for three years. He made two expeditions, one eastwards and one northwards from the Cape, returning to France in 1784 with a collection of 2,000 bird skins, a huge number by the standards of the day. However, he found that there was now no-one in the capital interested in purchasing his collection, so he was obliged to sell a good many of them to collectors in Holland, particularly Temminck, and set about writing the memoirs of his journeys. The resultant book (1790) was a tremendous success, and subsequent English and German translations made this flamboyant, charismatic ladykiller the toast of Europe. Encouraged, he began to think of the first of his series of huge

François Levaillant

illustrated works, the *Histoire naturelle des oiseaux d'Afrique*, (Levaillant 1796–1808) of which he published the first volume in 1796, simultaneously in three editions, folio, quarto and duodecimo, all sumptuously illustrated in colour. Although highly priced, this work had a wide sale. The second and third volumes appeared in 1799 and 1802 respectively.

Drunk with success, Levaillant now launched out on three works at the same time, to be illustrated by the popular artist, Jacques Barraband. These were *Histoire naturelle d'une partie d'oiseaux nouvaeux et rares de l'Amerique et des Indes* (Levaillant 1801–82), *Histoire naturelles des perroquets* (Levaillant 1801–85) and *Histoire naturelles des oiseaux des paradis* (Levaillant 1801–06). No-one before had attempted anything so ambitious, but at this stage of his career everything Levaillant touched seemed to turn to gold. He had the material at his command to produce these works, for he had been building up his collections of specimens again (he was extremely skilled in stuffing and mounting birds), and his travels through Europe meant that he had been able to study the rich collections, both private and public, such as those of Jacob Temminck, Gevers Arnz, Holthuyzen and the Prince of Orange. Previously he had studied the collections of Abbé Aubry and Mauduyt de la Varenne, from the former of which he had purchased a number of specimens, and the museum at Cape Town. He was also able to study extensively at the Muséum d'Histoire Naturelle in Paris, since the bird specialist there, L. Dufresne, was a particular friend. The Paris museum had grown much stronger, due to the efforts of Geoffroy Saint-Hilaire, and enriched by the French armies when they carried off the Prince of Orange's collection from Holland in 1785. By 1808, however, Levaillant's reputation started to decline, and after the sixth volume of *Histoire naturelle des oiseaux d'Afrique* interest began to fall away. In 1818, he was to be found living in an attic, and soon he left Paris and retired to a small country house, where he died in poverty on 22 November 1824, at the age of 71. Forgotten by the public, he was still revered by ornithologists. The doubts came later.

Levaillant's virtue for many ornithologists of the time was his championing of the Buffon school of writing, and his opposition to Linnaeus and the binomial nomenclature. His contempt for 'sterile' attempts to classify birds by squeezing them into the prewritten straightjackets of the Linnaean genera, echoed exactly those of his mentor, Buffon. He complained, for example, of ornithologists classifying honeyguides with cuckoos solely on the basis of similarity of foot structure, whereas in all other respects, the two birds had nothing whatever in common. One of those to be the butt of Levaillant's scorn was the Abbé Bonnaterre (1747–1804) and his *Tableau encyclopédique et méthodique: Ornithologie* (Bonnaterre 1790–92). Bonnaterre had been one of the first French ornithologists to abandon the Buffon school, and embrace the Linnaean. Levaillant had justification for his scorn. He had studied his birds in the field, which his opponents had not, and knew their habits. "All our usual classifications", he wrote, "will remain inadequate so long as we do not possess a more complete knowledge ... of the habits of each individual species. Exact observation of an animal's behaviour, which is without doubt the best way of discovering the true place that each species occupies in the natural order, has been completely neglected." In many ways, Levaillant was a man before his time.

Then the recriminations began. The narrative of Levaillant's second South African journey, north of the Orange River and into Great Namaqualand, contains a great many errors, particularly of dates, because it was edited by another hand, no doubt from imperfect notes and journals. For many years this safari was regarded as fictional, but in 1957, Claude Grant published a paper in *The Ostrich* in which he demonstrated convincingly that both Levaillant's itinerary and geography stand up to scrutiny, even if the chronology does not. In 1963, a set of hitherto unknown paintings made by and for Levaillant were found in the South African Library of Parliament. These proved that he did indeed reach the Orange River, a point which had long been doubted. Research has not yet proved his claim to have travelled far north of it. Then there was the matter of erroneous descriptions of his African birds. Johann Wagler in 1832 made a detailed study of the *Histoire naturelles des perroquets*, and complained that Levaillant had taken plates of parrots from Buffon and George Edwards without acknowledgement, but with a few alterations to conceal the fact. In 1857, C. J. Sundevall, in a criticism of *Histoire naturelle des oiseaux d'Afrique* claimed that Levaillant had created new species which were not part of the South African fauna, or that he had fabricated new species by gluing together feathers from several different species. The reason for the deception was partly to give credence to the 'fictional' journey across the Orange River. Although some have considered the possibility that Levaillant may have been deceived by others, no-one seems to have considered that at least some of the species may be genuine but now extinct.

It is, however, useful to examine Sundevall's synopsis of Levaillant's species, which he assessed as follows:

Birds accurately described and recognisably from South Africa 134
Birds probably from South Africa, but with faulty descriptions 9
Birds not identifiable from the descriptions 10
Birds stated to have come from South Africa, but which almost certainly came from elsewhere 50
Birds said to have been obtained outside South Africa 71
Birds considered to be artefacts 10

Sundevall was one of those who considered that Levaillant may have been deceived, pointing out that on his return to Europe the latter may have been shown specimens of birds said to have come from South Africa, and felt obliged to say he had seen them there. Sundevall also claimed that Levaillant had obtained many specimens from a Paris taxidermist who was skilled in the production of composite birds, made of heads, bodies, wings and tails of different species. For these birds, therefore, Levaillant would have invented habits which he claimed to have observed in the field. So Levaillant remains an enigma. On the one hand much that he wrote may be false, but there is much more that is genuine and of abiding worth. He was an excellent observer of birds and their habits, and exceptionally gifted in expressing his observations in a way that was both vivid and elegant. It was he who first noticed the Peach-faced Lovebird *Agapornis roseicollis* breeding in the communal nests of the Sociable Weaver *Ploceus socius* though this fact was long unrecognised. But he was the last of his race, and, much as he despised binomial nomenclature, was forced to watch all his discoveries named binomially by others, who reaped the credit as the first describers.

Chapter 6
The Germans and The Dutch

During much of the 18th century, Germany had lagged behind England and France in the building up of large reference collections of specimens for taxonomic purposes. The first attempt proved to be tragic. About 1780, Duke Charles II of Pfalz-Zweibrücken set up a cabinet of specimens at his great castle at Karlsburg, by purchasing two large collections which had been used by Levaillant: those of Jean-Baptiste Bécoeur (1718–1777) and Mauduyt de la Varenne (1730–1792). But on 28 July 1793, republican revolutionaries attacked and burnt down the castle with all its contents. It was not till 1810 that the Zoological Museum of the University of Berlin was founded, and its chequered early career was bound up with the name of the most eminent German ornithologist of the turn of that century, the tragically short-lived Carl Illiger (1775–1813). His father was a merchant in Brunswick, and at college Illiger showed such promise that he was actively encouraged by his tutor, the famous entomologist J. C. L. Hellwig. Through the latter, Illiger obtained an introduction to Count Johann Centurius von Hoffmannsegg (1776–1849), who, in 1801, had returned from an expedition to Portugal with large botanical and entomological collections. So impressed was Hoffmannsegg with the young man that he had his own zoological collections sent from Dresden to Brunswick so that Illiger could work on them. From 1802 to 1807, Illiger edited an entomological journal which made his name well known to specialists. However, his health was never robust, and he suffered from weak lungs which spasmodically interrupted his work. When in 1809 the founding of Berlin University was discussed, Count Hoffmannsegg recommended that a collection for the study of natural history should be established, with Illiger as curator. All Hoffmannsegg's zoological collections were transferred to Berlin and for the next few years Illiger devoted himself to their study. The collections were particularly rich in Brazilian material – the first bird collections that had been sent back from that country since those of Marcgraf nearly 150 years before. Further collections were received in 1812, including much material from Brazil, North America and Australia. But Illiger's weak body had not the determination of his soul, and in May 1813 he died. Towards the end of his life, during which he never ceased his feverish activity, he became weary of the sheer volume of entomology and turned himself entirely to the study of higher vertebrates. The main work to concern us was his *Prodromus systematis mammalium et avium* (Illiger 1811), a vast overhaul of the Linnean system. Illiger divided birds into 7 orders, 41 families and 147 genera. The concept of the 'Family' had been first used by Daudin in 1800 but its general adoption by ornithologists was solely due to the influence of Illiger's *Prodromus*. This significant little book rejected the dictatorship of Linnean classification and cleared the way for unrestricted studies of relationships (see Appendix 16). Although the actual order is very different to present-day arrangements, most of the major groupings are recognised. And though the differences between the Oscines and sub-Oscines are not appreciated, and rollers were still placed with the crows, and swifts with the swallows, the arrangement was a great advance.

Two years after Illiger's death Germany began to develop an interest in museum collections. By 1825 there were five more: at Frankfurt under Cretzschmar; at Darmstadt under J. J. Kaup; at Munich under Spix and Wagler; at Dresden under Reichenbach; and at Halle under Nitzsch. There was the

Philipp Jacob Cretzschmar

Edward Rüppell

important private museum of Prince Maximilian zu Wied-Neuwied, who had conducted an expedition to Brazil. More expeditions were planned. Hemprich and Ehrenberg went to the Red Sea, Rüppell to Kordofan, and Kittlitz to the Pacific. Philipp Jacob Cretzschmar was born on 11 June 1786 in Sulzbach in Germany, and at an early age showed an interest in natural history, tending live animals and acquiring a private museum. He studied medicine at Würzburg and then Halle. Because of the Napoleonic wars he was forced to return to Würzburg where he graduated, and was recruited into the French army as a doctor, serving in military hospitals in Germany, Vienna, Paris and in Spain. Later, he settled in Frankfurt where, on 22 November 1817 he was one of the founders of the Senckenberg Natural History Society, with which he was active for nearly 30 years, building up its collections. He died on 4 May 1845, at the age of 59. *Emberiza caesia*, which he described in 1826 is generally known as Cretzschmar's Bunting. Edward Rüppell was born in Frankfurt on 20 November 1794, and is best remembered for his travels in northeast Africa, which resulted in the *Atlas of Rüppell's Travels in Northern Africa*, (Rüppell 1826–30). The ornithological section, published in 1826, was compiled by Cretzschmar while Rüppell was still in Africa, and was the former's major contribution to ornithology. However, it resulted in disharmony between the two men. Rüppell was originally intended to be a merchant, but in 1817 he made a trip to Sinai, which aroused in him such an interest in natural history that he decided on scientific travel. As a result, he attended lectures on zoology and botany at the Universities of Pavia and Genoa. His first great expedition took place in the years 1821 to 1827 through Sinai and much of Egypt. After his return and on reading a copy of the *Atlas*, Rüppell wrote crossly to Cretzschmar, criticising the descriptions and some of the plates, and blaming him for the choice of his assistant, who had proved less than satisfactory. In old age, Rüppell became an irritable recluse. He died on 10 December 1884 at the age of 90, having donated all his collections to the Senckenberg Natural History Society, of which, like Cretzschmar, he was a founder member.

Friedrich Wilhelm Hemprich was born on 24 June 1796 at Glatz (now Klodzko in Poland) the son of a surgeon. He studied at Berlin, where he met Christian Gottfried Ehrenberg (1795–1876). The latter had been born near Leipzig, where he studied medicine. The two men formed a close friendship, and Ehrenberg accompanied Hemprich on his expedition to Egypt, Sinai and Lebanon. The expedition set off in 1820. Hardship after hardship afflicted the pair, and on a number of occasions they were struck down by illness. Hemprich died on 30 June 1825, less than a week after his 29th birthday. Ehrenberg buried his friend and set off homeward with the huge collections of zoological, botanical and minerological specimens which they had accumulated and which were deposited in Berlin Museum. Ehrenberg, who later became professor of medical history in the city, published an account of the expedition, *Symbolae Physicae*, with both their names as authors. From a scientific point of view, the expedition had been a great success, the whole collection filled 114 chests, each with a capacity of 20–30 cubic feet. Many of the birds were described by Lichtenstein, the then director of the Berlin Museum, but he disposed of some of the skins before they could be properly worked on, and, in addition, removed some of the original labels and notes, replacing them with his own, and thus confused the origin of many of the specimens.

Baron Friedrich Heinrich von Kittlitz

Baron Friedrich Heinrich von Kittlitz was born on 16 February 1799 in Breslau, the son of a colonel in the Prussian army. He began by following his father into the army, but he is remembered for his round-the-world voyage, and the many ornithological discoveries he made on the way. This voyage began in 1826, as a result of meeting Rüppell, who became a firm friend and stirred his interest in natural history. Kittlitz returned to Germany in 1829, and began to write up his voyage. But he made slow progress owing to poor health suffered since visiting the Pacific. In 1831 he joined Rüppell in Egypt, but his health was so poor that he had to return home within a few months. He continued to contribute articles to journals, and gained a reputation as a writer, but his health did not permit him to go on any further expeditions. However, he lived to be 75, dying on 10 April 1874.

Coenraad Jacob Temminck

The emphasis in world ornithology now shifted to the Netherlands. Because of Coenraad Jacob Temminck (1770–1858) Leyden Museum was able to take its place as one of the important world bird collections. Temminck's father, Jacob, in his capacity as treasurer of the Dutch East India Company, had been a close friend of Levaillant who stayed frequently at the house. There was a fine menagerie where Temminck senior kept exotic finches and many other kinds of birds in large flight aviaries. It was Levaillant's books which fired the interest of the younger Temminck, although he had no initial scientific training and at the age of 17 followed his father into the East India Company where he became an auctioneer. His spare time was spent in an occupation which would now be called 'birding' and one of his companions on such excursions was Dr J. P. A. Leisler, who in 1812, named a new species of stint, *Calidris temminckii*, in Temminck's honour. In 1800, Temminck's employment came to an end when the East India Company was dissolved. In 1807, he took over his father's collection of birds, and published a catalogue of nearly 1,100 species. This hobby was to occupy him for the rest of his life. There were many fine and rich private Dutch collections and live menageries at this time. Levaillant had already described the zoological gardens of Arnoldus Ameshoff as containing such birds as Mandarin and Wood Ducks, pelicans, Lappet-faced Vulture, several species of curassow, Demoiselle, Whooping and Sarus Cranes, flamingos, Scarlet Ibis, Crowned Pigeons, Secretary Birds, bustards, Agami Herons, several species of peacock pheasant, jacanas, Purple Gallinules, Hoazins, and many more. Since 1804, Temminck had been collecting information for a planned monograph on gallinaceous birds of the world. This work was set aside for a time, when he came into contact with the gifted artist Pauline de Courcelles (1781–1851), later Madame Knip, a pupil of Barraband. She was working on a book on pigeons, and, wishing to have a specialist to write the text, settled on Temminck. It proved to be a sumptuous work, but Temminck was swindled, because the last instalments were issued with his name omitted, and hers as sole author on the title page (Temminck and Knip 1808–11). So the two parted in anger, but Temminck took his full text on the pigeons and added to it his text on the gallinaceous birds to produce a three volume octavo work *Histoire générale des pigeons and des gallinacées* (Temminck 1813–15), which firmly established his ornithological reputation. Immediately afterwards his *Manuel d'ornithologie* (Temminck 1815) appeared, the first comprehensive book on European birds. It was such a success that Temminck began work on a second, greatly enlarged, edition. The first two volumes of this appeared in 1820, but the last two had to wait until 1835 and 1840, owing to Temminck's preoccupation with other matters.

For a long time Temminck had wanted to make his collection the most complete in the world and in 1819 his chance came when the collection of his greatest rival, William Bullock, came up for sale. Of the early life of this extraordinary Englishman, nothing is known. The first indication of his existence was in 1799, when he published the first of a number of catalogues of the contents of his private museum, then housed in Sheffield. In this, he described himself as a silversmith and jeweller, almost the only existing clue to his former occupation. In 1801, he moved the museum to Liverpool, and in 1809, to London, where it was housed in Piccadilly, at first in temporary lodgings, but later (from 1812) at the Egyptian Hall, built specially to house it. Bullock became a fellow of the Linnean Society in 1810 and remained on the membership list until 1827. He seems to have been an astute businessman with a collecting mania. In May 1807, he visited the Bass Rock where he collected two shags, and in 1812 made an expedition to Orkney to try to catch a Great Auk which had been frequenting the island of Papa Westray for some years. He was unsuccessful, but the following year the bird was killed and sent to him. It is now in the Natural History Museum at Tring. It has never been clear why in 1819 Bullock decided to dispose of his entire collection. It was certainly not a case of failed business, as was the case with the Leverian Museum. However, he retained the Egyptian Hall, and later held other exhibitions there, including, we are told, a family of Laplanders tending a herd of live reindeer! In 1822–23, accompanied by his son, he visited Mexico, where, he was the first ornithologist since Hernandez to visit the country. This visit resulted in an exhibition of specimens and artefacts in 1824, entitled "Ancient and Modern Mexico", a collection which was dispersed by auction a year later. Bullock visited Mexico again in 1826, and North America in 1827. He remained for some years in the Americas, living, according to one author, "for many years far up the Orinoco". He was said to have been seen again in London in 1840, and known to have died in 1849.

By the time of its dispersal, Bullock's museum is said to have contained some 32,000 items, and to have contained the "entire" collections made by Banks and Cook on their voyage. This statement is quite erroneous, though the collection undoubtedly contained a considerable number of Banksian specimens. Bullock's sale attracted buyers from all over Europe. Lord Stanley was there, William Leach from the British Museum, Fector from Vienna, Lichtenstein from Berlin, Baron Mieffren Laugier de Chartreuse from Paris. The sale lasted 26 days, and consisted not only of natural history specimens, but also of objets d'art, guns, jewellery, costumes, and "The Military Carriage of the Ex-Emperor of France". In spite of fierce competition, Temminck was able to obtain 536 specimens of 363 species for which he paid the high price of £445. Most of his specimens survive, which is more than can be said of those bought by Leach for the British Museum. William Elford Leach (1790–1836) was Assistant Keeper of Zoology at the British Museum, but was forced to resign early owing to ill health. Leach's Petrel *Oceanodroma leucorhoa* is named after him, but his contribution to ornithology is small, and he earned a certain notoriety for the periodic bonfires he held in the museum grounds to dispose of 'unwanted' or badly preserved specimens. Many unique and historic specimens probably perished in this way.

Temminck now had the materials at hand to commence his systematic work. In collaboration with his young assistant, Heinrich Kuhl, he launched into a more ambitious project, the *Planches coloriées* (Temminck and Laugier 1820–39), a continuation of Daubenton's *Planches enluminées*. This was the idea of, and was largely financed by, the wealthy Baron Laugier de Chartreuse. The first issue appeared on 1 September 1820 and further parts quickly followed. The artistic merit of the work is, however, indifferent, and the accompanying text not always accurate. But Temminck's success was not to go unchallenged. He had a bitter rival in Louis Jean Pierre Vieillot (1748–1831), who was born at Yvetôt, France, on 10 May 1748, held a small clerical post in his youth and devoted himself wholeheartedly to

Louis Jean Pierre Vieillot

the study of ornithology. Being dissatisfied with his life in France, he emigrated to Santo Domingo in the West Indies where he remained, engaged in business pursuits. However, on the outbreak of the French Revolution he was one of those proscribed for military duty, so he moved to the United States, where he remained for some years. He was a most discerning ornithologist, and one of the most underrated in view of his published output. His particular skill lay in his ability to recognise generic similarities, but his work was always overshadowed by that of Buffon, and later by Cuvier. He left America for France with his family in August 1798, but his wife and three daughters died of yellow fever on the voyage. Eventually arriving back in France, almost destitute, he remained poor for the rest of his life. Alone in Paris, he consoled himself with writing, but many of his publications were costly, and he received little or nothing from them. He lived out an eccentric old age, often depriving himself of the necessities of life, in order to indulge his special interest. He died in poverty in Rouen and as he left no papers, biographical information is scanty. One of his first productions was in collaboration with Jean Baptiste Audebert (1759–1800), the *Oiseaux Dorés* (Vieillot and Audebert 1802), a forerunner of the gorgeous and sought-after monographs of the 19th century, but when Audebert suddenly died in 1800, Vieillot continued the work alone. His ideas on classification were utilised in the voluminous articles (almost all those on birds) he contributed to the multi-volumed *Nouveau Dictionaire d'Histoire Naturelle*. In his *Analyse d'une nouvelle Ornithologie élémentaire* (Vieillot 1816) he provided a serious competitor to Temminck's work. But his many new names aroused the resentment of his contemporaries. Scorned by ornithologists during his lifetime, since his death he has been more highly valued by systematists. No fewer than 26 generic names established by him are still in use. Vieillot's *Analyse* of 1816 (see Appendix 17) exhibited a system which differed much from that of Illiger, being very conservative in the use of orders. By contrast, Temminck's system (1820, see Appendix 18) recognised more orders, but is much less satisfactory.

In 1820, Temminck was appointed director of the Imperial Museum in Leyden. With the existing material there were incorporated Temminck's own rich private collection and that of the Royal Museum of Natural History in Amsterdam. Temminck proved to be an excellent administrator, organising expeditions, arousing enthusiasm in others, corresponding with people all over the world. From this point on, it was this quality, rather than his scientific performance, which distinguished him. He lacked formal education, knew no Latin or Greek, and as a zoologist, he never progressed beyond the stage of collecting and describing. But his excellent memory enabled him, for many years, to hold his reputation as one of the most knowledgeable authorities on birds. Under Temminck's influence, exploration began for the first time of the Dutch East Indies. Carl Reinwardt was sent to Java in 1815, where he founded the Bogor Botanic Gardens, and in 1821 undertook an expedition to the Celebes (Sulawesi), Amboina and the Lesser Sundas. He sent back rich collections of spectacular new species, which were described and published by Temminck, although at least four of the consignments failed to reach home. Furthermore, Temminck did not make proper use of Reinwardt's notes. Although based in Java, Reinwardt had travelled through Indonesia, visiting Sumbawa, Timor, Banda, Amboina, Ternate and northern Sulawesi, returning to Holland in 1822, where he took over the chair of natural history at Leyden.

Before Reinwardt's return to Holland, Temminck's young protegé Heinrich Kuhl had sailed in 1820 to Java. Born on 17 September 1797 in Hanau, at the age of 20 Kuhl had published a monograph on bats. This was the culmination of the work of his mentor and Temminck's friend, Leisler, who had died on 8 December 1813. Leisler had devoted his later years to the study of bats, but relied on memory and never wrote anything down. All his results would have been lost had not Kuhl taken the information in hand. Two years later, Kuhl's greatest contribution to ornithology, the *Conspectus psittacorum*, appeared. He had travelled to London with Temminck in 1819, and the following year spent some time studying in Paris. Before leaving for Java, he hurriedly finished off and published several other papers including a revision of shearwaters and petrels, and a treatise in which he listed and identified all the birds depicted by Daubenton in the *Planches Enluminées*. From December 1820 to August 1821 Kuhl and his assistants explored the mountains of Java, sending back enthusiastic letters predicting that soon the birds of western Java would be as well known as those of Europe. On 14 September 1821, he died of a liver infection brought on by the climate and overexertion. He was only 24 years old. Kuhl's death was an indescribable loss to science. His teacher, Professor van Swinderen, had compared him to Pallas and said that had he lived he would have been the greater of the two. Certainly he achieved a staggering amount in his tragically short life. The material sent back to Leyden by Kuhl and his associates from Java consisted of 200 skeletons, 200 mammal skins of 65 species, 2,000 bird skins, 1,400 fish, 300 reptiles and amphibians, many insects and molluscs, manuscripts and 1,200 drawings. Unfortunately, Temminck did little with the material except lock it away, and publish a few descriptions of the birds. Heinrich Boie wrote up the reptiles, while Cuvier and Valenciennes attended to the fish. The manuscripts and drawings vanished; fortunately they had been copied, but the copies remained mislaid and unused for many years, until Finsch commented on them in 1906. In 1835, Friedrick Boie named a bird in Kuhl's honour, *Puffinus kuhli*, based on a specimen from Corsica. But unfortunately the name proved to be pre-dated. It is the bird we now know as Cory's Shearwater, *Calonectris diomedea*. But Kuhl's name still lives. In 1824, Vigors described a new lory from the Austral Islands, which he named *Psittacula* (now *Vini*) *Kuhlii* with these words: "I have named it after the late M. Kuhl, who has left us a valuable disquisition upon the present family, and whose early loss cannot be sufficiently lamented by the lovers of ornithology." It is still known as Kuhl's Lory.

Temminck by now showed signs of senility, although he had many more years to live. It was not until 1823, over two years after Kuhl's death, that he agreed to send successors to Java to continue the latter's work, and it was a further two years before they were able to depart. They were Heinrich Boie (Friedrick's more famous brother), Macklot and Salomon Müller. Heinrich Boie was born on 4 May 1784, and studied law, much against his will, first at Kiel, and then at Göttingen. When at university, he found the natural history lectures of Blumenbach rather more interesting than those on law. In 1814, he met Kuhl in Offenbach where the latter was examining a bird collection, and moved to Heidelberg the following year for further legal study. Here he attended natural history lectures by the anatomist Tiedemann and could no longer resist the urge to become a zoologist. Tiedemann did all in his power to help him, and obtained a curatorial post for him at a newly founded zoological collection. After Temminck had published his *Manuel d'Ornithologie* in 1815, he received from Boie a friendly criticism as thorough as it was outspoken. This resulted in correspondence between the two, and in Boie's appointment as Temminck's assistant. Heinrich Christian Macklot was born in Frankfurt am Main in 1799 and studied medicine and natural history at Heidelberg. There he met and became friendly with Boie, and was appointed by Temminck in 1822 to the charge of the osteological collections at Leyden. Salomon Müller was the son of a saddler in Heidelberg, and met the other two by accident when they stopped at an inn in the town to discuss the results of a collecting trip. They noticed that the young waiter was listening to their conversation. They discovered that in his spare time Müller was a dab hand at shooting and stuffing birds. Boie, Macklot and Müller, together with a draughtsman, Pieter van Oort, left Holland in December 1825, taking six months to reach Java. On the way, they were becalmed at Cape Town for 15 days, and took the opportunity of going on shore to collect, and to visit Andrew (later Sir Andrew) Smith (1797–1872), then director of the museum there. Administrative problems with the governor awaited them in Java, but an expedition to Sumatra was accomplished.

On 4 September 1827 after only a year on Java, Boie died of gall fever. He left behind a voluminous quantity of papers, including a draft for a projected complete ornithology of western Java, several monographs, and descriptions of new genera. Macklot took over the material, intending to prepare it for publication. But it was not to be. Five years later he was caught up in a revolt by Chinese workmen who burned the houses of all the Europeans. Macklot's house was completely destroyed, and in the inferno perished not only all his instruments and books, but all the manuscripts of both himself and Boie. A few days later, Macklot himself died in hand-to-hand fighting. Sometime later, van Oort succumbed to malaria, and Salomon Müller requested to be repatriated. Selfishly, Temminck refused even to consider this, in spite of the fact that the six years stipulated in Müller's contract were long over. Reinforcements, however, were obviously necessary. Ludwig Horner was sent out in 1835, and as a result of Boie's death Pierre Médard Diard was now head of the Commission. Diard was an incomparably efficient collector, but lacked the ambition to publish his findings. He was, therefore, just the sort of man Temminck liked. Some years later, his name was commemorated in the name of the Siamese Fireback Pheasant, *Lophura diardi*. Horner, physician and geologist, was born in Zürich in 1811. Soon after arrival, he went to Sumatra where he worked for two years before, on 7 December 1838, he too fell victim to the unhealthy climate. Salomon Müller was at last permitted to return to Holland. This left Diard the only scientist remaining in Java. He had been in the Dutch Indies for over 20 years, having travelled to the east in 1817. There he met Sir Stamford Raffles, then governor of Benkulen on the west coast of Sumatra.

After Horner's death in 1838, the Dutch Commission nearly came to an end, but in the meantime, Temminck had found another recruit, Eltio Alegondus Forsten, who was sent out in 1838, but although he visited a number of islands, he, too, died prematurely in 1843. He was succeeded by C. A. L. M. Schwanner who survived from 1841 to 1851, before he, in his turn, was carried off by fever. At that point, by Royal Decree, the Natural History Commission was disbanded. Temminck had failed to understand the excellent work done by all these men in surveying the natural history of the area, for he was really only interested in spectacular specimens for his museum. Had he been a man of greater intellect and vision, he might have been able to interpret the zoological riches sent home, but he was in truth little more than an experienced amateur. Whereas he had once been praised for his conservatism when he attacked Vieillot for introducing new family and generic names, the world now favoured Vieillot's approach. In 1836 William Swainson attacked the *Planches coloriées*: "The figures are stiff and formal, and they are all put into nearly the same attitudes. The descriptions of the birds are meagre, and for the most part related to the mere colour of the plumage." In his old age, Temminck became insanely jealous of his assistants, keeping all the cupboards locked and refusing to allow anyone access to his collections.

He wanted no-one to work on them but himself, but at the same time lacked the energy and drive to do more than dabble. In disgust, the assistants (those who remained alive) all left; the only survivor was Hermann Schlegel, who became Temminck's successor.

During this period, the French had turned their attention to New Guinea. In 1817–19, the ships *L'Uranie* and *La Physicienne* had visited the New World under the command of Claude Desaulses de Freycinet (1779–1842) and included scientists Louis Isidore Duperrey (1779–1856), Jean Rénée Constant Quoy (1790–1869) and Joseph Paul Gaimard (1790–1858). *L'Uranie* was wrecked on the Falklands and the naturalists stranded there for three months. But they survived to produce one of the volumes in the report of Freycinet's voyage (Appendix B). Duperrey set off again in 1822–25 as commander of the ship *La Coquille*, calling at the Falklands where the crew found traces of the earlier camp. In his account in the zoology volumes of Duperrey's *Voyage Autour du Monde*, Dr Prosper Garnot (1797–1836) mentioned two dozen birds of the islands, including two penguins, gulls, terns, frigatebirds, cormorants, two kinds of geese and four of ducks. René Primavera Lesson (1794–1849), the assistant to Garnot on board the *Coquille* had been collecting birds and mammals at Doré in the west of New Guinea. Soon after the return of the *Coquille*, the first mate, Dumont d'Urville, set out again, this time on a round the world voyage with Quoy and Gaimard. As a result of these voyages, Temminck commissioned Macklot and Müller to explore New Guinea.

Stamford Raffles

Stamford Raffles was born at sea on 5 July 1781, the son of a ship's captain. As a young man he entered the service of the East India Company. As an addition to his political career he was an enthusiastic zoologist, botanist and ethnologist. His first collection, together with all his notes, grammars and vocabularies of local languages, a large map of Sumatra, 2,000 drawings of zoological and botanical subjects, the work of over 20 years, perished when the ship on which he was returning home to Europe caught fire; he barely escaped with his life. Nothing daunted by this experience, on his return to England he became involved in the founding of the Zoological Society of London, but did not live to see the fruits of his labours, for he died of a stroke at the age of only 45, on 6 July 1826. It was left to N. A. Vigors to oversee the progress of the Society. The latter engaged the young John Gould (1804–1881), a gardener who was skilled at taxidermy, as keeper and superintendent of the Society's collection. The origins of the Zoological Society of London are uncertain, but it is likely that Raffles took an active part in its establishment. In 1817 he had been introduced to the Prince Regent and his court, and had expressed an interest in founding such a Society. London was, at that time, one of the few capitals in Europe where no collection of live animals existed. On 26 February 1826, a committee chaired by Raffles was set up, which included Vigors (secretary), Sir Humphrey Davy, Dr Horsfield (assistant secretary), and Joseph Sabine (1770–1837, treasurer). Raffles died between the second and third meetings. After some discussion, the plot of land in Regent's Park on the south side of the Outer Circle was purchased. The architect Decimus Burton was appointed to design the gardens, but few of his designs were used. In 1827, a request went out:

> Living specimens of all rare animals, and particularly of such as may possibly be domesticated and become useful here, will be much valued by us; and above all, varieties of the deer kind, and of gallinaceous birds; but beyond this, preserved insects, reptiles, birds, mammalia, fishes, eggs and shells will be gratefully received (Chalmers Mitchell 1929).

The gardens first opened to the public on 27 April 1828.

When Vigors's father died in 1832, the son resigned from the secretaryship to become MP for Carlow. After several secretaries, it was decided to make the post a full-time paid office, and in 1847, David William Mitchell was appointed to the post. Mitchell was an Oxford graduate, a fellow of the Linnean Society, and a zoological artist. At the time of his appointment, he was engaged in creating the plates for G. R. Gray's *The Genera of Birds*. He remained secretary until 1859, and died the same year. His successor was P. L. Sclater, who served from 1859 till 1902. In 1903, David Seth-Smith, author of a standard monograph on parakeets, was appointed Curator of Birds.

Chapter 7

The Beginnings of American Ornithology

William Bartram

Since the early days of colonisation, birds had been commented upon in travelogues of settlers and visitors to North America. However, most of these accounts were cursory, erroneous or of little value from an ornithological point of view. One of the first really useful accounts was that of the Swedish botanist and economist, Peter Kalm (1716–1779), who had come to study and collect plants suitable for cultivation in his native country. He had been a student of Linnaeus at Uppsala, graduated from Åbo (now Turku), in 1735, and had travelled in Sweden and Russia. On the recommendation of Linnaeus, he was sent to the English colonies in America, where he remained from 1747 to 1751, visiting parts of Ontario, New York, Pennsylvania and New Jersey and writing an account of his visit (Kalm 1753–61). Although scrupulously accurate in some ways, he was unscientific and credulous in others: like many naturalists of his day, he believed in the hibernation of swallows, but was observant enough to point out errors in Catesby's plate of the Bluebird. Kalm's ornithological observations are in no particular order or system, but are scattered throughout the book, in among a mass of other miscellaneous information. However, a number of his accounts seem to have formed the basis for Linnean names. The next writer of importance was William Bartram (1739–1823). He was the son of the farmer and botanist John Bartram (1699–1777). Already, by the age of 14, he loved to draw and observe birds and his father decided to apprentice him as engraver to a printer, while learning the trade of merchant. These efforts coming to nothing, however, his father arranged to take him on a major expedition through the Carolinas, Georgia and Florida. This resulted in the major work for which William Bartram is remembered: *Travels through North and South Carolina* (Bartram 1791). His special contribution to ornithology lies in his Catalogue of the Birds of North America in which he lists 215 species, together with observations on migration, song, nesting and general habits. In this work many of Catesby's claims are challenged, and modern research has shown Bartram to be right. He was a pioneer in the study of migration, and several pages in his *Travels* are given to a description of this. Yet, incredibly, nearly a century later, some reputable American scientists still believed in the hibernation of swallows. It was not only for his writing that Bartram was important. The Bartram family home, with its botanical gardens, was a Mecca for all travellers interested in natural history. Peter Kalm was one of the visitors who stayed with the Bartrams in the course of his travels. Bartram also corresponded with such writers as George Edwards and Thomas Pennant, and sent them specimens to illustrate in their books. John Abbot (1751–?1840) was primarily an entomologist, and his work on birds, mainly coloured drawings, remains unpublished. He was born in London, the son of an attorney, and was stimulated by the books of George Edwards, Mark Catesby and Eleazar Albin. In 1773 he emigrated to America where he worked as a planter, and collected birds and insects, many of which were sold to collectors in England.

Charles Peale

It was with the setting up of the private museum of Charles Peale in Philadelphia, that American museum ornithology began. Charles Willson Peale (1741–1827), born at Chestertown, Queen Anne County, Maryland, was one of the fathers of American painting, an inventor, naturalist, scientist, portrait painter and friend of Thomas Jefferson. His dedication to art is shown by the fact that he named several of his sons after famous painters such as Titian, Rubens and Rembrandt. Among many things he was an inspired taxidermist who invented a technique for preserving specimens using arsenic and mercury. When he began to suffer the effects of arsenic poisoning he invented a partial antidote. He was so disappointed with, or jealous of, the talents of his eldest son Raphael, that, it has been alleged, he allowed him to die of arsenic poisoning without informing him of the antidote (Macintyre 1993). Peale's museum originally opened in 1784, consisting mainly of paintings by Peale himself; later artefacts and other curiosities were added, and the premises were changed several times as the collection grew. For a private collection of those times it was large and rich. Not only were the series of birds, mammals and reptiles comprehensive, but the collection contained two mounted skeletons of a mastodon, the sort of exhibit that would have been exceptional anywhere. In the early days, Peale attempted to open it on Sundays; not surprisingly this brought condemnation from the press, to combat which the resourceful Peale had a sign painted for display on Sundays, that read: "Here the wonderful works of the Divinity may be contemplated with pleasure and advantage. Let no-one enter today with any other view." In 1821, the museum's name was changed to the Philadelphia Museum under the management of a company consisting mainly of Peale's sons. Unfortunately debts resulted, and after vain attempts to boost the receipts by adding musical entertainments and other attractions, it finally closed in 1844. Even more unfortunately the dispersal of its contents was largely undocumented, and many important bird specimens vanished for ever. Some of the natural history material was purchased by the travelling showman P. T. Barnum and later destroyed by fire in New York.

Alexander Wilson

Alexander Wilson

Peale's museum provided the Father of American ornithology with his mission in life. Alexander Wilson (1766–1813) was born in Paisley, Scotland, but emigrated to America at the age of 20. Several birds have been named after him, including Wilson's Petrel *Oceanites oceanicus*, and Wilson's Phalarope *Steganopus tricolor*. Unfortunately, all his type specimens were housed at Peale's museum, and, except for two, were lost when the contents were dispersed. In their absence, one or two of Wilson's species cannot now be identified. Wilson had become interested in birds before leaving Scotland, but the flowering of his interest occurred after his arrival in America, where he began his career as a schoolteacher. By a fortunate coincidence, his residence was on one of the great flyways for migrating birds, and he began to draw and paint them. His dream of painting all the American species began to become a reality when, in 1806, the publisher Samuel Bradford offered him a job as assistant editor of an encyclopaedia, and Wilson showed him his proposed book, which Bradford agreed to publish. So Wilson was able to resign his teaching post and devote himself to these new projects. The first volume of *American Ornithology* (Wilson 1808–14) duly appeared, and in the autumn of 1808 Wilson set off on foot to 'peddle' copies of his book in an attempt to find subscribers. In Louisville he visited a storekeeper who gave painting lessons, and tried to sell him a copy. The man was Audubon who was initially sufficiently impressed by the book to agree to buy it; but when it was pointed out to him by his partner that his own paintings were far superior to Wilson's, he changed his mind. In Philadelphia, Wilson continued work on the later volumes of his book, dying soon after the completion of the eighth. The last volume and later editions were prepared by his friend and disciple, George Ord (1781–1866), who saw Audubon's work as a threat to Wilson's memory and did his best to discredit it.

George Ord

John James Audubon

John James Audubon (1785–1851) was born on Hispaniola, the illegitimate son of Captain Jean Audubon by a servant girl. The girl died soon after, and Audubon spent his childhood with his father and childless wife in France. In 1803, Audubon was sent by his father to oversee his plantation in Pennsylvania, but showed little aptitude and spent much time sketching birds. The chance meeting with Alexander Wilson launched Audubon on his life's work, which culminated in the production of the world's largest and most expensive book, the elephant folios of *The Birds of America* (Audubon 1827–38). Although Audubon has been much praised, he is important chiefly as a bird artist; his contribution to the development of ornithology is comparatively slight. The rivalry between Wilson and Audubon was unfortunate, and was probably caused by the jealousy of the latter, who accused Wilson of plagiarism in copying some of his pictures, when it was quite obvious that he, Audubon, was guilty of the same misdemeanour. Ord stoically defended Wilson from these slanders. Although unquestionably a more charismatic personality and a finer artist than Wilson, Audubon was far from being his equal as an ornithologist. Elliott Coues was later to write of Audubon "... he liked to exaggerate and embroider ... he had no genius for accuracy, no taste for dull dry detail ...". Coues, however, also pointed out that Wilson's original paintings were vastly superior to the published engravings, which are loud and garish when compared with the delicacy of the originals and give a poor idea of the quality of his work. Very much a Wilson supporter, Coues added that science would lose little, but gain much, if every scrap of pre-Wilsonian writing on American ornithology were destroyed (Burns 1908–09).

Audubon's Turkey

Charles Lucien Bonaparte

In the meantime, a new figure had appeared on the ornithological horizon, a whirlwind named Charles Lucien Bonaparte (1803–1857), Prince of Musignano and nephew of Napoleon Bonaparte. A small, black-eyed and talkative man, he had been fascinated by natural history from an early age, and had used Temminck's *Manuel d'Ornithologie* as a beginner's guide. There was one bird, however, which he had failed to identify, so he sent it to the author in Leyden and it was featured in the *Planches coloriées*. It was indeed new to science; it was the Moustached Warbler *Sylvia* (now *Acrocephalus*) *melanopogon*. In 1823, Bonaparte left Europe for America. He settled near Philadelphia and began to work on the local avifauna, consulting the specimens in Peale's museum, in effect continuing the work which Wilson and Ord had begun. Since Wilson's death, the collections at Philadelphia had been augmented by an expedition to the Rocky Mountains on which the zoologist was Thomas Say and his assistant Titian Peale, the 19-year-old son of Charles Peale. The result of this period of Bonaparte's life was *American Ornithology* (Bonaparte 1825–33). Nominally a continuation of Wilson's work, it was very different in scope, for Bonaparte was a systematist, not a field man as Wilson had been. Illustrated with superb plates by Titian Peale, and, reprinted through a number of editions, it became a popular handbook for much of the remainder of the century.

Titian Ramsay Peale was born in 1800, his mother was Charles Peale's second wife, and he was named after a half-brother who had died young. He is believed to have derived as much of his education from his father's museum as from formal schooling. At the age of only 17 he was elected a member of the Academy of Natural Sciences of Philadelphia, and in the autumn of that year visited the Sea Islands of Georgia and east Florida with Thomas Say, George Ord and William McClure. The Rocky Mountain expedition was two years later. Among the discoveries of this expedition, the Lark Sparrow *Chondestes grammacus* was found near the mouth of the Missouri. Later at Omaha, the first specimens of the Orange-crowned Warbler *Vermivora celata* and the Yellow-headed Blackbird *Xanthocephalus xanthocephalus* were collected. Later discoveries included the Dusky Grouse *Dendragapus obscurus*, Lazuli Bunting *Passerina amoena*, Cliff Swallow *Petrochelidon pyrrhonota*, House Finch *Carpodacus mexicanus* and Say's Phoebe *Sayornis saya*. The collections consisted of 60 skins of animals, several thousand insects, 500 plants, shells, minerals and 122 drawings by Peale, all of which were deposited in the museum.

In 1824, Bonaparte engaged Titian Peale to visit Florida to obtain novelties to be published in his *Ornithology*. Although many interesting specimens were obtained, the only 'novelty' was Peale's Egret, which is now identified as a colour phase of the Reddish Egret *Egretta rufescens*. Bonaparte, realising that the collections in Philadelphia were inadequate to his scheme for a grand world ornithology, now began a life of globe-trotting, visiting many countries of Europe, both for ornithological and political reasons. As a member of a famous family he could not but be involved in politics, but that side of his life is not relevant to this study. However, his political views had resulted in his expulsion from both his fatherland of France and the Austrian Empire and so after a visit to England and Scotland in 1849, he took refuge in Leyden. He was a restless and volatile worker, and his publications were many, but his main contribution to ornithology was the *Conspectus generum avium* (Bonaparte 1850, 1857), the first volume being published in 1850. It was hailed as the most significant work since Latham's *General Synopsis of Birds*, 80 years previously. Reconciled with his fatherland after Napoleon III's accession, Bonaparte now settled in Paris, where he entertained many foreign naturalists and penned vast numbers of small articles and papers. However, the second volume of his *Conspectus* beckoned, and before long he was off on his travels again, visiting collections and making notes for his work. In 1856, he addressed an ornithological conference in Köthen, in which he spoke on the identity of species, and hinted very strongly at an evolutionary process by which now-extinct forms gave rise to different, but obviously closely related extant species. This address was published in the *Journal für Ornithologie* (Bonaparte 1856). In 1857, the second volume of the *Conspectus* appeared, but Bonaparte died the same year and the work was never finished. It appears that he knew of his frail condition, and that time had never been on his side. This was the reason for his volatile, feverish activity – a determination to try to get as much as possible done. He is said to have been found, on at least one occasion, writing in his bath. Bonaparte's chief contribution to ornithology was his talent for systematic arrangement, and as a worker with knowledge of both Old and New World species, he was particularly well equipped in this way. The remarkable aspect of Bonaparte's learning is that it embraced not only ornithology, but the whole of zoology. His perspicaciously critical 'Observations on the State of Zoology in Europe' was translated by Strickland and published in 1845 in the Ray Society's series of 'Reports on the Progress of Zoology and Botany for 1841 and 1842' (Bonaparte 1845). His systematic arrangement (Bonaparte 1841) divided the class of birds into two subclasses and eight orders (Appendix 19).

Constantin S. Rafinesque was a colourful contemporary of Bonaparte's in the USA. He was born near Constantinople on 22 October 1783 of Franco-German parents, and was described as combining superabundant enthusiasm with egotism and vanity, which he made no attempt to disguise, and had remarkably advanced and radical views on bird classification. At about the same time as Wilson was commencing his studies on American birds, Rafinesque had formed the idea of a grand survey of the same subject, but, perhaps fortunately, this came to nothing. Rafinesque began to study birds in 1797, near Marseilles, and corresponded with the French ornithologist F. M. Daudin, to whom he sent accounts of rare birds. But he was an unwilling hunter; his first specimen (a titmouse) upset him so much that he could never become an enthusiastic collector, and eventually devoted much of his time to botany. However, his brother, who travelled with him to Philadelphia, where they arrived on 18 April 1802, was a sportsman and provided him with many specimens. Rafinesque's first few descriptions of birds, from specimens he found in Peale's museum, were communicated to Daudin for publication in the Bulletin of the Philomathic Society of Paris, in 1802. These were commendably full and painstaking, but later he lapsed into brevity and imperfection, which not only caused criticism, but insuperable difficulties to those who endeavoured to identify his species and give him the posthumous credit that was his due. In 1805, he moved to Sicily, where he remained for about ten years, married, was deserted by his wife and children, and published several works, including a Natural History of Sicily. On 2 November 1815, on returning to America, he was wrecked on the Race Rocks off Long Island, escaping only with his life; his entire luggage, including all his specimens and manuscripts being lost. Undaunted, he returned to Philadelphia in 1818. At this

Constantin S. Rafinesque

time he had published 269 books and papers, but only seven of these contained anything on birds. Many of his publications were privately printed pamphlets, of limited circulation, and are extremely rare today. Of the birds described by Rafinesque, a number are quite unidentifiable, and on at least a couple of occasions, he seems to have been the gullible victim of practical jokes by Audubon, who palmed off artificial specimens of birds on him, to be duly described by Rafinesque as new. But not all of the descriptions believed to be erroneous may in fact be so. In 1832, he described a new species, *Aquila dicronyx*, based on a captive eagle brought from Buenos Aires and held in captivity in a garden in Philadelphia. It has been dismissed as a Bald Eagle with aberrant coloured toes and a band of brown staining the ends of its tail feathers, but if it really was found in the claimed locality this seems unlikely, for the Bald Eagle is a purely North American species. Rafinesque died in a garret in Philadelphia either in November 1840 or in 1842 (accounts differ), alone and in poverty, a sad end to a remarkable life. Samuel N. Rhoads (1911) summed him up as follows:

> With the increase of his publications on all sorts of subjects, many of them relating to hobbies undreamed of by his most versatile fore-bears in biological science, the enemies and pecuniary troubles of Rafinesque increased. He had some friends in Philadelphia, who were willing to overlook his faults and stand by him in his worst extremities, but perhaps no man of his talents suffered more keenly in his closing days the ingratitude and neglect of the world which he had so actively endeavoured to benefit and enlighten by his researches. Anyone reading his autobiography and willing to overlook the many egoisms and exaggerations of it, will be impressed with the thought that here was a man striving

> after truth, a real lover of nature, sincerely endeavouring to impart his interpretation of the cosmos to his fellowmen. At the same time he was sadly handicapped by the necessity of making a living out of his discoveries, his peculiarities depriving him of that patronage and encouragement of wealthy friends which is so essential to the best success in the career of a scientific man addicted to poverty (Rhoads 1911).

Turning now to South America, we find that apart from Marcgraf's researches in Brazil, already discussed in a previous chapter, one of the first South American countries to receive attention was Chile, but some would regard Father Juan Ignacio Molina as something of a mixed blessing. He was a native of the province of Maule, and his *Saggio sulla Storia Naturale del Chile*, published in Bologna (Molina 1782), included 33 species of Chilean birds. The descriptions, however, are sometimes imprecise, and a number have never been satisfactorily identified with any known species. This is believed to have been due to the fact that Molina, having been forced to leave his country, wrote the book from memory or incomplete notes. Although several critics have claimed that the 'unidentified' birds of Molina were mythical, it is possible that at least some of them represent species that formerly occurred in Chile, but are now extinct. Molina was born at Guaraculen on 24 June 1740, and died at Bologna on 12 September 1829. He received his early education at Talca, and when 16 entered the Jesuit College at Concepción, where he studied languages and natural sciences. He entered the Jesuit order and became librarian of the college, but was forced to flee Chile in 1768 when the Jesuits were expelled. He settled in Bologna where he became professor of natural sciences, and where he wrote most of his works. Some of his lectures maintained the theories of evolution and were censured by his superiors, but he remains the classic author on the natural history of Chile. The next writers on Chilean ornithology were voyagers who visited the shores briefly, made a few collections and described a few new species. These included R. P. Lesson (1823), Baron von Kittlitz (1827), Eduard Poeppig, Captain Parker King (1830), Charles Darwin on the Voyage of the 'Beagle' between 1832 and 1836, and Titian Peale on the United States Exploring Expedition of 1838 to 1842.

Poeppig (1798–1868) was born in Germany and was in Chile in 1827 and in 1832. His two-volume account was published in 1835–36. The first important work was by the French botanist and explorer Claudio Gay (1800–1873), who spent 12 years in the country from 1830 to 1842 and founded the Museo Nacional at Santiago. His multi-volume work *Historia fisica y politica de Chile* (Gay 1844–71) is in 28 volumes, of which eight are devoted to zoology. The mammals and birds form volume 1 of these eight, but the birds were assigned to O. Des Murs, and constitute the first catalogue of the birds of Chile. Unfortunately, Des Murs's work on classification was not good, several species are included that do not occur in the country, and some species appear more than once, under different names. Gay added notes on distribution and habits to Des Murs's classifications. His most important discoveries were two species of seedsnipe, and the remarkable Des Murs's Spinetail *Sylviorthorhynchus desmursii*. Philip Parker King (1793–1856), a Fellow of both the Linnean and Royal Societies, surveyed the Straits of Magellan and nearby islands in 1826 to 1830 and sent back specimens to London Zoo and the British Museum. In 1859, the eccentric Alphonse Boucard (1839–1904) saw his first hummingbirds in Chile while on a voyage to California. He later became an expert and dealer in this group of birds, suggesting in his writings that collecting hummingbirds was a very suitable occupation for ladies – more varied and beautiful than jewels, and much cheaper! Ludwig Landbeck and Rudolph Amandus Philippi (1808–1904), who arrived in Chile in 1852 and 1853 respectively, inaugurated another phase, and they greatly advanced the knowledge of birds of the country by a series of detailed papers which exhaustively treated various genera. Philippi was the son of a Prussian courtier. He studied medicine in Berlin, and geology and conchology in Sicily. After his arrival in Chile, he took employment at the University of Chile, and of his 150 published papers, 23 were devoted to birds. However, some of his work has been criticised for creating new species on the basis of very slight anatomical differences. Harry Berkeley James (1846–1892) a wealthy businessman after whom James' Flamingo is named, was born at Walsall in Staffordshire. At the age of 21 he was employed as a clerk in Valparaiso. He soon rose to become the manager of a large nitrate mine, and his wealth and leisure were employed in collecting birds and their eggs.

Paraguay benefited from one of the greatest ornithologists to visit the continent in the early days. Don Felix d'Azara (1742–1821) was born at Barbunales, near Balbastro, Aragon, Spain, the younger of two sons of Alexander d'Azara and Marie de Perera, country landowners. Azara's elder brother, Don Joseph Nicholas, was 15 years his senior, and the two rarely met. Nevertheless, Azara was devoted to his elder brother, who had a distinguished political career in Europe. At the age of 18, Azara commenced a military career, received rapid promotion to the rank of captain, and, around 1775, was nearly killed when shot during a Spanish attack on Algiers. In 1781, he was ordered to sail to South America to act as commissioner in territorial and boundary disputes between Spain and Portugal in regard to their colonies. He was dispatched in great haste, before he had time to send for his books or belongings, and arrived in Buenos Aires little realising that it would be 20 years before he returned. It soon became apparent that the Portuguese had no interest in settling the boundaries, but only in prolonged procrastination. Azara was posted to Assumption, the capital of Paraguay; to relieve the inevitable boredom, and finding himself in a continent almost totally unknown scientifically, he determined to find out as much as he could about the geography, biology and ethnography of the lands around him. Azara was a modest man, deeply conscious of his lack of preliminary qualifications, and he greatly regretted the lack of reference books available to him. Nevertheless, he pursued his task with dedication and perseverance. Azara described 448 birds in his work, *Apuntamientos para la natural, etc.* (Azara 1802–05). The number can be reduced to 381 when duplications of age, sex and plumage are considered, and eight remain unidentified. A total of 178 are type descriptions on which the scientific names are based. Most of these scientific names are from Vieillot, who had supervised the illustrations for the French edition. The work exists in Spanish and French but the Spanish edition is very rare. It is more accessible in the French translation of Sonnini, who, unfortunately, was very hostile to Azara and his work, and often rewrote or omitted sections of it. This appears to be, at least partly, because Sonnini was a loyal disciple of Buffon, whose work Azara had often (correctly) criticised. As a result, Azara has received less than his due, and his work has been neglected, although not in Latin America or Spain. The lack of reference books during his years in Paraguay makes Azara's achievement the more remarkable. He did not merely describe species, but his book is full of observations of the birds' habits. He kept a number of birds in captivity, either in cages, or roaming freely indoors. A young Burrowing Owl for example, had the free run of his room for a long time. He would feed it raw meat, which it seized and carried behind the clothes trunk where it was swallowed. It climbed on everything, but as soon as Azara moved about it ran to hide itself. Azara's achievement is that he worked as a dedicated amateur, and probably published his book at his own expense, whereas Spain had sent out two official scientific expeditions to South America at this time, the reports of which were never published. Azara wrote the most complete natural history account of Paraguay, either before or since. He was followed by Johann Rudolph Rengger (1795–1832) a Swiss pharmacist and naturalist, who was primarily a mammologist. Little further happened in Paraguayan ornithology until in the late 19th century John James Kerr (1869–1957), a student at Edinburgh University, was offered an appointment as naturalist on an expedition to the Gran Chaco, but the greater part of his collections was lost.

Prior to Marcgraf, the French missionary André Thevet (1503–1592) had visited Brazil. He arrived in Rio de Janeiro in 1555, returning to France a year later. An account of his travels appeared in 1557 or 1558. While in Brazil he had collected Indian artefacts, birds and insects. The work stirred up considerable debate amongst those of the orthodox establishment who were not prepared to accept that opossums had pouches or that the bodies of natives were not densely furred. The scholar-missionary José de Ancieto (1534–1597) came to Brazil from Portugal in 1553 and remained there for the rest of his life. His writings mention birds and other animals common in the eastern part of the country. After Marcgraf (1610–1644), little was added to Brazilian ornithology until Count Hoffmannsegg, the mentor of Illiger, met the wealthy Brazilian Francisco Agostinho Gomes when the latter visited Portugal where Hoffmannsegg was temporarily living. The latter taught Gomes how to skin and preserve birds. After Gomes's return to

Brazil, he collected and sent many specimens to the Count. Hoffmansegg then sent his servant Friedrich Wilhelm Sieber to the eastern Amazon for 11 years where he made large collections. He returned by way of London where he traded many of his duplicate specimens for skins from Australia and North America. In 1807, the Royal House of Portugal fled to Brazil to escape an insurrection, and King John VI relaxed travel and trade restrictions. By 1813, naturalists had started to visit the country in numbers. One of the first was Georg Friedrich von Langsdorff (1774–1852), who held a medical degree from Göttingen and had served as consul in Prussia and Russia. He settled southwest of Rio de Janeiro where he collected birds and insects, and hired natives to collect for him. Visiting naturalists were entertained, but in 1828, Langsdorff had a breakdown and returned to Germany.

Prince Alexander Philipp Maximilian zu Wied-Neuwied

Prince Alexander Philipp Maximilian zu Wied-Neuwied (1782–1867) arrived in Brazil in 1815. Born in Prussia, he was the second son of the ruling prince, and the last of the line. Wied studied biological sciences under Blumenbach, and in 1802 joined the Prussian army in which by 1814 he had risen to the rank of Major General. After the Napoleonic wars he retired and from 1815–17 led an expedition to south-east Brazil. His itinerary is difficult to track because many of his journeys were circuitous and many of the localities he visited were not plotted on any existing map. Wied was interested in all aspects of nature, and his species accounts are said to be models of precision. On returning to his castle on the Rhine, he spent the next few years writing his memoirs and compiling his *Reise nach Brasilien* (Wied-Neuwied 1820–21) and *Beiträge zur Naturegeschicte von Brasilien* (Wied-Neuwied 1825–33). Then, in 1832, he went to North America for two years. He was an industrious and enthusiastic worker, and, when well into his eighties, was still writing ornithological papers. His collections of Brazilian birds, numbering nearly 4,000 mounted specimens, were bought, after his death, by the American Museum of Natural History. The *Ibis*, in reporting his death, said of his collection that: "Many tourists on the Rhine will remember to have seen [it] at his unpretending residence at Neuwied below Coblenz; for its doors were always open with the greatest liberality to visitors" (Wied-Neuwied 1867). Wied described 160 species as new, and the types of about three-quarters of these were still in the collection when it was received by the American Museum, and discussed at length by J. A. Allen in 1889. For his time, Wied was an exceptionally careful worker, recording the colour of eyes, bill and feet, etc., and making careful measurements. Unfortunately, between the time of collecting and the publication of his books, many of his new species were redescribed by Vieillot and Lichtenstein based on Azara's descriptions. Thus he was pre-empted in rather more than half of his new species.

Soon after Wied, Johann Baptist von Spix (1781–1826) and Karl Friedrich Philipp von Martius (1794–1866) arrived in Brazil in the course of their long journey. Spix was born in Hochstadt, Bavaria on 9 February 1781, and originally studied for the priesthood, but after two years his attention was turned to medicine and natural history, and he obtained his doctorate in 1806. The same year he became curator and organiser of the zoological collections of the Munich Zoological Museum, at the Academy of Science. In 1816, he was sent to Brazil by the King of Bavaria to undertake a three-year scientific survey together with Martius, a Bavarian botanist. This expedition was as a result of the marriage of the Archduchess of Austria, Caroline Josepha Leopoldina, to the Crown Prince of Portugal, Don Pedro D'Alcantara, the court of Portugal being at that time removed to Rio de Janeiro. Spix was detailed to survey the entire animal kingdom of Brazil. The expedition left Munich on 6 February 1817, arriving in Rio in July. It had originally been intended that the Archduchess would sail with them, but in the end she travelled later. Spix and Martius stayed in Brazil till 1820, travelling through southeast Brazil to Belém and then up the Amazon to Tabatinga and Avaracuara. On their return they collaborated on *Reise in Brasilien* a 3-volume work on the natural history of the country (Spix and Martius 1824–25). Spix also published *Avium Species Novae* in two volumes in 1824 and 1825. He died in Munich on 13 March 1826. Among the many new species he discovered was the now severely threatened macaw which bears his name, *Cyanopsitta spixii*.

Johann Natterer (1787–1843), born near Vienna, studied biology, modern languages and illustration. He was the son of an imperial falconer, and learned from his father how to preserve specimens. In 1816, he was employed as assistant at the Vienna Museum and the following year appointed a member of the expedition to Brazil. He collected many items, but was principally a bird collector, and his collections of 12,293 bird skins of about 1,200 species, now in the Vienna Museum, remain of prime importance. He remained in Brazil until 1835, returning to Europe where he was recognised as an authority on South American birds. Francis Louis Nompar de Caumont, Comte de la Porte de Castelnau, was born in London on 25 December 1812 and died in Melbourne on 4 February 1880. In 1843, Castelnau led a French government-backed expedition to central South America. He landed at Rio de Janeiro, crossed the southern grasslands to Lima and then descended the rivers to Pará. Accompanying him were the geologist Eugene d'Osery and the young ornithologist Deville, the former was assassinated on the journey and many of his notes were lost. A number of species were named after Castelnau as a result of this journey. His monumental multi-volumed report, *Expédition dans les parties centrales de l'Amerique du Sud, de Rio de Janeiro à Lima, et de Lima au Para; executée ... pendant ... 1843 à 1847, sous la direction de F. de Castelnau* was published between 1850 and 1859. As in the case of Gay's work on Chile, the birds were written up by O. Des Murs. Castelnau travelled to Australia in 1862, as Consul General for France in Melbourne. After his death his ornithological collections were returned to the Paris Museum.

The Danish frigate *Galatea* had as its principal naturalist, Wilhelm Friedrich Georg Behn (1808–1873), who since 1837 had been director of the Kiel Museum. However, he left the expedition in northern Chile and crossed the mountains into Brazil, to São Paulo and Rio de Janeiro. His collections included a considerable number of birds. Behn appears to have published no account of the voyage. Several species are named after him, but though these are probably based on specimens collected by him, they were named long after the event. On the *Galatea* was Dr Johannes Theodor Reinhardt (1816–1882) whose father, Johannes Christopher Reinhardt (1776–1856) was a professor of zoology, specialising in the birds of Greenland, also left the ship at Rio on the way home and travelled north to visit Dr Peter W.

Lund (1801–1880) at his home at Lagoa Santa near Belo Horizonte. Lund had made large collections of fossils, birds and insects which he sent back to Copenhagen. Reinhardt visited him on several occasions, in 1850–52 and 1854–56, publishing many papers on ornithology, based on Lund's specimens at Copenhagen.

In 1850, Hermann Konrad von Burmeister (1802–1892) arrived in Brazil as a political refugee, and remained for two years. He had been a professor of zoology at the University of Halle, and the author of a number of books, including a history of creation. During his period in Brazil, he collected at the Swiss Colony at Nova Friburgo. After his return to Germany in 1852, he published a five-volume work on the fauna of Brazil, two of these dealing with birds. Carlos Euler (1834–1901) from Basel in Switzerland, came to Rio de Janeiro in 1853 and bought a farm near Cantagallo. He became Swiss vice-consul, and in his spare time collected birds and made observatons on their habits. His results were published in a series of papers in the *Journal für Ornithologie* between the years 1867 and 1893. For some years before his death in 1855, Jean Théodore Descourtilez collected birds in the forests of Brazil. For a time he was on the staff of the National Museum in Rio, and is believed to have died as a result of his experiments with poisonous chemicals for preservation of specimens.

Herman von Ihering (1850–1930) proved to be one of the most important figures in Brazilian ornithology. He had served as a doctor in the German army when about 20, and took degrees in medicine, zoology and geology. In 1880, he reached Brazil to practice medicine and make collections of birds and spiders which he could sell to museums in Europe. In 1883 he was appointed travelling naturalist to the National Museum in Rio de Janeiro and in 1893, he was appointed director of the State Museum in São Paulo. He remained for 23 years during which time he greatly enlarged the museum and created a botanical garden. In his monograph on the birds of Brazil (1907) Ihering pointed out that the avifauna was very poorly known when he first arrived. Although many thousands of skins had been sent back to Europe, they had been taken from only a few limited localities.

The priest Jacintho Carvajal (1567–1647) served in San Domingo and at Bogotá in Colombia. His journals contained a number of observations on birds, some of these were described as tiny emeralds and were evidently hummingbirds, which are still called 'esmeraldas' in the country. Serious systematic ornithology in Colombia got off to an indifferent start with the so-called 'Bogotá' collections. About 1838 a French collector, resident in Bogotá, began to send large numbers of bird skins to Paris, where they were examined and described by Boissoneau, Lafresnaye, Des Murs and Bourcier. The natives soon began to learn how to prepare skins in increasing numbers. Many of these species were of wide distribution or came from areas far removed from the capital, but they arrived in Europe with only the locality 'Bogotá' attached to them. This trade was fed by the fashion for using small birds to trim hats, and it continued for close on a century, but probably reached its peak about 1855. Justin Goudot, who was in Colombia from 1822 to 1843, made large collections of birds which were sent to dealers in Europe. Thomas Knight Salmon (1841–1878) was an engineer who was also a nature lover and collected birds and eggs in the state of Antioquia, which were sent to the dealer Gerrard in London. Another early collector, the 'indefatigable' Adolphe DeLattre, about whom little is known, collected in Colombia around 1847.

Alcide Charles Victor d'Orbigny (1802–1857) belongs in a section by himself, for his travels and researches extended over several South American countries. French-born, his travels in South America were under the auspices of the Paris Museum. After a period of study with Cuvier, de Blainville, Geoffroy Saint-Hilaire and others, he arrived in Rio de Janeiro in 1826 and stayed in South America until 1833, exploring Patagonia south to Cape Horn, Uruguay, Chile, Paraguay and Bolivia. He first arrived in the middle of the Argentine-Brazilian war, and was captured by Brazilian soldiers, but managed to make some brief expeditions round Montevideo. His itinerary is impossible to track because of the inaccuracies of the then existing maps. His complete account is contained in seven large tomes (d'Orbigny 1835–47). These include a memoir outlining his theories concerning the distribution of approximately 400 species of passerine birds he encountered. He proposed three zones of latitude from Lima to Chubut and three zones of altitude from sea level to 1,700 m, 1,700 to 3,700 m, and above 3,700 m. The proportions of species in each zone were worked out. Other influences considered were temperature, habitat and exposures between east and west of mountain ranges. Thus 219 species were found in scrub, 125 in forest, 26 on plains, 14 in swamps and 11 on cliffs and buildings. The great collections of specimens brought back by d'Orbigny, were studied by himself and several colleagues including Isidore Geoffroy Saint-Hilaire and Baron de Lafresnaye. Their descriptions of new species appeared in the *Revue et Magasin de Zoologie*.

The earliest study of the fauna of Peru seems to have been that of José de Acosta (1539–1600) a Jesuit priest who lived in the country for 17 years as rector of the College in Lima and assistant to the Viceroy. He was able to travel extensively, particularly to Lake Titicaca, a centre of Inca agriculture. The resultant book was *Historia natural y moral de les Indias*. Acosta was acquainted with a manuscript copy of Hernandez's work on the fauna of Mexico and used this extensively in his own work. Garsilasco de la Vega (1539–1615) was the son of a Spanish settler in Peru and an Inca princess. He is remembered as a historian, particularly of the Incas, but his writings also contain incidental notes on birds. After 1600, the Spanish became secretive about the activities of their representatives in South America. The Jesuit Bernard Cobo (1562–1657) was not allowed to publish the information he had gathered about the natural history of Peru. It was not till the late 19th century that ten of his 43 manuscript books were found and published. Johann Jacob von Tschudi (1818–1889) travelled through much of Peru from 1838 to 1843, and his results, the *Avium Conspectus*, were published (Tschudi 1844). He also published a book of his travels, followed by an English translation. His taxonomic arrangement, though it covers only South American birds, is of interest (see Appendix 20). William Nation (1826–1907) was trained as a botanist at Kew Gardens, and at the age of 23 emigrated to Lima where he taught botany and languages. He sent a number of bird skins to the Zoological Society of London, and at least two bird species are named after him. During the later part of the 19th century a number of Polish ornithologists were active, either in Peru or in writing up the results of field explorers. These included the wealthy Count Constantin von Branicki (1824–1884), Ladislaus Taczanowski (1819–1890), Constantin Jelsk (1837–1896), Jan Stanislaus Sztolcman (1845–1928), Jean Kalinowski (1860–1942) and Dr Josef Siemiradski (1858–1933). Taczanowski was an expert on raptors and in 1865 became custodian of Warsaw Zoo and director of the zoological museum at Warsaw University. In 1884–86, his four-volume monograph of Peruvian ornithology appeared. Jelsk had been curator at the Kiev Zoo, but in 1867 was sent by Count Branicki to collect birds in Peru and became attached to the museum in Lima.

The most important early explorer in Venezuela was Friedrich Heinrich Alexander von Humboldt (1769–1859). He had studied at the University of Göttingen and the mining academy of Freiberg. After some years in the service of the Prussian Mining Department, an ample inheritance enabled him to devote himself to natural history without having to earn his living. He was rigorously trained in the natural sciences and developed into one of the most versatile of scientific investigators of his time. He was a botanist, zoologist, anthropologist, ecologist, geologist, cartographer, biogeographer, physicist, chemist, astronomer, demographer, historian, mountaineer, poet, artist and linguist (how many people today can claim to encompass all those disciplines, even in the most perfunctory manner?). He arrived in Venezuela in 1799 and explored the headwaters of the Orinoco, confirming its connection with the headwaters of the Rio Negro, which flows into the Amazon, one of the few examples of river division. In the caves at Caripe, he studied the Oilbird, then unknown to science, while in the mountains at the head of the Rio Magdalena he drew pictures of the Andean Condor and described its anatomy. From here he went to Peru where he noticed the thousands of birds on the guano islands, and was told of the use of this material as fertiliser by the Incas. The Humbolt Penguin *Spheniscus humboldti* was named after him in 1834. He remained in South America for five years, making vast collections. After working on his material in Paris, he published a best-selling account of the voyage. Most of his money had been spent on the journey and the publication, which filled 30 volumes; but the King of Prussia offered him a well-paid post as Chamberlain. In 1827, he settled in Berlin where, except for a short expedition to Russia and Siberia, on which he was accompanied by C. G. Ehrenberg, formerly Hemprich's companion on the Red Sea Expedition, he spent the rest of his life. Humboldt was responsible for elevating geography to the level of a science. He pioneered the study of climatology and the use of isotherm lines on maps was invented by him. In biological matters his chief contribution was in plant geography.

Pierre Barrère (1690–1755) lived for five years (1722–27) in Cayenne, (French Guiana), with instructions to prepare a detailed report, but the resultant work, published in 1741, is no more than a checklist. A number of early books and travelogues mentioned birds in passing, but the information they contained was of little or no value. Charles Nicholas Sigisbert Sonnini de Manoncourt spent three years in Cayenne, from 1772 to 1775, returning to France with skins representing 160 species of birds and with detailed notes on their behaviour and habitat. Edward Bancroft (1744–1821) lived for three years in Surinam, and produced his *Essay on the Natural History of Guiana* (Bancroft 1769). Philippe Armand Fermin (1720–1790) was the author of two accounts of the fauna of Surinam published in 1765 and 1769 respectively, but was described as a European who never penetrated far into the woods. Sir Robert Hermann Schomburgk (1804–1865) and his brother Richard (1811–1891) travelled in Guyana in 1835 and as a result of these expeditions, Guyana advanced from a totally unknown country to the then best known for its geography, fauna and ethnology. Robert Schomburgk was born at Freiburg, Saxony, on 5 June 1804, and in 1829 went to the USA. A year later, aged only 26, he left for the Virgin Islands, which he surveyed at his own expense, so successfully that the Royal Geographic Society in London, on the basis of his report, appointed him to conduct an expedition to Guyana. He was knighted on his return to England. In 1848 he was appointed British Consul to San Domingo and in 1857 British Consul in Bangkok. He was the author of books on Guyana and Barbados. Eugene André (1861–1922) was born on Trinidad of French parents, travelled extensively in Guyana and the Orinoco delta, and his book *A Naturalist in the Guianas* (André 1904) describes the trade in bird feathers (the so-called 'osprey-plumes' of commerce). Unfortunately, many of the notes and collections from his first expedition (1897–98) were lost through accidents on the river. A later trip (1900) was more successful.

Robert Schomburgk

As early as 1736, well-known travellers such as Charles Marie de la Condamine (1736), Pierre Bouguer (1744) and Humboldt (1802) visited Ecuador. Condamine (1701–1774) was a French scientist and traveller, who travelled the Amazon from 1735 to 1743. He had been sent by the Bourbon king, Philip V, to Quito to attempt to determine the length of a degree of meridian at the equator. The expedition was led by Condamine, already a distinguished scientist at the age of 34, and the Buff-tailed Hummingbird, *Eutoxeres condamini*, was named after him. He was accompanied by Jorge Juan y Santacilia (1713–1773) and Antonio Ulloa (1716–1795). As well as doing odd jobs for the Spanish Viceroy, they compiled voluminous notes on the ethnography, geography and natural history. Ulloa's account of his travels was published in 1748 and later translated into English by John Adams as *A Voyage to South America*. (Ulloa 1760). This was evidently a popular book, as it ran through at least five English editions. In describing a visit to the Juan Fernandez Islands, he recounted a story that 'quebrantahueso' birds (*?Buteos*) would take shelter in flocks in the lee of ships on the shore before local north winds developed into storms, and wait until the storm had passed. In 1745 Ulloa left for home, but was intercepted by the British and taken to London where his accomplishments were recognised and he was elected a Fellow of the Royal Society. He is remembered in Spain as the most distinguished Spanish scientist of the 18th century. Bouguer (1698–1758) was a French mathematician and hydrographer, who also visited Peru from 1736 to 1742. During the years 1789–94, a Spanish voyage consisting of two ships commanded by Alejandro, Marqués de Malaspina (1754–1810) spent five years surveying the coasts of South America, particularly Peru and Ecuador. The naturalist on this expedition was Antonio Pineda y Ramirez (1753–1792), born in Guatemala but educated in Spain. After school, he joined the army in which he served with distinction in the 1779–83 war with England. Retired from the army he studied natural history and foreign languages, and selected the zoology books for the voyage. When at Buenos Aires he met Azara and obtained a copy of the latter's manuscript, as a result of which he was able to make useful suggestions to Azara regarding the duplicating of species. Ramirez spent five months in Ecuador collecting birds. On the return of the voyage to Spain in 1795, Malaspina expressed unpopular opinions on political policy in South America and was imprisoned. Consequently, the expedition was never written up, though several crew members wrote partial accounts. It was not until 1844 that Lesson described four new species from Guayaquil and accounts of the voyage were not published for over a century, by Iris Engstrand in Buenos Aires in 1974 and Barbara Beddall in Connecticut in 1979. The first scientific investigation was by William Jameson, a Scottish professor at the University of Quito. He was principally interested in botany, and published a two-volume work on the plants of Ecuador. He explored the eastern and western Andean slopes over a considerable area and made large collections of birds. Several species, including the Andean Snipe, *Gallinago jamesoni*, were named after him. From 1849 to 1850, Jules Bourcier (1787–1873) was the French consul to Ecuador, resident in Quito. He was a noted expert on hummingbirds, and collected and described many new forms either alone, or in partnership with other French ornithologists such as Mulsant and DeLattre. For quite a long time, most of the specimens from Ecuador (other than those collected by Jameson) were taken by natives, and traded, probably passing through the hands of more than one dealer before reaching Europe, and thus lacked precise data or locality. Louis Fraser (1819–1866), who was in Ecuador from 1857 to 1859, inaugurated the scientific period in Ecuadorean ornithology. His notes and specimens were written up by P. L. Sclater in nine separate papers between 1858 and 1860. They were the first comprehensive accounts of Ecuadorean bird life. That expert on the country, Frank Chapman, said of him admiringly: "Fraser was one of the first collectors of his type to visit South America. There had been explorers like d'Orbigny and Schomburgk, expedition leaders like Darwin, and general collectors and field students like Wallace and Bates, but I do not recall anyone who more nearly approaches the modern collector than did Fraser" (Chapman 1926). Exploration continued with a huge, but haphazard collection by Charles Buckley, containing many new species which were described by Sclater and Salvin in 1880. Much better was the work of the Polish scientists Josef Siemiradski (1858–1933) and J. Stolzmann (1854–1928), who collected in Ecuador from 1875 to 1883; their work forms one of the most important series of papers on Ecuadorean birds.

Argentina was the last South American country to attract the attentions of ornithologists, although the work of writers such as Molina and Azara contained a number of species which occur in the country. For many years, the only writers on the birds to occur here were members of passing expeditions, for whom Argentina was a port of call on a circumnavigation of the world. Commerson, on Bougainville's expedition, visited the country and his notes later incorporated in the work of Buffon, include accounts on the Rufous Hornero *Furnarius rufus*, the Checquered Woodpecker *Picoides mixtus* and the Shiny Cowbird *Molothrus bonariensis*. Other visitors include James Cook; Louis de Freycinet; Duperry in 1822 (Garnot and Lesson wrote on the birds in a series of papers in French scientific journals); the Fitzroy expedition in the 1830s with Charles Darwin aboard; Dumont d'Urville between 1837 and 1840 (his physicians Jacquinot and Hombron collected about 800 birds of 400 species; this work was published in 1853); and Enrico Giglioli, the naturalist on the Italian ship *Magenta* which explored the South Pacific Ocean in the 1860s. Hermann Burmeister, professor of zoology at Halle, is the first great name in the ornithology of Argentina. He visited several South American countries during the 1850s, which resulted in a small number of publications. In 1862, he became curator of the Museo Publico at Buenos Aires. He transformed the little museum into an institution of importance, building up the collections, particularly of fossils, insects and birds. He sent out collectors to all parts of the country, and by the time of his death in 1892, the bird collections had increased from 445 to 3,500 specimens. He also bought some collections, such as one of hummingbirds from the dealer Deyrolle of Paris, and the Chubut collections of Henry Durnford. However, his work on palaeontology, geology, and other branches of zoology occupied a great deal of his time, and thus his contribution to ornithology, though considerable, was not as great as it might have been. As a result of Burmeister's efforts, the study of natural history was now established in Argentina. In 1878, two young Argentinians, Enrique Lynch Arribálzaga and Eduardo Ladislao Holmberg, founded the magazine *Naturalista Argentino*.

William Henry Hudson, son of American immigrants, was born in Buenos Aires on 4 August 1841. As a young man, he lived as a gaucho in the pampas, learning about its animals and plants. Although not a scientist, he made many useful observations on bird behaviour, which he confirmed by repeated observation year after year. In 1867, he sent a collection of skins to the Smithsonian Institution, which was eventually studied in London by Sclater and Salvin, who found 14 species not mentioned by Burmeister. After this, Hudson continued to send specimens to Sclater (see Chapter 9) who published accounts of them. In 1874, Hudson visited London and in 1887 received a grant from the Royal Society to publish his Argentinian notes. This resulted in a book *Argentine Ornithology*, published jointly with Sclater (Hudson and Sclater 1888–89). It was a special limited edition of only 200 copies in two volumes with 20 colour plates. When newer agricultural practices transformed his beloved pampas out ofa all recognition, Hudson emigrated to England, where he died on 18 August 1922. Ernest Gibson was the son of an Englishman whose family owned an estate near Buenos Aires called 'Estancia Los Yngleses'. He was a keen bird observer and collector, and published a series of ornithological notes on the birds of his area in the *Ibis* in 1879 and 1880. Henry Durnford was a businessman who came to Buenos Aires in 1875 from London. When his business failed, he turned to ornithology, making several visits to the Chubut River and elsewhere in Patagonia. In 1878, he died of a heart attack while on safari. His specimens were bought by Burmeister for the museum.

Chapter 8

The Quinary and other Nineteenth Century Systems

The idea of the mutability of species was gradually developed, and to understand it, it is necessary now to retrace our steps. The first man to query the truthfulness of the story of Noah's Ark was Sir Walter Raleigh, who, while imprisoned in the Tower of London, wrote a *History of the World* in which he pointed out that no Ark could possibly have been large enough to contain two of every known animal. The animals of the New World, he concluded, must have "evolved" independently from those of the Old. In 1685, the Lord Chief Justice of England, Matthew Hale, decided that even the Old World species were too numerous to have fitted into the Ark. Noah could only have been able to save a few, and these must have been the original ancestors of all present-day species. The physicist Robert Hooke (1635–1703) suggested that in the course of the earth's history many animals had been exterminated, while others had varied and changed. He also believed that new varieties of some species had developed as a result of changing climates, soil conditions and food. In 1700, the philosopher Gottfried Wilhelm von Leibniz suggested that when the ocean covered everything, the animals which now inhabit the land must have been marine; when the sea retreated they must have become amphibian, and finally completely terrestrial. An infant prodigy, he had read almost all the classics as a boy and received his doctorate at the age of 17. Born at Leipzig in 1646, the son of a professor, he died at Hanover in 1716. Mathematics and law were his chief subjects, and his energy throughout his life was remarkable. In the course of his travels in most cultural countries he had met all the eminent men of his time, and corresponded with many statesmen on matters of culture.

In 1715, Benoit de Maillet proposed a theory as to how fish could have changed into birds:

> Under the influence of the air the fins split, the supporting spines were transformed into quills, the drying scales into feathers. The skin was covered with down, the ventral fins became feet, the whole body assumed a different shape. Neck and beak lengthened, and at last the fish was transformed into a bird.

These ideas were ridiculed at the time, but Maillet's writings were seen and noted by the sagacious Buffon. In 1746, Jean-Etienne Guettard had noticed that in geological formations, the oldest fossils lie in the deepest layers. This observation led him to the conclusion that there have been a number of periods in the earth's history, and that each period had possessed its own animal life which later became extinct. His theories and writings were so contrary to the biblical account of the Creation that he was forced to recant, but other writers were more adept at evading the attentions of the censors. Buffon was one of the cleverest of his age in this respect. Some 120 censors sifted through his *Histoire Naturelle* and other writings, but his heretical views were so skilfully concealed behind his scintillating presentation that only seven of them discovered a few passages that might give offence. These were insufficient to warrant a ban on the book, so in despair they gave up. Thus in 1750, Buffon was able to write a *Theory of the Earth* which went far beyond Guettard's hypotheses. In this, he proposed the theory of Catastrophism, that each era of the earth and the life forms associated with it, had been destroyed by a catastrophe, of which Noah's Flood was merely the last.

Twenty years after Buffon's death in 1788, his disciple Lamarck (who is credited with the invention of the term 'biology') carried the matter further. Although poor all his life, Lamarck's talent was recognised early, and he became an assistant to Buffon at the Jardin du Roi. Jean Baptiste Pierre Antoine de Monet, Chevalier de Lamarck (1744–1829) trained as a priest, served as a soldier, then as a literary hack, but always had a love of natural history, and through Buffon obtained a place at the Academy of Science. He was commissioned to travel in Europe as companion to Buffon's young son, and eventually became an assistant in the botanical department of the natural history museum. On the outbreak of the French Revolution, a year after Buffon's death, Lamarck went to the government and persuaded them to preserve the precious collections under the new name of Jardin des Plantes. Probably as a result of this, Lamarck obtained a 'safe' position; the National Convention wanting to reform everything, instituted new professorships including two in zoology, which were offered to Lamarck and Geoffroy Saint-Hilaire. Thus, at the age of 50, Lamarck began research in the discipline in which he was eventually to achieve fame. He was retiring and modest, but his domestic life was far from happy. He was married four times, all his wives and most of his seven children predeceasing him. He was 57 when he first published his repudiation of the immutability of species (Lamarck 1802).

Jean Baptiste Pierre Antoine de Monet, Chevalier de Lamarck

Lamarck was not the only writer to be influenced by Buffon. Charles Bonnet, born in Geneva in 1720, was a pupil of Réaumur, and died in 1793. His writings are diffuse and difficult to read, but his chief contribution to biological history was the discovery of parthenogenetic reproduction (as in aphids). Bonnet formed a link between Buffon and Lamarck. He saw living beings and heavenly bodies as developing towards a common goal. The earth underwent a series of progressive developments, each of which was terminated by a catastrophe. All living creatures were destroyed, but germs out of which new life forms could arise, survived. The geological theory of catastrophes is evidently based on Buffon and the elaborated catastrophe theory of Cuvier is too similar to be a coincidence. Bonnet proposed that changes take place in species, foreshadowing the views of Lamarck. Also, in agreement with Buffon, Bonnet maintained no sharp divisions in life forms; all pass into one another. Petrus Camper (1722–1789) was born and studied in Leyden where he took degrees in philosophy and medicine. He was described as brilliantly gifted, but quick-tempered and despotic. He was an expert in many subjects: anatomy, surgery, gynaecology, hygiene, medical law and veterinary surgery. He was also a leading art critic and connoisseur. Primarily interested in anatomical investigation of rare animals, he published monographs on the elephant, the rhinoceros and the reindeer. His major contribution to ornithology was his study of avian bone structure, being the first to describe how the bones are filled with air to assist in flight. Felix Vicq d'Azyr was born in 1748 and died sometime during the French Revolution. He was personal physician to the King, having applied, unsuccessfully, for a position at the Jardin du Roi in 1774. His main contributions were in comparative anatomy, in pointing out for the first time the essential differences between the growth of crystals and of living creatures, indicating that of the three great kingdoms, animal, vegetable and mineral, a better division is between organic and inorganic objects.

Lamarck's theory attracted little attention in his lifetime, and was dismissed in the immediately succeeding period as fantastic speculation. He classified vertebrates on the basis of certain features. Mammals ranked highest, for they were viviparous and milk-secreting. Birds came lower for they lay eggs and have an incomplete diaphragm. Below these came reptiles (with which batracians were then grouped), which were cold blooded, and lowest of all, fish, which have lost lungs and extremities. (Lamarck seems to have assumed that fish were degenerate reptiles rather than that reptiles developed from fish.) There is little doubt that he was greatly influenced by Buffon; from the master came his assertion that only individuals occur in nature, and that all categories in a classification system exist only in the mind. But Lamarck had a far keener eye than Buffon for actual classification, though less of a capacity for understanding the difference between hypothesis and fact. For instance, he believed that waders have long legs and long necks through stretching these parts in their search for food, while swimming birds have webbed feet through stretching out their toes in the water. If a group of young animals had their left eye put out at birth and were allowed to interbreed, he asserted, these would eventually develop a race of one-eyed creatures. He also believed that spontaneous generation occurs under the influence of heat and light, the lowest invertebrates and plants being continually reproduced out of inanimate matter; including the interesting suggestion that freshwater polyps freeze to death each winter and regenerate spontaneously each spring. He considered that life evolved continuously, not as a result of catastrophes, and that no animal species has ever died out save those exterminated by man. Fossil species no longer extant have evolved into now existing forms. Birds, he claimed, were descended from tortoises, and mammals from crocodiles. It was his assertion that fossils represent the remains of ancestors of present-day animals, which called down on him a declaration of war by one of the most formidable brains of the time, Baron Georges Cuvier.

As well as being one of the most fertile minds of the 19th century, Cuvier seems to have been one of the most odious of men. Georges Léopold Chrétien Frédéric Dagobert Cuvier was born at Montbéliard on 23 August 1769, and died during the cholera epidemic of 1832. A delicate child not expected to live, he studied at Stuttgart under the eccentric biology tutor Karl Friedrich Kielmayer (1765–1844). While still young, his erudition came to the knowledge of Geoffroy Saint-Hilaire, as a result of which he obtained a position as professor of comparative anatomy in Paris. Étienne Geoffroy Saint-Hilaire (1772–1844) was born at Étampes, Seine-et-Oise, and had been destined for the Church. He entered the college of Navarre in Paris where he studied natural philosophy under Brisson, and subsequently under the latter's friend, the mineralogist Abbé Haüy, whom he later helped to escape from prison during the revolution. In March 1793, Saint-Hilaire was appointed sub-keeper at the cabinet of natural history, and later that year was appointed to the chair of zoology at the newly constituted museum of natural history. In 1798, he took part in an expedition to Egypt, returning to Paris in 1802. In 1807 he was asked by the emperor to visit the museums of Portugal, in order to obtain collections from them. In the face of competition from the British, he was able to obtain these collections for France.

Cuvier's energy was inexhaustible and he could discharge many duties at the same time without overlooking a single detail. Very early in life he became a follower of Buffon. He was bright, he attracted the attention of those in high places, and eventually became an important figure in the French

Georges Léopold Chrétien Frédéric Dagobert Cuvier

Government, pursuing an exacting political career as well as his studies in natural history. In 1818 he travelled to London where he met King George IV, recommending to him that if the British private collections of zoological specimens could be amassed as one, they would form a great national museum which would surpass any other. Cuvier was extremely widely read; his researches are alleged to have embraced all historical and philosophical science. He was a most systematic and clear-thinking man, and his opposition to Lamarck's speculations was inevitable.

His interest in fossils is believed to have been inspired by the fact that Paris is at the centre of a calcareous area, and the stone used for building is particularly rich in fossils. Indeed, Cuvier's work in this field was of such importance that he virtually created the science of palaeontology, as we now understand it. Cuvier was a supporter of Catastrophism, but not in quite the same way as Buffon. He did not claim that life had been totally exterminated at the end of each era, but that local catastrophes had wiped out animals in a particular area, to be replaced by immigrants from elsewhere. Thus, elephants once occurred in France but no longer did so. Towards the end of his career, Lamarck lectured in a room adjacent to that of Cuvier, but was now almost blind and his lectures did not attract anything like the audiences those of the scintillating Cuvier could command. On one occasion, Cuvier went into Lamarck's lecture room where he used his own superior powers of oratory to publicly humiliate the old man. Lamarck lost his students, his last source of support, his daughters had to beg for food, and when he died he was buried in a pauper's grave. Cuvier wrote an obituary that was so vicious and vindictive that the Académie Française, which published a memorial volume, refused to include it.

Cuvier, in his *Régne Animal*, divides birds into six orders (for details see Appendix 21):

1. Birds of prey and owls. Beak crooked, with a cere, talons crooked, three toes forward, one behind.
2. Passeres. All such as cannot be classed in the other orders.
3. Climbers (inc. parrots). Toes two forward, two back.
4. Gallinaceae. Upper mandible vaulted, nostrils covered by a cartilaginous scale, anterior toes united at base by a short membrane.
5. Grallae (inc. ostriches, ibises). Thighs naked.
6. Palmipedes. The only birds whose neck exceeds the length of their legs, palmated feet, set far back, plumage imbued with an oily juice (Cuvier 1817).

Alfred Newton in his *Dictionary of Birds* considered that Cuvier's arrangement was not much better than those it superseded. Although Cuvier claimed to have dissected more than 150 species of birds to determine the structure of the vocal organ, he seems to have used neither this, nor skeletal information in his classification, which is based entirely on external features. So little regard did Cuvier pay to bird osteology that for many years a skeleton of a fowl with a hornbill's skull was exhibited in the Museum of Comparative Anatomy in Paris! Although severely judged in modern times, and no doubt despotic in his manner, he was highly regarded by his contemporaries, and his kindness to younger scientists is documented. In 1794, Geoffroy Saint-Hilaire entered into correspondence with Cuvier, and collaborated with him on a number of papers. But later, his views found an opponent in his erstwhile friend, though these disagreements were conducted with unfailing dignity. Geoffroy Saint-Hilaire agreed with Johann Wolfgang Goethe (1749–1832) that organs which are superfluous in a species are retained as rudiments if they have played an important part in other species of the same family – thus testifying to the permanence of the general pattern of creation. Cuvier, on the other hand, maintained the absolute invariability of species. Just as the speculations of Lamarck were almost bound to arouse Cuvier's opposition, so were those of Geoffroy Saint-Hilaire. The latter studied the bone anatomy of vertebrates, and was able to draw parallels between certain bones in mammals and fish. He asserted that hearing is most highly developed in birds because they are so musical. Such wild flights of fancy combined with sober observation indicate the enigma of this remarkable man, and his wealth of both productive and unproductive ideas. He was the very antithesis of the strait-laced Cuvier.

Among Cuvier's followers, Johann Friedrich Meckel (1781–1833) was known as 'The German Cuvier'. He carried out extensive investigations into the anatomy of the platypus and the cassowary, resulting in monographs on these species. His studies also included the brain of birds. Progress in the study of the development of the hen's egg was made by Karl Ernst von Baer (1792–1876) who virtually created modern embryology. Martin Heinrich Ratke (1793–1860), a pupil of Blumenbach, and Baer's successor, investigated the respiratory organs of birds and mammals. Heinrich Christian Pander (1794–1865) also carried out work on the embryo of the chicken. His results were published in 1817 in collaboration with Baer.

By the early 19th century, zoologists had become dissatisfied with the situation that species, genera and orders were separated by gaps, and that these might be small or great. For long it had been assumed that these gaps would eventually be closed by the discovery of new forms, but this proved not to be the case. In 1635, Eusebius Nieremberg in his *Historia Naturae*, had described nature as a continuous weft, without gaps. In 1687, Leibnitz had argued that gaps in Creation could not occur, because this would point to the absence of a possible living creature, a contradiction of the completeness of the Creation. Charles Bonnet, in his *Contemplation of Nature* (Bonnet 1764), postulated life as a series of rungs on a ladder, in which each form or group is linked to those above and below. Thus water birds were placed on the rung above fish, birds which lived both on water and land came above that. Bats and flying squirrels represented bird-like mammals, while the Ostrich, which had goat's feet, was a mammal-like bird. Carnivorous quadrupeds corresponded to birds of prey, while quadrupeds that fed on grass and grain corresponded to birds with a similar diet. Thus, without realising it, Bonnet had almost stumbled on the principle of the ecological niche. Donati (1750) carried the matter further. He conceived nature as a web of knots, a species' resemblance to other forms being compared to the threads running between the knots. Although some pointed out that osteology flatly contradicted any such mystical relationships, the theory gained a considerable following. One of its leading exponents, Johann Baptist von Spix, reiterated Donati's views, but elaborated the theory, seeing nature as a series of concentric circles and figures with various angles. He opposed any idea that chance, rather than unchangeable order, should be the cause of the natural order of things. The characters of each species, he believed, were conditioned by other taxa in analogous categories. Vigors, the editor of the *Zoological Journal*, took the matter yet further, and proposed a theory which came to be known as the Quinary.

Nicholas Aylward Vigors was born in 1785 at Old Leighlin, Co. Carlow, Ireland, the son of Nicholas Aylward Vigors (1755–1828). In 1803 he matriculated from Trinity College, Oxford and in 1806 became a law student at Lincoln's Inn. From 1809–11 he served in the Peninsular War and was severely wounded. In 1811, he resumed his studies at Oxford, graduating with a BA in 1817. In 1826 he assisted in founding the Zoological Society of London and was its first secretary, a post he held until 1833. He became a fellow of the Linnean Society in 1819 and was elected a fellow of the Royal Society on 23 February 1826. He also held Fellowships of the Royal Society of Antiquaries, and of the Geological and Historical Societies, and was a Member of the Royal Irish Academy and the Royal Institution. He succeeded to his father's estate in 1828 and shortly afterwards embarked on a parliamentary career. He represented Carlow from 1832 until his death, unmarried, at his house in Chester Terrace, Regent's Park on 26 October 1840. He was buried in the nave of the Cathedral at Old Leighlin. Between 1825 and 1836 he was the author of 40 papers, mostly on ornithology. The day after his death, the *Proceedings of the Zoological Society* announced:

> In consequence of the lamented death of N. A. Vigors Esq., one of the founders of the Society, and during the first years of its existence its active and zealous Secretary, whose reputation and influence had materially increased its numbers, as his liberality augmented its collections, the Society adjourned to November 10th (Vigors 1840).

Vigors obtained many specimens at his own expense, and these were donated by him to the Society's collection. The collection was disbanded in 1855, the first choice of specimens being given to the British Museum, which acquired a number of types, but some were overlooked by G. R. Gray, the officer involved. In the new Quinary System, Vigors divided birds into a series of five circles, within which were five more, and so on. Thus each of the five orders of birds, Raptores, Insessores, Rasores, Grallatores and Natatores may be divided into five tribes, each tribe into five families, each family into five genera, each genus into five species, and each species into five varieties. This gives 3,125 species, not an unreasonable estimate for that time. Vigors's system was read before the Linnean Society on 3 December 1823 and published in volume 14 of their *Transactions* (Vigors 1825). It is based entirely on superficial similarities; thus the mottled plumages of the owls and nightjars means that they must be related, but the disparity of bills and feet keeps them apart. The hoopoe is placed near the treecreepers on account of the similarity of the shape of its bill. Vigors was puzzled by the overlapping of structural characters found in birds (of which he cited many examples), thus many species clearly belonging in one order, nevertheless possess characters which occurred in others. We now know that these similarities are due to genetic inheritance, or to convergence, and we should not sneer at men like Vigors who were trying desperately to find explanation for phenomena which were beyond he comprehension of the era in which they lived.

The concept that everything in the living world occurred in circles had first been proposed by William Sharpe MacLeay in his *Horae Entomologicae* (MacLeay 1819). At that time, before the acceptance of the concept of evolution, it offered a convincing explanation for adaptive characters found in otherwise different groups of birds. Vigors, a friend of MacLeay, drew attention to the danger of taking any single character on which to classify, for birds present a mosaic of interdigitating characters. Each of the five families of the Insessores apparently correspond to one of the five orders of birds. Thus the Dentirostres, consisting of thrushes, warblers, flycatchers, but also manakins and shrikes, correspond to the Raptores (birds of prey); the Conirostres consisting of crows, starlings, finches etc., correspond to the Insessores (passerines); the Scansores consisting of toucans, parrots, cuckoos, woodpeckers and treecreepers, correspond to the Rasores (gallinaceous birds); the Tenuirostres (hummingbirds, sunbirds, honeyeaters etc.) correspond to the Grallatores (wading birds); and the Fissirostres (swallows, bee-eaters, nightjars, kingfishers and todies) correspond to the Natatores (web-footed birds). Such analogies seem incredibly strained to us today, but they were accepted unquestioningly at the time. The theory was seized upon with fanatic zeal by William Swainson. Not content with Vigors's simple circles, Swainson elaborated the theory, seeing the circles as three-dimensional. Each bird had an affinity with one in another circle

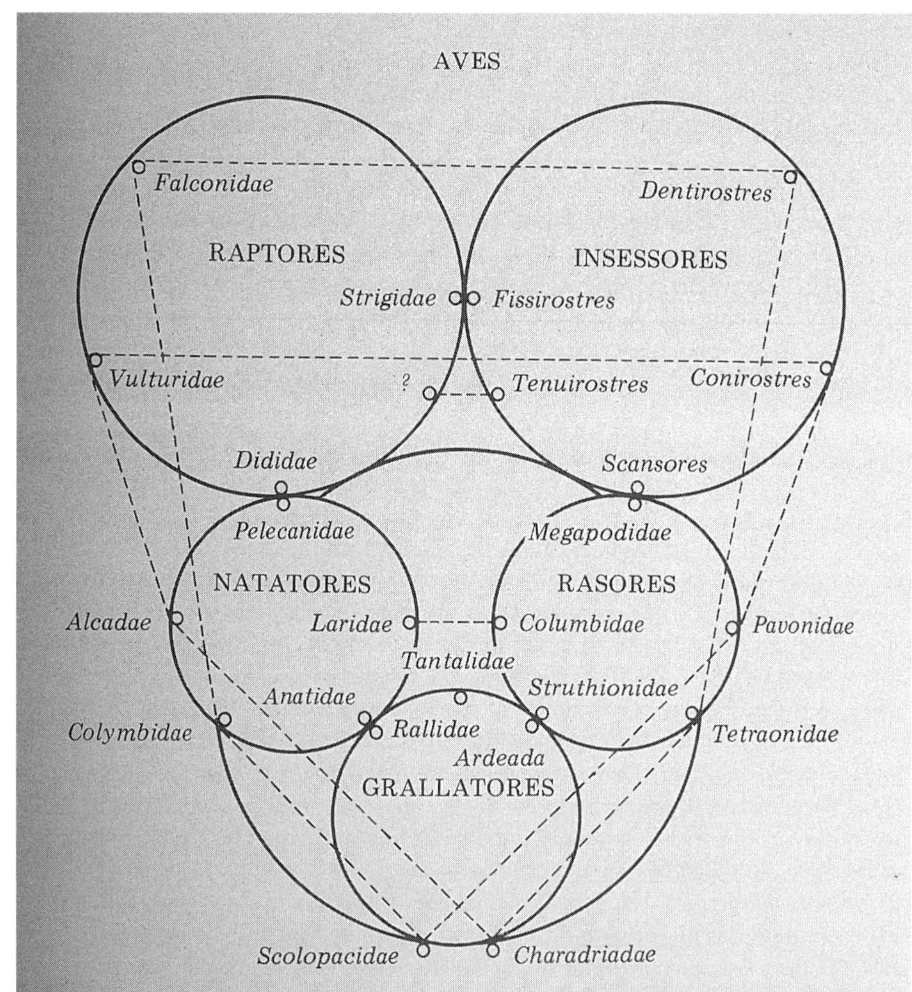

Figure 1. The classification of birds proposed by Swainson in 1837. The five orders are represented by three large circles, of which one, that for aberrations, is subdivided into three circles. Each circle includes five families, and family "affinities" join it to the two neighboring circles. The arrangement of the three remaining families is determined by "analogy" (broken lines).

Swainson's diagram

where these touched. Swainson (1789–1855) had been born at Dover Place in St Mary Newington, London, on 8 October 1789. His mother died before he was three years old, and at the age of 14 he began work as a customs clerk in his father's office in Liverpool. Fired with an interest in natural history by his father's collection of shells and insects, he became so knowledgeable that by his late teens he was asked by Liverpool Museum to write the text for *Instructions for Collecting and Preserving Subjects of Natural History* (1808). Soon he was drafted to Sicily in an army post and there was able to study natural history subjects. In 1815, at the end of the Napoleonic Wars, he retired from his army post owing to ill health and returned to England. The following year he made a collecting expedition to Brazil, where, at Rio de Janero, he met Prince Maximilian zu Wied-Neuwied and other explorers. He returned to Britain in 1818 with many specimens of birds and plants. His friend W. E. Leach, then Assistant Keeper of Zoology at the British Museum, encouraged Swainson to experiment with lithography for the illustration of his

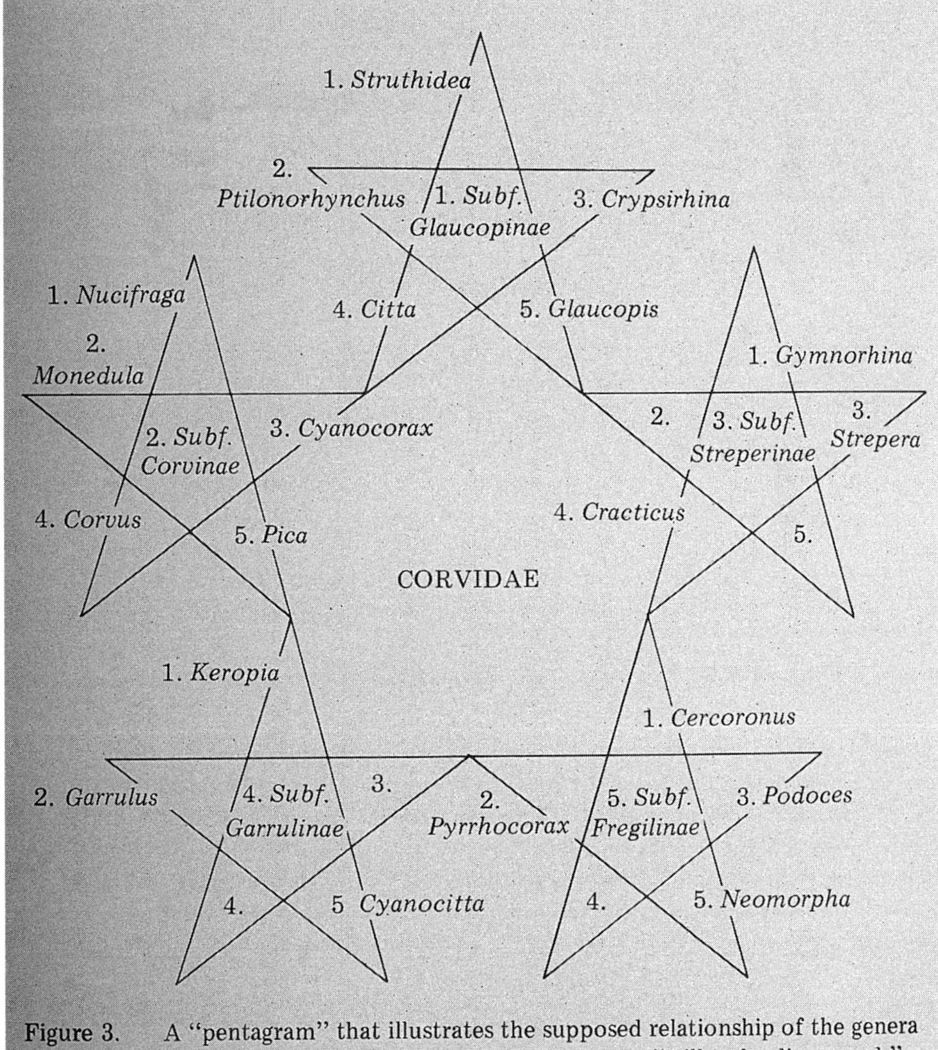

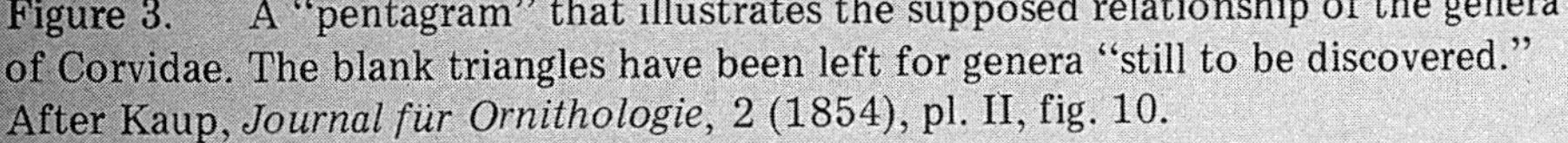
Figure 3. A "pentagram" that illustrates the supposed relationship of the genera of Corvidae. The blank triangles have been left for genera "still to be discovered." After Kaup, *Journal für Ornithologie*, 2 (1854), pl. II, fig. 10.

Kaup's diagram

book *Zoological Illustrations* (Swainson 1820–23), generally considered his finest work. About this time, Leach was forced to resign owing to ill health, and Swainson applied for the vacancy, but, although there was a great deal of support for him, J. G. Children was appointed instead. Swainson's later years were something of an anticlimax. He emigrated to New Zealand in 1841, but the farming venture in which he engaged, proved a costly failure. He took a botanical post in Australia in 1851, but this was not a success either, and he returned to New Zealand in 1855, where he died on 6 December.

In Germany, Lorenz Oken had been impressed by the idea that man was the ultimate aim of creation and considered that all other creatures were stages in this direction. Oken (1779–1851), born on 1 August 1779 in the village of Bohlsbach, Baden, studied medicine and natural history at Würzburg, and later at Göttingen. It was Oken's stated opinion that animal classes were nothing else but a representation of the sense organs, and he recognised only five classes: invertebrates; fishes (in which the tongue appears for the first time); reptiles (in which the nose opens for the first time and inhales air); birds (in which the ear for the first time opens externally); and mammals (in which all the senses are present and complete). Oken was important in the history of German philosophy and natural history, his influence on the development of cultural matters was considerable, but his contribution to exact science was of little importance. In 1816, he commenced publication of his periodical *Isis*, which he continued until 1848 (*Isis*; Jena 1817–48). This contained essays on all branches of natural history, and other subjects of interest, including poetry and even comments on the politics of various German states. The first paper of Sundevall published in 1835, was an unsuccessful attempt to arrange birds according to Oken's theory. Reichenbach, the director of Dresden Museum, met with more success. He arranged birds in ascending order, starting with penguins ('fish-birds'), these were followed by swamp-birds, and so on up the scale to the highest evolved, 'mammal-birds' i.e. ratites.

Even more intricate was the system of Johann Jacob Kaup (1803–1873), the Director of the Natural History Museum at Darmstadt, who transformed Swainson's circles into pentagrams. His arrangement was published in 1854 in the *Journal für Ornithologie* (Kaup 1854). According to Kaup, the number five totally dominated all creation. Thus there are five senses, five anatomical systems (brain, lungs, bones, muscles and skin), five parts of the body (head, chest, trunk, abdomen and pelvis), and so on (see Appendix 22). Although some of his contemporaries laughed at him, these various systems had their admirers. As late as 1862, T. C. Jerdon wrote in *Birds of India* that he saw more virtue in an ordered system than in a system of evolution:

> That species were created at hap-hazard, without any reference to others, either of the same group, or more distant ones, is a doctrine so opposed to all the affinities and analogies observed throughout the animal world, that the mind refuses to accept it, and intuitively acknowledges the evidence of design (Jerdon 1862–64).

It was H. E. Strickland who led the reaction in a famous article *Report on the Recent Progress and Present State of Ornithology* in 1845, in the 'Report of the Fourteenth Meeting of the British Association for the Advancement of Science' (Strickland 1845). He pointed out that comparative anatomy showed all beings to have general types of structure, modified solely with reference to external circumstances, and the purpose of these modifications must be sought in the conditions under which each being is destined to exist. Irregularity, not symmetry, is what should be anticipated in the natural system. In his classic work *The Dodo and its Kindred* (Strickland 1848), Strickland correctly classified the Dodo as a large, flightless pigeon; it had been defined by Vigors in 1823 as a link between the ostrich and the curassows; by de Blainville in 1835 as a bird of prey; by Owen in 1845 as a vulture; and by Brandt in 1847 as a pigeon-like wader. Hugh Edwin Strickland was born at Righton in Yorkshire on 2 March 1811, and died on 14 September 1853 as a result of a tragic accident in which he was struck by a train while examining the geology of a railway cutting. Next to geology, which was always his first interest, came ornithology, and might have become the more important one had he had more years of life for it to develop. He began collecting birds early, and by his death his collection totalled 6,006 specimens, but

Hugh Edwin Strickland

only 301 had been actually collected by him; he received many from his cousins and his brother, all of whom were connected with the navy. Other specimens were purchased from dealers. Strickland's widow presented his entire collection to the museum of the Department of Zoology, Cambridge, and his father-in-law, Sir William Jardine, published a posthumous collection of his papers.

Perhaps the most eloquent summary and obituary of the Quinary System was by Alfred Newton in the Introduction to his *Dictionary*:

> The purely artificial character of the System of Linnaeus and his successors had been perceived, and men were at a loss to find a substitute for it. The new doctrine, loudly proclaiming the discovery of a "Natural" System, led away many from the steady practice which should have followed the teaching of Cuvier (though he in Ornithology had not been able to act up to the principles he had laid down) and from the extended study of Comparative Anatomy. Moreover, it veiled the honest attempts that were making [*sic*] both in France and Germany to find real grounds for establishing the improved state of things, and consequently the labours of de Blainville, Etienne Geoffroy Saint-Hilaire, and L'Herminier, of Merrem, Johannes Müller and Nitzsch – to say nothing of others – were almost wholly unknown on this side of the Channel [i.e. in Britain], and even the value of the investigations of British ornithotomists of high merit, such as Macartney and MacGillivray, was almost completely overlooked. True it is that there were not wanting other men in these islands whose common sense refused to accept the metaphorical doctrine and the mystical jargon of the Quinarians, but so strenuously and persistently had the latter asserted their infallibility, and so vigorously had they assailed any who ventured to doubt it, that most peaceable ornithologists found it best to bend to the furious blast, and in some sort to acquiesce at least in the phraseology of the self-styled

> interpreters of Creative Will. But, while thus lamenting this unfortunate perversion into a mistaken channel of ornithological entry, we must not over-blame those who caused it. Macleay indeed never pretended to a high position in this branch of science, his tastes lying in the direction of Entomology; but few of their countrymen knew more of Birds than did Swainson and Vigors; and, while the latter, as editor for many years of the *Zoological Journal*, and the first Secretary of the Zoological Society, has especial claims to the regard of all zoologists, so the former's indefatigable pursuit of Natural History, and conscientious labour in its behalf – among other ways by means of his graceful pencil – deserve to be remembered as a set-off against the injury he unwittingly caused (Newton 1896).

But Strickland had not been the first to see the folly of the Quinary System. As early as 1826, Friedrich Boie had adhered to it as well as he could, but, if he thought about it at all, must have realised its unworkability. Boie divided birds into five orders, and of these, four consist of only five families, as in the Quinary System; however, the remaining order is very large indeed (see Appendix 23). The first (Raptors) contains the birds of prey and owls, but also, rather surprisingly, the nightjars and allied groups. The third order (Rasores) contains the game birds, pigeons and ratites; the fourth (Grallatores) the waders and most of what are now the Gruiformes and the Ciconiiformes. The fifth and last (Natatores) contains all the seabirds and the Anatidae. Everything else is in the second order (Insessores) the perching birds. This is divided into 30 families, the composition of which seems somewhat eccentric, but serves to demonstrate how relationships which seem obvious to us were by no means so in these early years. Thus, the pipits are grouped with the larks not with the wagtails, and the latter group also contains the forktails and some chats, presumably those with long tails. The ioras are placed with the titmice; while the fairy-bluebirds, today in the same family as the ioras, were placed in a group which includes the orioles, broadbills and rollers. Most of these assemblages are arranged on the grounds of differences and similarities

Edward Blyth

which we now know to be superficial. Separation of the ioras from the fairy-bluebirds is interesting, as some authorities have doubted that they are closely related. In 1839, Keyserling and Blasius pointed out that the larks differ from all other passerines which have 'fine songs' in that the tarsus is scutulated, agreeing in this feature with 'non-singing' passerines and the hoopoe. Much importance was attached to this superficial character for years before its irrelevance was recognised. In 1838, Edward Blyth had attempted a classification in which he recognised that geographical distribution was of importance in understanding the relationship and differences between taxa. He may well have been the first to recognise this important point, but unfortunately his published classification covered only one group, the Insessores. Perhaps for this reason, Blyth's conclusions were largely overlooked or ignored at the time, and when later rediscovered by others were regarded as new. The difficulties under which researchers at this period worked must be understood. Anatomists such as Blyth had no access to specimens of a full range of species, but relied on such bodies as could be obtained from dealers and zoological gardens. Their work does not seem to have received much encouragement, for although the minutes of the Zoological Society of London reveal that Blyth submitted a complete classification of birds to the Society, this was never published in full. Soon after this, Blyth's transfer to India, and the inevitable lack of both time and books of reference, would have prevented him from carrying out any further studies in this line. Although Blyth was for some time unjustly neglected, since the time of Alfred Newton at the end of the 19th century his importance to ornithological history has been recognised. Johann Friedrich von Brandt (1802–1879) published a series of short papers on classification in the *Mémoires de l'Académie Impériale Sciences de St Petersbourg* (Brandt 1836–39). This followed closely the arrangement of Illiger, but with a number of important variations. Brandt failed to appreciate the differences between the penguins (which Illiger had placed as a separate family) and the auks, divers and grebes, and placed the finfoots and coots with the webbed-footed swimming birds. But on the other hand he noticed Illiger's error in placing the phalaropes with these birds and correctly placed them with the waders.

Johann Friedrich von Brandt

Brandt was born near Brandenburg, Germany on 25 May 1802, the son of a doctor who wanted his son to follow the same profession. However, Brandt's uncle had stimulated the young man's interest in natural history by presenting him with a copy of Humboldt's *Travels*. Brandt studied first at Wittenberg and later at Berlin University, where he met and became friendly with Lichtenstein, then the head of Berlin Zoological Museum. Brandt graduated in 1826, but six months in the medical profession was quite enough for him, and in 1831 he took up a post as director of the Zoological Museum at the Academy of Sciences in St Petersburg. After the death of Catherine the Great in 1796, exploration had languished in Russia. Under her successors, the tyrannical Tsar Paul I (who mercifully was assassinated after only four years) and the mystic Alexander I (1801–1825) the country was concerned with defeating Napoleon and other political matters, and had little time for science. Brandt, however, was able to initiate a period of intensive study under the more enlightened reigns of Nicholas I (1825–55) and Alexander II (1855–1881). He began to build up the library at the Academy and encouraged the collection of native animals, most of which were not represented in the museum's collections. After 1840, much material began to filter back from the second phase of empire expansion, with the expeditions of Severtsov and Przevalski, Middendorff, Schrenck and Radde. Brandt's publications amounted to about 300 papers of which two-thirds were mainly concerned with birds and mammals. Probably owing to overwork, many of his projects remained unfinished, but Brandt's Cormorant, *Phalacrocorax penicillatus*, and the Spectacled Eider, *Somateria fischeri*, both of which he named, remain as tributes to his memory.

In 1833 Nitzsch in his *Pterylographiae Avium pars prior* (Nitzsch 1833, translated into English in 1867 as *Nitzsch's Pterylography*) showed almost for the first time that feathers were not distributed uniformly over the whole surface of a bird, but concentrated into areas or tracts. This had indeed previously been noticed by John Hunter, but not published until 1836 in Richard Owen's Catalogue of the Museum of the College of Surgeons, and by Macartney, whose comments were published in *Rees's Cyclopaedia*, 1819 (vol. 14, article 'feathers'). This article had, however, been largely overlooked, and many celebrated bird painters of the time, including Landseer, show a completely unnatural depiction of the plumage. This, though detectable to an ornithologist, does not detract from the artistic value of such work even to the discerning public, who see only the correct attitude and expression. But the spectre of the feet, which had haunted writers like Barrère, was still to haunt writers as late as 1846, when the distinguished amateur, John Hogg, produced a treatise based entirely on this feature, though of his two main divisions he carefully pointed out that his *Constrictipedes* make, in general, well-built nests in which they rear altricial young (born naked and helpless), while the *Inconstrictipedes* make a poor nest on the ground and rear precocial young (born feathered and mobile). There are numerous errors and anomalies in this arrangement. The herons, for example, are in fact much more like *Constrictipedes* than larks or kingfishers, and indeed almost all the claimed characteristics of the two subclasses break down under careful analysis (see Appendix 24). In 1850, Edward Newman contributed a not dissimilar arrangement, which was read before the Zoological Society of London on 12 March, and later published in the *Proceedings*.

1. Hesthogenous Birds (i.e. precocial)
    1. Gallinae game birds
    2. Brevipennes ostriches
    3. Pressirostres plovers
    4. Longirostres snipe
    5. Macrodactyli rails
    6. Plongeurs divers
    7. Lamellirostres ducks
2. Gymnogenous birds (i.e. altricial)
    1. Totipalmes pelicans
    2. Longipennes gulls
    3. Accipitres birds of prey
    4. Cultrirostres herons

5. Passeres sparrows
6. Grimpeurs climbing birds
7. Columbae pigeons (Newman 1850)

In 1856, Professor Gervais returned to the study of the sternum, which had been used by de Blainville (1815) and L'Herminier (1827), and based his arrangement on this feature. The practicalities of his system were to place the tinamous with the rails, the bustards and the sheathbills with the waders, and divers, auks and penguins together. This work was largely negated, when, in 1859, Charles Émile Blanchard finally demonstrated that the sternum alone was not a reliable feature; it was necessary to study the skeleton in all its details. Thomas Eyton had for many years collected bird skeletons. His *Osteologia Avium* (Eyton 1859–75) contains a great many illustrations of skeletons and detached bones, but these are of rather poor quality. A similar work by A. B. Meyer, *Abbildungen von Vogel-Skeletten* (Meyer 1879–97) made the mistake of using photographs, a process incapable of distinguishing between diagnostic and trivial features. Similarly, the systematic arrangement of Des Murs in *Traité général d'Oologie ornithologique au point de vue de la Classification* (Des Murs 1860) based entirely on eggs, resulted in some ill-assorted unions, due in part to the misidentifications which were then rife, but also because although oology can be of use in classification, it cannot be used as a *basis* for classification, since many unrelated birds can produce virtually indistinguishable eggs. In practice, eggs are more useful for indicating differences than similarities. The anonymous review of Des Murs's work which appeared in *Ibis* (1860, pp. 325–335) makes clear the misidentifications and gaps in knowledge which fatally flaw this work.

In Britain during the 19th century, there were writers less interested in systems than in factual accounts of birds, and their contribution to the progress of ornithology should not be underestimated. William MacGillivray was born in Old Aberdeen on 25 January 1796. His parents were unmarried. At the age of three, he was sent to Harris in the Outer Hebrides, where he was brought up by his uncle, Roderick MacGillivray, on a farm at Northton on the southwest tip of the island. The ruins of the cottage still stand. Before he was 11, MacGillivray was already capable of shooting eagles, one such incident is vividly described in his *Descriptions of the Rapacious Birds of Great Britain* (MacGillivray 1839). At 12 years of age, he went to Aberdeen University, and would walk home for the long vacations – a distance of 180 miles, including a boat crossing of the Minch. After completing his MA degree he began to study medicine, but gave this up in favour of natural history. He was handicapped by the fact that the only textbooks available to him were Linnaeus and Pennant. In 1817, aged 21 and with no plans for a career, he spent a year based at home visiting parts of Lewis and Harris; his diary for this period records observations on the local plants and animals, especially the birds. He must have earned a living as a peripatetic medical man, and also taught for a time in his old school at Obbe. After leaving Harris, he went back to Aberdeen, where he taught botany, and worked as assistant dissector in the anatomy class. In 1819, he decided to visit the British Museum in London to see the bird collections. He reached London by walking, deliberately taking a roundabout route, a distance of 837 miles, so as to see as much of the country as possible. A week was spent in London, sightseeing, and visiting the Museum's collections. He returned to Aberdeen by sea, his mind and career now decided – to become an ornithologist.

From 1820 to 1841, MacGillivray lived in Edinburgh. Of this period of his life little is known, but in 1830 he met J. J. Audubon, who also lived there for a period. The two became firm friends and it is believed that a great deal of the text of Audubon's *Ornithological Biography* (Audubon 1831–39) was actually written by MacGillivray. In 1831, MacGillivray became conservator to the museum of the Royal College of Surgeons Edinburgh. In spite of his enormous literary output, poverty plagued MacGillivray all his life. He painted a series of superb pictures intended to illustrate his five-volume *History of British Birds* (MacGillivray 1837–52), but could never afford to have them engraved, and they remained unpublished. Audubon is said to have admitted they were the equal of anything he himself had painted.

Audubon left Edinburgh in 1839, and, in 1841, MacGillivray also left and returned to Aberdeen to the Regius Chair of Civil and Natural History at Marischal College, Aberdeen. *The History of British Birds* was a commercial failure, partly because his work was highly idiosyncratic. His classification system, founded upon the structure of the digestive system, had no real taxonomic basis. The book received devastatingly vicious reviews, and, as a result, did not sell. Yet in 1909, W. H. Mullens wrote that MacGillivray's work was far superior to that of any of his contemporaries (Mullens 1908–09). One of the most important reasons for its failure had been MacGillivray's personality. He was irritable, sensitive, ambitious, plagued with ill health and a sense of neglect, tactlessly frank with a contempt for fools and pedants.

No such problems affected the life of William Yarrell (1784–1853), who was born on 3 June 1784 in London, the son of a newspaper agent. He was educated at Ealing and began employment as a bank clerk, but soon left this vocation to join his father in the newspaper industry. His hobbies were shooting and fishing, which probably aroused in him his interest in zoology. In 1823, he began to note the occurrence of rare and interesting birds. Two years later he was elected a Fellow of the Linnean Society. He was also a founder of the Zoological Society of London. Already in 1829, he had published a treatise on the syringeal muscles of birds. He followed this with the *History of British Fishes* (Yarrell 1836) and a year later the first of the three volumes of *History of British Birds* (Yarrell 1837–43). The latter work was a great success, largely because it was elegantly written. Yarrell understood the type of book that would be popular with the public. He realised that they did not want the sort of anatomical treatises which made up MacGillivray's more scholarly work. MacGillivray's masterpiece was undoubtedly hampered by the appearance of Yarrell's work, and the latter had the advantage that it was completed quickly. But a more important reason may be found in the state of British ornithology at the time. There were three types of student: anatomists, museum naturalists and field naturalists. Museum workers directed their attention to classification of taxa and discussion of different systems. Field naturalists were despised, their works regarded as non-scientific. Suddenly, a Scotsman who had studied birds in the field had proposed to combine this expertise with that of an anatomist, and the fact that his position as an anatomist was unassailable, identified him to the museum naturalists as an enemy. Presumption combined with merit must be crushed, so word went forth that MacGillivray's work was choked with anatomical details, and this repelled the public. In a bitter indictment of the ornithological opinion of the time, Mullens commented: "That they had incidentally broken the heart of the greatest ornithologist this country has ever possessed, that they had nearly prevented the completion of one of the greatest books on British birds, was to them of course, not a matter of the least importance." The book never recovered, and although it subsequently commanded high prices in the salesrooms, it was never reprinted. On the other hand, a second edition of Yarrell's book was published in 1845, only two years after the last volume of the first, and a third edition appeared in 1856, after Yarrell's sudden death at Great Yarmouth on 1 September 1853. The popularity of the book is shown by the fact that further editions appeared through the remainder of the century, edited by Alfred Newton and Howard Saunders. It was the standard popular bird book for the 19th century, probably filling much the same functions as T.A. Coward's *The Birds of the British Isles* did in the 20th.

Although John Gould contributed little to the development of ornithology other than a number of de luxe books and a considerable quantity of new names, he cannot be ignored. He was born on 14 September 1804, the son of a gardener. Nothing is known of his early education, if any; certainly the quality of the spelling and grammar in his correspondence does not indicate schooling of anything more than a very rudimentary kind. When he was 14, his father obtained a post at the Royal gardens at Windsor, and about this time Gould seems to have been introduced to the art of taxidermy – an art in which he was to become very proficient. He was the first practitioner to receive the patronage of the British Crown, and in 1830 was responsible for stuffing a giraffe for King George IV. Around the same time he was appointed 'curator and preserver' to the recently formed Zoological Society of London. Less than four years later, Gould published his first book, having lost no time in building up a network of correspondents and

John Gould

friends who would enable him to realise this ambition. At face value, Gould was not the right person to publish illustrated books. He was not an artist, was unlettered, and had no obvious talents other than a keen business sense and an ability to ingratiate himself to anyone he thought could be of any use to him. However, in January 1829 he had the fortune to marry a very competent artist and later that year to engage the services of a dedicated secretary, Edward Prince. The marriage was probably one of convenience, and it lasted for 11 years, when Elizabeth Gould died of an infection following childbirth. Nearly 3,000 plates were published by Gould in his multifarious works, but, although he claimed authorship for a number, not one was actually drawn by the 'author'. His 'rough sketches', from which his draughtsmen made the completed masterpieces, are almost embarrassing in their crudity. He was to apologise for, and seek to conceal this point, throughout his life. His first work, *A Century of Birds from the Himalaya Mountains* (Gould 1831), made Gould's name. It also earned him a reputation for ruthlessness, as a number of the birds illustrated were from the collection of Brian Hodgson, who received no credit for the material.

Among those treated with callous unscrupulousness by Gould was the genial and charming Edward Lear, author of the famous nonsense rhymes, and an artist of considerable importance. Lear's one excursion into publishing, the *Illustrations of the Family of Psittacidae or Parrots* (Lear 1832), ensured him a toe-hold in ornithological history, since some of his plates are the first representations (and therefore descriptions) of the species in question. After this work, Lear was an illustrator to Gould for many years, but all his plates, although signed by Lear, were credited by Gould to himself. Lacking skill as an engraver, Lear had turned to the newly developed style of lithography. This involved drawing directly on to a slab of limestone with fine wax or crayon, which attracted the thick ink used. After the excess ink was wiped off, a sheet of paper pressed over the stone took an impression of the inked wax drawing. It was a much more flexible and easier process than engraving, and it was to prove the focal point of Gould's process of reproduction.

Edward Lear

It had been used very little before Lear introduced it, but, typically, he received no credit for it from Gould. Reviews cited Lear's plates in *The Birds of Europe* (Gould 1832–37) as proving Gould to be one of the greatest bird artists of the day. Gould described many new 'species', but a considerable number of them have subsequently proved to be no more than colour varieties, aberrant specimens, or simply examples falling within the normal range of variation of a species. Anything that differed in any degree from the large, but inevitably inadequate, series of specimens then available was automatically named as new. This enabled a plate of it to be painted, and helped swell Gould's publications. Of course it should be realised that the naming of invalid species was by no means confined to Gould: many authors of the time made similar errors, but they had less to gain by it. Gould probably lacked the ability to understand the significance of speciation, and those species which proved to be valid were probably arrived at fortuitously rather than perspicacity on his part. When Charles Darwin returned from the voyage of the *Beagle*, bringing specimens of the Galapagos finches, it was Gould who, by virtue of being on the spot at the right time, identified and described the various forms as separate species, for which he has been given much probably undeserved credit. Gould's work sparked off Darwin's train of thought, but, although Gould had identified the finches, he did not, apparently, appreciate the significance. It was left to Darwin to do this. Later, Gould strove to distance himself from the evolutionary theory, because to be seen to espouse such a controversial idea might easily lose him subscribers to his books. His naivety is demonstrated by the fact that he bought a Great Auk egg for £10 in a toyshop in Regent Street, London and, thinking it was a coloured model, sold it again a few days later.

In 1838, Gould set off with his family for his fabled trip to Australia. Almost nothing had been published on the birds of that continent, so almost everything he found would be new. By this ambitious stroke, and the subsequent monumental publication, *The Birds of Australia* (Gould 1840–48), Gould secured himself the position of pioneer of Australian ornithology. He spent about 18 months in the country, making vast collections, many of which he sent back to England ahead of him, though a good many of these deteriorated badly on the way, or after arrival. He obtained many more specimens from other sources, such as several of his wife's brothers who had settled permanently on the continent, and from his collector, John Gilbert, who remained behind, collecting for a total of five and a half years until he was murdered by Aborigines. Callous and calculating as ever, Gould saw Gilbert's death merely as the loss of a source of material and revenue. He did not even trouble to find out the cause of the tragedy.

Project followed project, but the next one for which Gould's name is particularly remembered is his *Monograph of the Trochilidae or Family of Hummingbirds* (Gould 1849–61). He had thousands of these glittering creatures sent to him, and there is no doubt that a high proportion of those now known were described and named by him. In 1758 only 18 species were named in the tenth edition of the *Systema Naturae* by Linnaeus, of which several later proved to be sunbirds, and several others are based either on artefacts or pictures of unidentifiable species. In 1829, R. P. Lesson began his several illustrated volumes devoted to the family, and described about 110 species (today over 330 species are recognised), while in 1833 Sir William Jardine published a *Monograph of the Trochilidae*. In these early days the male and female of dimorphic forms were frequently mistaken for separate species, and sometimes the reverse also occurred with the alleged male and female proving to be distinct. One of the greatest collections of hummingbirds at the time was that of George Loddiges, primarily a botanist, and it may have been this which inspired Gould to begin his own rival collection. A number of new species were named by Gould from specimens in the Loddiges Collection – the most spectacular being the marvellous Loddiges' Spatuletail *Loddigesia mirabilis*. Loddiges returned the compliment by naming a new species *Lesbia Gouldii*. The Loddiges Collection remained in his family till 1933, when it was bought by the Natural History Museum, London (now in Tring).

Of the many bird specimens which Gould mounted or had mounted during his lifetime, almost none now seem to survive except some of the spectacular octagonal cases full of hummingbirds which he set up for London's Great Exhibition of 1851. All of the vast number of Gould specimens in the Natural History Museum are flat study skins, and moreover possess a disappointing lack of accompanying information. Gould, for all his professed scientific dedication, had little interest in data, being mainly concerned with exciting novelties with which to illustrate his books. The volumes of the *Monograph of the Trochilidae* were innovatory in their use of gold leaf combined with varnish and colour to depict the glittering iridescent patches. This has proved impossible to reproduce in the various facsimile editions made in the late 20th century. There is, however, reason to believe that this gold-leaf method was borrowed from William Loyd Baily of Philadelphia without acknowledgement. Baily, a hummingbird enthusiast, had developed the technique for a planned book of his own (never published), and although Gould knew of this process, he claimed that it was of no use to him, and that the process he used was different. Baily's nephew took a very different view, and after the death of both men, expressed the opinion that Gould had taken unfair, if not dishonest, advantage of his uncle's good nature.

In 1862, Gould's last completed project, *The Birds of Great Britain* (Gould 1862–73), was begun – several others were left unfinished at his death. *Birds of Great Britain* proved to be another triumph. Gould died on 3 February 1881, aged 76.

Chapter 9

New Theories, and Explanations in the Far East

On 30 January 1858, Temminck died, and his place at Leyden was taken by a man who had spent 33 of the best years of his life working under and sacrificing his ability to the capricious old man. Herman Schlegel (1804–1884) was 55 when he finally came into his own. He had been born in Thuringia, the son of a brassfounder. His father had had a butterfly collection, which first stimulated Schlegel's interest in natural history. A chance discovery of a buzzard's nest led him into ornithology, and he was encouraged to visit Pastor Christian Ludwig Brehm. Schlegel had begun in his father's business, but soon tired of it, and set off on his travels. Armed with a letter of recommendation from Brehm to Joseph Natterer, the brother of the famous explorer, he secured a small post under Natterer at the Vienna Museum. In addition to his scientific interests, Schlegel had a love of music, and it was a great thrill to him, on several occasions, to be able to watch Beethoven walking in the park. After a year in Vienna, the museum director, Schreibers, received a letter from Temminck asking if there was a young scientist suitable to be his assistant. Schlegel was recommended, and left for Leyden almost at once. It was intended that Schlegel should be sent out to Java to join the Natural History Commission, but after the death of Boie, already designated by Temminck as his successor, the latter begged him not to leave the museum. During his first years, Schlegel worked mainly on reptiles, but soon extended his interests to other groups. At about this time, he met Philipp Franz von Siebold, the famous explorer of Japan. Siebold was born in Würzburg on 7 February 1796, graduated in medicine in 1820 and the following year was called to Holland as personal physician to King William I. However, he disliked court life, so transferred to a position as an army surgeon and was posted to Batavia, where he arrived in 1822. There the Governor persuaded him to visit Japan as doctor to the head of the Dutch trading station on an island off Nagasaki. Because doctors were in demand, he was permitted more freedom of movement than was normally accorded to foreigners in Japan at that time. He was therefore able to send back many crates of specimens to Java, but because he had made and kept maps of Japan he was accused of spying, and permanently banished from the country. He returned to Holland in July 1830 with his collections. These were bought by the Dutch Government in 1837.

The Japanese themselves had been fairly well acquainted with birds from their earliest times, and they figured frequently in Japanese mythology. However, the doctrines of Buddhism, which prohibited hunting, and of Chinese literature which slighted natural science, discouraged the Japanese from studying natural history, and for many centuries no progress was made in ornithology. The first truly Japanese ornithologist was Atsunobu Kaibara, whose book *Yamato Honzo* (The Natural History of Japan) was published in 1709. Volume 15 of the 21 volumes is devoted to ornithology. He considered 99 species of birds, dividing them into water birds, mountain birds (eagles, nightjars, White's Thrush etc.), small birds (most passerines, woodpecker), domestic birds (fowl, geese, ducks, pigeons), miscellaneous and foreign birds. A number of the species have never been satisfactorily identified. After this book, a number of others were published during the same century, the *Wakan Sansai Zuye* (Illustrations of the Natural History of Japan and China) by Ryoan Terashima (1713) and *Ko Yamato Honzo* (Enlarged Natural History of Japan) by Kosai Naoumi (1755). However, the most comprehensive was Ranzan Ono's *Honzokomoku Keimo* (1803), which comprised 48 volumes, of which volumes 32 and 33 were on ornithology. The author classified birds into four groups: water birds, plains birds, forest birds and mountain birds. The book is an improvement over the *Yamato Honzo*, giving a synonymy and describing birds more accurately. E. Kämpfer visited Japan in 1690, and published his *Historia Imperii Japonici* in 1727. This contains a few accounts of birds and other animals, but, for the most part, during the long period of time when Japan was closed to foreigners, no study of the ornithology could take place. It was the 19th century before any scientific work was done. The first person seriously to collect birds in Japan was Siebold, who sent specimens to Temminck in Leyden, and these were figured in the *Planches Coloriées*. Temminck also wrote them up as the *Aves* section of Siebold's *Fauna Japonica* (1845–50). A total of 175 species of native Japanese birds are described, but all are from the southern part of the country, chiefly round Nagasaki. The first scientific description of birds from other parts of Japan was by John Cassin, in a report on a collection of birds made by Commodore M. C. Perry who visited the country from 1852 to 1858, and Captain J. Rodgers in 1856 and 1862. After the Meiji Restoration of 1868, foreigners were able to travel freely and collect specimens. Sharpe (1870) described specimens collected by T. H. Bergman, and Finsch (1872) collected at Hakodate, Yokohama and Nagasaki. The first *General Catalogue of Japanese Birds* was published in the *Ibis* (Blakiston and Pryer 1878), and this was later revised and enlarged in the *Transactions of the Asiatic Society of Japan* (1880 and 1882). After Pryer's death, his collection was obtained by Henry Seebohm, who published *The Birds of the Japanese Empire* (Seebohm 1890), a milestone in Japanese ornithology. It discussed all previous important writings on the subject, the geographical distribution, and gave a diagnosis of all known species. It was around 1890 that Japanese writers began to write scientifically on their native avifauna. In 1891, Nobutoshi Okada published the vertebrate (including birds) section of the *Catalogues of the Animals of Japan*. Shortly before this, the zoological department of the Imperial University of Tokyo began a zoological magazine, in which papers on birds were published (op. cit). Dr Isao Ijima was one of the first leaders of this movement. Labouring under the handicap of scanty specimens and poor reference books, his achievements were remarkable, and he succeeded in laying the foundations of modern Japanese ornithology. Besides writing a number of important papers and books, Ijima was one of the founders of the Ornithological Society of Japan in 1911, and its first president.

Schlegel and Siebold became firm friends and collaborated on the *Fauna Japonica* (Siebold 1845–50). Although officially holding only a subordinate position in the museum, Schlegel was quickly regarded as the leading light in the establishment, and visitors applied to him and sought his advice, rather than approach Temminck. Schlegel was one of the first ornithologists to adopt the concept of the geographical race proposed in 1825–26 by Friedrich Faber. Faber was born in 1796 in a village near Henneberg in Denmark. His father was a lawyer, and Faber studied law at Copenhagen University. But he must already have been a keen zoologist, for, as soon as he had passed his final examinations in October 1818, he was given a grant to conduct research in Iceland. He remained on the island till 1821, collecting birds and fish, and keeping copious notes on his collections. The following year, he published his *Prodromus der isländischen Ornithologie*, but because he was technically a lawyer not a zoologist, it attracted little attention at the time. Faber had been of the opinion that the geographical forms seen in many species (and which many ornithologists would have considered separate species) were due to differences in climate. Schlegel, however, showed that this was not necessarily the case. He therefore took the view that geographical races had existed since the beginning of creation, and were immutable. This corresponded to the theories long held by his old friend Brehm.

Christian Ludwig Brehm

Christian Ludwig Brehm was born on 24 January 1787 in Schönau-vor-dem-Walde, and probably became interested in birds through the influence of J. M. Bechstein, who was a friend of his father, and a frequent visitor to the parsonage. When he was a scholar at Gotha, Brehm had already begun to collect birds, prepare and stuff them, and observe their habits. After a short period of service at Drakendorf (1812–13), Brehm became pastor at Renthendorf, where he remained until his death on 27 April 1864. He wrote over 200 papers and several large volumes; the first paper, which appeared in 1820 was entitled 'Whooper Swans in the Osterland', and the last, published only a few weeks before his death, 'Some Species of Birds which are distinguished from their Allies by their Males wearing a similar dress to those of the Females'. The first of his major works was *Beiträge zur Vögelkunde* in three volumes, (Brehm 1820–22). A total of 104 German species of birds are described in minute detail, the plumage, dissected bodies, and much information on their habits, all from personal observation. By 1823, Brehm's great collection already contained 4,000 skins, and after he had discovered two species of treecreeper and two of goldcrest, which had previously been unrecognised, he became obsessed by minor variations and began describing species by the dozen and then by the hundred. Brehm became notorious for binomially naming every minor variation he could find in birds so that they appeared as species. Only his notes, not his names, made it clear whether he regarded a form to be a species, a subspecies or merely a variety. All of these forms, according to Brehm, were immutable. For instance, he discovered three 'subspecies' of the Red-backed Shrike breeding together, and believed that only examples of the same 'subspecies' regularly paired together. After the modest 104 species in his first book, in 1831 he published *Handbuch der Naturgeschicte aller Vögel Deutschlands* (Brehm 1831), in which over 900 species and subspecies were described. Brehm's collection finally totalled about 15,000 skins. During his lifetime he had offered it to the Berlin Museum, as he feared it might be lost if lightning were to strike the house. But this sale never materialised, and it remained for many years in the garret after Brehm's death, till Otto Kleinschmidt had occasion to consult a type specimen in it. He persuaded Lord Rothschild to buy it and in 1900 it was transferred to Tring, but later sold again to the American Museum of Natural History with the rest of Rothschild's specimens.

Brehm's books were read only by ornithologists; they were eclipsed by the more popular *Natural History of Birds in Northern Germany* by Johann Friedrich Naumann (Naumann 1795–1803). The

Naumanns were a family of farmers who lived in the village of Ziebigk near Köthen, and were zealous birdcatchers. Johann Andreas Naumann (1744–1826) was passionately devoted to ornithology, and when his eldest son, Johann Friedrich (1780–1857), showed a talent for painting, his father had him take lessons. Whenever Naumann senior shot an interesting or rare bird, his son painted it for him. Naumann began to make descriptions of all the birds which passed through his district, and this was published in four volumes (1795–1803) at the author's expense, entitled *Detailed description of all birds of wood, field, and stream that reside in or pass through the principality of Anhalt and some of its neighbouring districts*, by Johann Andreas Naumann, illustrated with engravings by Johann Friedrich (Naumann 1795–1803). In 1816, Naumann junior received a letter from Temminck, asking him to paint a series of plates for a second edition of *Manuel d'ornithologie*. He agreed under certain circumstances, pointing out that he had to run his farm as a full-time occupation, doing virtually all the physical work himself as the family was too poor to employ servants or farmhands. All his ornithology was done in his spare time, mainly in winter, when there was less farm work.

While endorsing many of Brehm's ideas, Schlegel was not happy with the vast increase of binomials, and adopted a suggestion proposed in 1828 by Carl Friedrich Bruch (1789–1857). This was trinomial nomenclature, first adopted by Schlegel for geographical races as early as 1844, though the practice did not come into general use until many years later. Brehm's multiplicity of names was criticised by many, particularly Gloger, and later Brehm himself realised the confusion he had caused by these binomials, and recommended Schlegel's suggestion of trinomials. One point on which Brehm and Gloger were in agreement, however, was the significance of coloration in birds and their eggs (Stresemann 1975). They both stressed the point, already noted by Zorn, that the eggs of birds which breed in the open on the ground are inconspicuously coloured. Eggs are only conspicuous when this is not disadvantageous. White or pale-coloured eggs are found in aggressive species, those which never leave the nest unattended, or those, like ducks and grebes, which pull a covering of down or other nesting material over the eggs before leaving. In many birds, egg colour matches the surroundings, or even the colour of the nesting material. Gloger noted that cuckoo eggs are not white, but failed to realise the significance, declaring merely that white eggs would be too noticeable by foster parents and predators alike, and therefore: 'they always preserve a mean between the extremes of all those eggs with which they will be incubated'. It was not till 1850 (in the journal *Naumannia*), that G. H. Kunz (Stresemann 1975) noticed that many cuckoo eggs actually match the colour of the eggs in whose nest they are laid. He thought that the female cuckoo was stimulated by the colour of the eggs in the nest, whereas Opel in 1858 believed that the type of food fed to the young cuckoo created the shape and colour of the egg, and forced her to seek out the same host species in which to lay.

❧

1858 was a significant year. Temminck died, and in England, Philip Lutley Sclater published a paper in the *Proceedings of the Linnean Society* setting up six zoological regions which he called the Palaearctic, Aethiopian, Indian, Australasian, Nearctic and Neotropical. Although some of the names have now been altered, these zoogeographic regions are still in use today. Sclater was persuaded to write this paper because of the confusion previously existing in regard to geographical regions in the Natural Atlases of his time. These divided the world very arbitrarily into regions with little regard to differences in animal life. He wrote:

> The world is mapped out into so many portions, according to latitude and longitude, and an attempt is made to give the principal distinguishing characteristics of the Fauna and Flora of each of these divisions; but little or no attention is given to the fact that two or more of these geographical divisions may have much closer relations to each other than to any third, and, due regard being paid to the general aspect of their Zoology and Botany, only form one natural province or kingdom (as it may perhaps be termed) equivalent in value to that third (Sclater 1858).

Philip Lutley Sclater

One of the works about which Sclater complained, A. K. Johnston's *The Physical Atlas of Natural Phenomena* (1848), divided the world into 16 areas: 1. Canada and Greenland northwards, 2. USA, 3. Mexico, the West Indies and Central America, 4. South America south to Bolivia and mid-Brazil, 5. From this line, south to the Bay of Conception and the Rio Colorada, 6. Southwards from this line to the South Pole, 7. Europe, 8. North Africa from the Mediterranean to the southern edge of the Sahara, 9. From this line to Lower Guinea and Mozambique, 10. From Lower Guinea and Mozambique to the Cape, 11. Asia, from the River Ob and the Sea of Okhotsk northwards, 12. From this line south to the Black Sea and the Sea of Japan, 13. From the Black and Japan Seas south to Arabia and the China Sea, 14. From here south to Indonesia, 15. New Guinea, islands to the north and east, and Oceania, 16. Australia and islands to east and south (presumably this includes New Zealand, though this is not stated). The author of this extraordinary work indicated the number of species to be found in each of the geographical regions, using the six avian orders of Cuvier: Rapaces (raptors), Scansores (climbers), Oscines (songsters), Gallinaceae (game birds), Grallatores (waders), Natatores (swimmers). These figures are totally meaningless today since so many more species have been described. The author pointed out, for example, that Europe possessed more species than any area other than tropical America. This, of course, only means that at the period in question the avifauna of Europe was better known than that of any other area.

Sclater (1829–1913) studied birds for over 50 years and wrote over 1,300 papers, either alone or as co-author. He was born at Tangier Park, Hampshire, and lived in the south of England for most of his long life. Although he practised law for a few years, he was soon stimulated by H. E. Strickland into an interest in natural history. He often visited European museums, and in 1856 spent some time in the eastern USA. Three years later, he went on a collecting trip to Tunisia and Algeria, and later that year (1859) became secretary of the Zoological Society of London. For many years he was editor of the *Ibis*, the journal of the British Ornithologists' Union. However, Sclater was a Creationist, and he believed that different creations of beings had occurred in each of the zoogeographic regions.

Alfred Russel Wallace

Temminck had lived too long. Schlegel, having a free hand at last, now began to try to build up the Leyden Museum's collections systematically according to his own principles, but he was already too late. The English had stolen a march on him. When the Natural History Commission had been disbanded in 1850, the Dutch had done nothing more about zoological research in Indonesia, and in the meantime another man had stepped in and won undying fame with his discoveries in Schlegel's own province. His name was Alfred Russel Wallace, and although he had slipped into zoology by chance, there can be no doubt that he was the right man at the right time. He was born on 8 January 1823 in Usk, Monmouthshire, the son of poor parents. Originally intended to be a surveyor, he became a teacher at Leicester, where he made the acquaintance of Henry Walter Bates, an obsessive collector of insects. Together the two young men hatched a fantastic and hair-brained scheme to make a collecting journey to the tropics on credit, and pay for it later by collecting vast quantities of zoological specimens, which they would sell. Wallace proved not only to be a genius, but to have an incredible slice of luck, that luck which fate had cruelly denied to equally gifted men such as Kuhl and Boie.

Wallace stayed in South America for four years, but, on the way home, the ship caught fire. He was lucky to escape with his life; but he lost the specimens he had collected in the last two years, and all his drawings. Undaunted on his return to England he set about writing two books of his travels. Fortunately he had sent some collections of specimens back to England ahead of him, and the sale of these financed his next expedition – to the Malay Archipelago. He was there from 1854 to 1862, during which time he visited and worked Sarawak, Sumatra, Bali, Lombok, Celebes (now Sulawesi), Amboina, Ceram, Buru, Ternate, Batjan, Waigeo, northwest New Guinea, the Aru Islands and Timor, while an assistant collected on Sulabesi, Morotai, Misoöl, Salawati and Flores. He continued to finance his journeys by the sale of duplicates from his zoological collections which were sent back to England. The entire collection consisted of 125,000 items, including 8,000 bird skins. Of the many spectacular finds, none is so popularly associated with his name than the extraordinary bird of paradise, *Semioptera wallacii*, which he found on the island of Batjan, still known as Wallace's Standardwing. But the greatest fruit of Wallace's journey was that it led to revolutionary insights into the reasons for animal distribution, and the origin of species. The studies of Wallace resulted in an evolutionary theory put forward virtually simultaneously by himself and Charles Darwin, and delivered in two papers read to the Linnean Society on 1 July 1858. Wallace was also one of the first to appreciate the principle of convergence. Thus similar-looking birds occurring in different parts of the world are unrelated, but because they have evolved to fulfil similar functions (or, as we would now say, to occupy the same ecological niche), have become similar in appearance. Such groups include the Old and New World warblers. The distinct avifaunas found on islands could only be explained by their having evolved in situ from ancestors which had subsequently died out. Almost as soon as Wallace and Darwin had proposed their theories of evolution, support was forthcoming in the form of a slab of Jurassic stone discovered in the Solnhofen quarries in Bavaria in 1861. The fossil, *Archaeopteryx*, caused T. H. Huxley to declare that "Birds are greatly modified reptiles", and proceeded to draw up the first avian genealogical tree.

Charles Darwin

A considerable influence on Darwin was that of his grandfather, Erasmus (1731–1802), a man of medicine and a poet. His poetry was well thought of at the time, but history has consigned it to the bin of ridicule. Erasmus Darwin believed that all living phenomena, both physical and psychical, are due to the contraction of irritable fibres, induced by 'idea'. The origin of an embryo is due to a filament, derived from the father, to which the mother gives nourishment. Organs which the animal needs are produced by 'irritation' in the respective parts of their bodies. Thus cocks acquire spurs by fighting for their mates. All living creatures, no matter how different, have their origin in the same primeval filament, but their offspring have been altered by the different conditions of life under which they exist. Here we see already a very primitive form of Charles Darwin's theory, but Erasmus was in no way interested in the origin of species. Another writer who influenced Charles Darwin took an opposing view to that of Erasmus. Charles Lyell (1797–1875) in *Principles of Geology* (1830–32) denied the possibility that the earth had previously existed in a molten state, and rejected Lamarck's theory that creatures of earlier times differed from those of today. He insisted that mammals and birds had existed from very early times. Lyell's criticism of Lamarck exposed the weakest aspect of the latter's theory, by pointing out that Lamarck never attempted to discover the origin of a single organ, but merely commented on the modifications to those already in existence. Lyell did not believe that species could vary beyond a certain limit – if one forced a species beyond this limit, it died, as happens if one tries to adapt species to unsuitable climatic conditions. Sterility of hybrids, he considered, is also proof of the immutability of species.

Charles Darwin was born in 1809 at Shrewsbury. He did poorly at school and failed to complete his medical studies at Edinburgh due to boredom with the subject. He then spent three years studying theology at Cambridge where he took a BA. He was prevented from pursuing a career in the Church, by a chance recommendation for the post of naturalist on the voyage of the *Beagle*, this voyage lasting five years. He suffered from sea-sickness for most of the journey, and this seems to have had a permanent effect, for he suffered from ill-health for the rest of his life – a life devoted to writing up the results of his voyage, and formulating (over a long period) his theory of natural selection. In 1842, he left London and settled in the village of Down (now Downe) in Kent. His health did not improve, and he suffered from insomnia and stomach problems. It was only the devoted nursing of his wife that enabled him to live as long as he did. He died in 1882. Brought up a devout Christian, Darwin's observations often conflicted with ecclesiastical dogma, and in particular the divine creation of every species. In observing island faunas he was deeply disturbed to find a different species on each island, differing only in minor points. Why should God have gone to the trouble to create so many different, but closely related, species? For nearly 20 years Darwin pondered these problems, seeking proof to give substance to his theory (Lamarck's theory he at once rejected as rubbish). Finally, he published *On the Origin of Species by Means of Natural Selection* (Darwin 1859), one of the most famous and controversial books ever written. Linnaeus had originally held that all species had been created 'in the beginning'; later he modified this view and suggested that a few species had been created, out of which others had evolved. By Darwin's time, however, 'creation' also meant 'immutability'. Various judgements have been made on Darwin. Even in his own day, opinions differed as to whether he was one of the world's greatest geniuses or an ignorant dilettante.

One of the first to espouse the theory of natural selection was Canon Tristram, who, in discussing the larks and chats of Algeria (1859) remarked that the theory could not be better illustrated than by these groups. Many of the desert species, he observed, seemed to be basically European birds adapted and altered in colour and structure to the conditions of the desert. Henry Baker Tristram (1822–1906) was born near Alnwick in Northumberland, and as a boy was often taken on field trips by friends of his father. After leaving school he went to Oxford where he graduated in classics in 1844 and spent six months travelling in Switzerland, studying the habits of wallcreepers, snow finches and other birds. He was ordained at the age of 23, but as he was delicate and suffered badly from tuberculosis, he was forced to spend a considerable amount of time abroad in warm climates while nominally a canon in Durham Cathedral. That he survived to such a ripe age was regarded as remarkable. North Africa and the Holy Land were visited regularly, and his *Natural History of the Bible* (Tristram 1867) became a classic on the subject. His large collection of skins went mainly to Liverpool Museum and his egg collection to the private collector, Philip Crowley, some of which ended up in the Natural History Museum; those which did not are now presumed lost.

T. H. Huxley acted as a public front, or bulldog, for the shy and retiring Darwin. Thomas Henry Huxley (1825–1895) was an eminent biologist of his day, and grandfather of both Julian and Aldous. He had received little schooling and was therefore largely self-taught. At the age of 15 he was apprenticed to a medical practitioner and was granted a free scholarship to Charing Cross Hospital Medical School. At the age of 21, he obtained a post as assistant surgeon on board H.M.S. *Rattlesnake*, on which he served for four years during an expedition to the southern seas. He studied marine life, sending back reports and observations from every port of call. These were published in leading journals. Returning to England in 1850 to find himself famous, he obtained three years' leave on full pay from the Admiralty to work up the specimens collected on the journey. He had originally intended to complete his studies and take a degree, but this was never completed. Later, he received honorary doctorates from a number of universities. He obtained a lectureship at the tiny School of Mines (which he later transformed into the Royal College of Science) and wrote on many biological subjects. He served as president of many learned societies, including the Royal Society and was a senator of London University.

Huxley's classification was to divide the class Aves into three 'orders': Saururae, Ratitae and Carinatae. The first (containing only *Archaeopteryx*), had well-developed metacarpals and numerous large vertebrae in the tail. A tail with bones in it is what principally distinguishes *Archaeopteryx* from all extant birds. The other groups correspond respectively to the Ratites and other birds. Huxley's subsequent arrangement 'On the Classification of Birds', *Proceedings of the Zoological Society of London* (Huxley 1867) was based entirely on the form and relations of the bones of the palate and base of the skull (see Appendix 25). One of his more interesting contributions to anatomy was to explode the absurd theory of Oken that the skull was composed of fused vertebrae. But the first attempts to discover the evolutionary relationships of birds relied almost entirely on structure, particularly osteology; the systematists (such as Huxley, Garrod and Sundevall) failed to realise that every structure is influenced by its function, and that study of form must proceed hand in hand with that of function.

Alfred Newton

Tristram soon rejected the theory of evolution, because it conflicted with Christian beliefs. A more constant disciple of Darwin's was Alfred Newton. Newton was born on 11 June 1829 while his parents were on holiday in Switzerland, and brought up on the family estate of Elveden, near Thetford, in Norfolk. Little is known of his childhood except that at the age of five he fell and injured his knee which resulted in permanent lameness. He took an interest in natural history from an early age, and, having graduated at Cambridge in 1852, visited Scandinavia in the company of John Wolley, an expert in the oology of that area. (Wolley died young, and Newton later published the catalogue of his friend's egg collection, as the *Ootheca Wolleyana* in four volumes, in 1864, 1902, 1905 and 1907 respectively). In 1857, Newton visited the West Indies and the USA, where he met the leading naturalists such as Baird and Coues, with whom he kept up a life-long correspondence. It was on this trip that he taught the Americans how to blow eggs with a single hole in the side by means of a drill and blowpipe, instead of the old method of two holes. He published *Suggestions for Forming Collections of bird's eggs* in the *Miscellaneous Collections of the Smithsonian Institution* in 1860.

Newton originally intended to enter the Church, but in 1862 decided against it, and eventually became professor of zoology at his old university. During his many years at Cambridge he gave regular lectures, which he hated, always reading from a carefully prepared manuscript. At intervals, in the margin, he had drawn the outline of a tumbler, and at these points, took a sip of water. Sadly, he destroyed all his lectures shortly before his death in case anyone should publish them. More illuminating were the informal Sunday evening discussions regularly held in his rooms. It was in Newton's rooms in Cambridge on 17 November 1858 that a meeting was held, as a result of which the British Ornithologists' Union was formed, and the *Ibis* journal founded. Newton was a regular contributor of ornithological articles to *The Record of Zoological Literature*, founded in 1864 by Albert Günther, who edited the first six volumes (after that the name was changed to *Zoological Record*, under which name it continues to this day). Newton also served on many committees which considered different aspects of zoology. He was particularly involved in conservation, fulminating against the then common practice of shooting seabirds (particularly Kittiwakes) during the breeding season, in the name of 'sport', thus condemning vast numbers of chicks to starvation.

One of the more ingenious attempts to explain what evolution actually meant, and how it differed from previous theories was by C. J. Sundevall in his *Tentamen* (Sundevall 1872–73, see Appendix 26). The much-used word at that time 'Affinity' can have two meanings, he explained. Firstly, that animals and vegetables are constant, that they were created in the beginning as they now are and will remain so for ever. This opinion, he pointed out, seemed to be in perfect accordance with experience, and had been held more or less from the beginning of time by naturalists. On the other hand, the theory that forms are in a constant state of change accorded well with the evidence of geology, which showed that each strata contains different forms, most resembling those in the immediately preceding or succeeding strata. This view seems to be in complete opposition to actual experience, but this could be explained if changes took too long a time to be observed. Sundevall then introduced a brilliant argument, by pointing out that the theory of evolution is actually much older than the theory of immutability. "It has always been adopted," he says, "or at least admitted, under more or less vague forms and almost unconsciously, by the majority of mankind ... The yokel, when he says that the fox is related to the dog, the pheasant to the domestic fowl, or even the eel to the serpent, means to say, without doubt, that they derive their origin from the same ancestor." It is a most subtle comment on the paradoxical working of the human mind, which finds it equally impossible to understand infinity, as to accept the lack of it. He continued: "We ask any one, if it is possible to imagine a new creation, out of nothing, for each geological period, or if one can suppose that men and animals could have subsisted in their present form under conditions so different of heat, of moisture, of atmospheric pressure, and of vegetation or sources of food which prevailed in the ancient epochs of the world's history." However, Newton believed that Sundevall's *Tentamen* betrays a lack of the sense of proportion. "In many of the large groups very slight differences are allowed to keep the forms exhibiting them widely apart, while in most of the smaller groups differences of far greater kind are overlooked." Although much praised in certain quarters, in Newton's view this system was a total waste of time and paper, or as he put it:

> When a man of Sundevall's knowledge and experience could not, by trusting only to external characters, do better than this, the most convincing proof is afforded of the inability of external characters to produce anything save ataxy. The principal merits it possesses are confined to the minor arrangement of some of the Oscines; but even here many of the alliances, such, for instance, as that of *Pitta* with the true Thrushes, are indefensible on any rational grounds, and some, as that of *Accentor* with the Weaver-birds and Whydah-birds, verge upon the ridiculous (Newton 1896).

An adaptation of Sundevall's system by W. Lilljeborg was read at the Zoological Society and published on 9 January 1866 in the Society's *Proceedings* (Lilljeborg 1866). The author provided a mere outline, but the attitude of classical correctness prevalent at that time is indicated by his comment in referring to other systems, that a system "that places the dirty Vultures highest, does not seem to us to indicate a correct idea of the nature of birds" (see Appendix 27). Newton (1885) pointed out that Lilljeborg's system was the first one to begin with the lower orders of birds and progress to the highest, instead of the reverse, and that, as such, it had a considerable effect on the opinions of American ornithologists. In the *Ibis* for 1880, P. L. Sclater attempted to create a general arrangement by summarising and harmonising all the views of the various systems which had been proposed in the fairly recent past. However, the result came out as very similar to that of Sundevall, the six main divisions being very similar to the main divisions of the latter, though with different names.

Needless to say, evolution was not without its opponents. As late as 1914, Anton Reichenow, still believing that systematics should be a purely artificial arrangement of convenience, divided birds into six groups: shortwings, swimmers, stiltbirds, skinbills, yoketoes and treebirds (Reichenow 1913–14). Most lamentably, this absurd system (see Appendix 28), already out of date when first proposed and never used by any ornithologist, was adopted by the decimalised library system worldwide, which must continue to cause much confusion. Another opponent of evolution, Richard Owen (1804–1892), one of the most influential of the anti-Darwinists, began life as an apothecary, later becoming a medical doctor. In 1860, he was made head of the natural history department of the British Museum, and subsequently supervised its move to the Natural History Museum, South Kensington. He is considered to be England's greatest comparative anatomist, among the birds he studied in detail being the moas of New Zealand, and *Archaeopteryx*. Much of his classification was based on the plan of comparing the same organ throughout all animal groups to determine the changes which the organ had undergone in different species. Most of his work, however, is not directly relevant to ornithology.

Bernard Altum, another anti-Darwinist, was born on 31 December 1824, in Münster, Westphalia, the son of an artisan and amateur naturalist. Altum's book, *The bird and its life*, published in Münster (Altum 1868), was more fiercely attacked than any other book of the 19th century. At an early age, birds had become his great passion; he learned to shoot and stuff them while still at school. After graduation, he entered the priesthood, but his passion for ornithology did not diminish, and he was later to combine the two callings. He worked from 1853 to 1856 as Lichtenstein's assistant in the Berlin Museum, before returning to Münster as curate at the cathedral. Two years later, he also became lecturer in natural history at the Royal Academy. Later still, he was active in the study of the zoology of forests, and in 1891 was elected president of the Deutsche Ornithologische Gesellschaft. He died on 1 February 1900. Throughout his life, he was one of the stoutest opponents of evolution. *The bird and its life* is a diatribe against Darwinism, which, as he had foreseen, gained him many enemies. But in spite of this, there was much in Altum's work that was new and important, as for example, his demonstration of instinctive behaviour. In the matter of camouflage, he exploded Zorn's theory by showing that while birds whose colour matches the ground crouch at the sight or sound of danger and do not fly away, they do not understand the function of their colour, because albinos behave in exactly the same way as normally coloured birds. He was the first to explain iridescent feathers as due to refraction of light by the feather structure.

Although Schlegel was unconvinced by the new theory, and remained all his life an ardent anti-evolutionist, it confirmed him in his opinion (inherited from Brehm) that the study of individual and geographic variation was important. On 29 March 1859 he wrote to the trustees of his museum that in the present state of science a duplicate meant something different to that which it had previously done:

> If one wishes to know a species thoroughly, one must possess specimens of different sexes, different ages, from different seasons, and from the most important localities (Stresemann 1975).

Poor Schlegel was destined to be flouted by the English. In 1857, he sent his son Gustav to China to collect for the museum; but he was just too late, an Englishman had got there first. This was Robert Swinhoe (1836–1877), generally considered the father of Chinese ornithology. As a result of Swinhoe's work, Gustav gave up ornithology, to become a renowned sinologist. His name still lives in ornithology, however, as the Pechora Pipit *Anthus gustavi* was named after him.

The Chinese were aware of birds from a very early period. The age of the *I-Ching* is not known, but the first pictorial representation of birds dates from the Shang-Yin Dynasty (1523–1100 BC), when they were used as decorative motifs on bronze vessels. These were very stylised, but owls and some other birds are recognisable. During the next Dynasty, the Chou (1100–256 BC) birds such as Mandarin Ducks (*Aix galericulata*) were modelled in bronze and pottery. The Mandarin Duck featured importantly in oriental literature and folklore; its fidelity became legendary. The earliest known paintings date from the Han Dynasty (206 BC–AD 220). From the end of the Sung Dynasty (AD 960–1280), oriental birds were beginning to reach Europe, but it was not till 1498 that Vasco da Gama discovered the main sea route to the east, and trade was established. Lacquer, silk paintings and porcelain decorated with birds flooded into Europe. By the early 18th century live Chinese birds began to reach the west, such as Silver and Golden Pheasants, and were kept on the estates of the rich, where they were seen and drawn by artists like Eleazar Albin. In 1741, Pierre Poivre reached Canton and made a series of drawings of Chinese birds, but unfortunately these were lost at sea when his ship was captured by the British on the return journey.

Robert Swinhoe

It was not till the 19th century that any scientific study of Chinese birds began, with the labours of Robert Swinhoe and Armand David. Swinhoe was born in Calcutta in 1836 and educated in England. He arrived in China in 1854 as interpreter for the Consular service, and devoted most of his spare time to studying and collecting birds, mostly in eastern China, though he made one expedition up the Yangtse to Sichuan. He collected extensively on Hainan and Taiwan, where every specimen was either a first record for the island or an undescribed species. Swinhoe was responsible for the first checklist of birds of China, published in 1863 in the *Proceedings of the Zoological Society of London*, with a revised list in 1871 (Swinhoe 1863, 1871). He also wrote many papers, both in this journal and in the *Ibis*. Swinhoe was a vice-consul in Amoy until his repatriation and death from syphilis at the early age of 41. While Swinhoe was in the east of the country, the Basque missionary, Père Armand David, was in the west. He had been born at Espalette in the Basses Pyrenees, and, in 1862, went to China, where he was in charge of the French school at Peking. In 1864, he made an expedition to what little was left of the Mongolian forests to the north of the city, and in 1868 made the first of a number of journeys through the western parts of the country, until his health broke down and he was obliged to return to France where he died in 1900. He made many discoveries, such as that of the wonderful Chinese Monal, *Lophophorus lhuysii*, and was the first to see the Blue Eared-Pheasant, *Crossoptilon auritum*, since Pallas had described it nearly a century before. His work culminated in the two-volume work, *Les Oiseaux de Chine*, written in collaboration with E. A. Oustalet and published in Paris (David and Oustalet 1877).

Swinhoe's Pheasant

In the 1870s, the Russians began to take an interest in the country and a number, including Kozlov, Roborovski and Berezovsky, explored Sinkiang, Tsinghai and Kansu. The most notable was Przevalski, whose travels became best-sellers and were translated into a number of languages. However, the first person to collect in the southwest of the country was Dr John Anderson, superintendent of the Calcutta Museum. He made two expeditions into the wild and lawless country, in 1868 and 1875. In spite of constant attack from unfriendly natives, the two expeditions collected 233 species of birds. Captain Nicolai Mikhailovich Przevalski made a total of four voyages through Mongolia and northern China, and was the first European to see Lop Nor since Marco Polo had passed it in 1275. On his first expedition (1871–73), he covered 7,500 miles and collected about a thousand birds of 289 species. These included about 20 that were new, including the Black-necked Crane, *Grus nigricollis*, and Severtzov's Grouse, *Tetrastes sewerzowi*. The second expedition (1876–77) explored Sinkiang, Tibet and the slopes of the Altin Tagh, where he collected about 500 specimens of 180 species. He had failed to reach Lhasa on this journey, but set out again in 1879 accompanied by V. I. Roborovski. On the fourth journey (1883), he was again accompanied by Roborovski, and by P. K. Koslov. Przevalski was preparing for a fifth journey when he died of typhus at the age of 49. Many bird species were named in his honour, as well as *Equus przewalskii*, the famous wild horse of the plains of central Asia. The fifth Russian expedition (1879–91) set off under the leadership of Colonel M. V. Pevtsof, accompanied by Kozlov and Roborovski, who made many important bird discoveries, and collected 234 species. Many of these were on the little-known Chang Tang plateau. Kozlov continued to explore western China, making three more expeditions between 1893 and 1909.

At the same time, the botanist G. N. Potanin and the ornithologist M. Berezovski were exploring Kansu, eastern Tsinghai and Szechwan during the years 1884–86. Between 1889 and 1890, the brothers Grum-Grzimailo travelled across Sinkiang to Kansu and collected many specimens. In east China, during 1893 and 1894 F. W. Styan, a tea merchant living in Shanghai, made many discoveries, including the White-eared Partridge *Arborophila ardens* from Hainan, and Styan's Bulbul *Pycnonotus taivanus*. C. B. Rickett was collecting further south, and John D. LaTouche, a British customs officer, made studies of the local birds culminating in his two-volume work *A Handbook of the Birds of Eastern China* (LaTouche 1925–37). Charles Boughey Rickett was born in Hong Kong on 10 December 1851, and died at Reading, England, on 8 April 1944. He was the son of John Rickett, an employee of the East India Company. C. B. Rickett joined an eastern banking company and served in such places as India, Japan and Java, being transferred to China in 1890. There he met LaTouche and Styan, and the three carried out a joint trip to the hills north of Fuchow where they made a number of interesting ornithological discoveries.

Schlegel was still interested in the East Indies, most particularly New Guinea, and in 1859 he sent out the collector H. A. Bernstein, a highly gifted and trained ornithologist. Heinrich Agathon Bernstein was born at Breslau on 22 September 1828, and studied medicine at the university there, graduating in November 1853. As a boy he had always been eager to get to the tropics, and decided to emigrate to Java, where he had been offered a medical post. He started to study birds intensively, resulting in a series of articles in the *Journal für Ornithologie* (1859–1861). Bernstein was the first to examine the salivary glands of cave swiftlets (genus *Collocalia*) and thus solve the problem of the construction of edible birds' nests. He also discovered that coucals (genus *Centropus*) have only one developed testis. He collected for some years, but died in 1865, of a liver abscess. He was succeeded by the collector Hermann von Rosenberg, who explored extensively in the area. Rosenberg was born on 7 April 1817, at Darmstadt, and began his career in the army. Posted to Java in the Dutch colonial service, he was commissioned to explore Sumatra, and remained in the army in Indonesia for the next 16 years. He did not, apparently, begin to collect birds till he was assigned on an expedition to New Guinea in 1858, Schlegel having succeeded in getting him a government grant to collect for Leyden Museum. In 1863 and 1864 he collected the Celebes. After this, he was sent to the Aru and other islands in the area. Until 1871, he travelled and collected all over the western part of New Guinea and its outlying islands, and Schlegel was delighted with the many new species he found and sent home. After his retirement to Europe, Rosenberg described his travels in a popular book *Der Malayische Archipelago* (Rosenberg 1878), which, however, does not match Wallace's similar travelogue *The Malay Archipelago* (Wallace 1869).

In the meantime, Schlegel had taken on a young assistant called Otto Finsch. At the same time he founded a new periodical *Notes from the Leyden Museum* and embarked on the 14-volume work *Muséum d'histoire naturelle des Pays-Bas* (Schlegel 1862–80). He was also the first to employ two talented illustrators, J. G. Keulemans and Joseph Smit, both later to achieve their greatest fame and reputation in England. But he was not to achieve much more in the East Indies. Schlegel's later years were troubled. His wife died in 1864, Finsch left to go to Bremen Museum, which offered better terms, and, under the vigorous direction of Bowdler Sharpe, the British Museum began to outstrip Leyden in terms of collections. Leyden's great days were over, and it remained a somewhat old-fashioned museum. Schlegel maintained all his collections in a mounted form, as he considered this was the only way in which they could be useful. The idea of skins kept in drawers was repugnant to him. He remained all his life a stubborn opponent of the theory of evolution, and regarded Darwin's theories as little more than speculation. The main objection, which was also held by such German ornithologists as Blasius and Cabanis, was that Darwin had proved nothing. Direct evidence for his hypothesis was impossible. Schlegel could not see the beginnings of future species in local races and geographical variations.

Otto Finsch (1839–1917) had been passionately devoted to birds from an early age. Having travelled in the Balkans in his teens, he was fortunate to secure an appointment at Leyden with Schlegel, moving to Bremen in 1864, where he obtained a post under Hartlaub. Gustav Hartlaub had been born in Bremen on 8 November 1814, and was the most distinguished German ornithologist for almost 50 years. He was described as a critic to be feared, but was himself seldom, if ever, attacked. He remained in Bremen for most of his life, but his interests embraced the world. He studied at Bonn and Berlin before graduating in medicine at Göttingen in 1838, but as his family were wealthy, he had no need to start work immediately, so he spent some time travelling in Austria, Holland, France, England and Scotland. In 1840 he decided to take up exotic ornithology, purchased an ample library of books on the subject at his own expense, and bought specimens from dealers to enlarge the trivial bird collection that already existed at the Bremen Natural History Society, which by 1844 had grown to 2,000 specimens. He soon began to find undescribed species among his purchases, and started publishing descriptions. In 1846, he was asked to begin a series of annual reports on the progress of ornithology for *Archiv für Naturgeschichte*. Not long afterwards, in 1852, Hartlaub was instrumental in setting up a new journal, the *Journal für Ornithologie*, which was, and is to this day, the leading German ornithological journal. Its first editor was Hartlaub's friend, and former assistant of Lichtenstein, Jean Cabanis (1816–1906). Hartlaub spoke and wrote English perfectly, and visited England on several occasions, making many English friends. He had an excellent library, but kept no personal collection of specimens, all those he obtained being donated to the Bremen Museum. To begin with, Hartlaub was interested in birds from all over the world, but soon began to specialise in those of West Africa, probably as a result of the stimulation of Carl Weiss, whose zoological collections reached Hamburg Museum, and resulted in Hartlaub's major work *System der Ornithologie Westafrica's* (Hartlaub 1857). Hartlaub also visited a number of other museums in Europe to study their collections. In 1856 he was in Paris, and in 1859 in London, where he forged links with the English ornithological fraternity. Another important piece of work was his 'Apuntamientos' on the Birds of Paraguay, in which he identified and indexed the species described by Azara. But it was more specially to Afrotropical birds that he directed his attentions. He was an honorary member of the American Ornithologists' Union, the British Ornithologists' Union and the Zoological Society of London, contributing frequently to their journals. He died in Bremen on 29 November 1900.

At the age of 30, Finsch published his monograph *Die Papageien* in two volumes, and while at Bremen developed a great interest in birds of the Pacific islands. Around this time, J. C. Godeffroy, the owner of a trading firm in Hamburg with principal interests in the South Seas, decided to set up a museum, the Museum Godeffroy, and had material sent to him from all parts of the Pacific. Finsch and Hartlaub worked on this material, and published the *Ornithologie der Viti, Samoa und Tonga-Inseln* (Finsch and Hartlaub 1867). The same team produced *Die Vögel Ost-Afrikas* (Finsch and Hartlaub 1870), based on the collections sent back by the traveller Baron Claus von der Decken. Finsch's feet started to itch and he desired to see the South Pacific. In the meantime, in 1872 he crossed North America. In the summer of 1873 he visited Lapland and in 1876 western Siberia. His frequent absences irritated the senate of Bremen Museum, and when he applied for leave in 1879 to explore the South Pacific islands, he was told curtly to choose between his journey and the museum. Finsch resigned without a moment's hesitation and set out. He travelled in the Pacific for several years, returning in 1882 laden with large collections of birds, which he delivered to the Berlin Museum. Shortly after this he was appointed by the German Government to conduct a secret mission to annexe much of New Guinea and the Bismarck Archipelago, which became known as Kaiser Wilhelm Land. This accomplished, he became disillusioned with politics and returned to natural history. In 1884, he had met the collector Carl Hunstein in Queensland, and bought the latter's collection of bird skins. For some years Finsch was a freelance writer, mainly studying geography and ethnology. But in 1899, he was offered a post at the museum in Leyden. In 1904, he returned to Germany to become curator of the ethnographic section of Brunswick Museum, which post he held until his death in 1917. He contributed over 150 papers to the *Ibis* and other journals, from which he emerges as a somewhat pompous and opinionated individual. A. O. Hume, in a 28-page review of *Die Papageien*, in which he demolished everything Finsch had to say on the genus *Psittacula*, remarked:

> We are all liable to error, but for a cabinet naturalist, on the strength of half a dozen wrongly sexed skins in some museum, to take upon himself to contradict the definite statements of trustworthy field naturalists ... appears to me to indicate a tone of thought incompatible with the philosophical investigation of any branch of physical science (Hume 1874).

Chapter 10

The Turn of the Nineteenth Century and the Introduction of Trinomials

After Bonaparte left the USA in 1828, American ornithology experienced a tremendous boom, largely due to the ambitious, and, by all accounts, somewhat abrasive, curator of the Academy of Natural Sciences in Philadelphia, John Cassin. Cassin intended to create a museum of world importance on the European model, and set himself to build up the size of the collections, acquiring, among others the great collection of the French nobleman Victor Masséna, John Gould's Australian birds, and the egg collection of Des Murs. The bulk of the Massena collection was purchased for the Philadelphia Museum, by Thomas Wilson, a wealthy member of the institution, and later its president. He approached J. E. Gray of the British Museum for advice as to a suitable collection to purchase for the institution. Gray recommended the purchase of the Massena collection. At the request of Wilson, Gray visited Masséna in Paris, and the deal was finalised in a few minutes. The collection consisted of about 12,500 specimens, and its acquisition by the Americans, over all the crowned heads of Europe, was described as greatly to the credit of American energy and enterprise.

John Cassin

John Cassin (1813–1869) was the son of a Quaker landowner. Although one of the most eminent American ornithologists of the mid-19th century, he was essentially a museum worker. In the back room of the library of the Academy, surrounded by books, he arranged, described and classified bird skins from all over the world. When the Civil War began in 1861, Cassin joined the army (at the age of 47) and was captured and imprisoned, which wrecked his health. He died on 10 January 1869, only four years after his release. A less pleasing side of Cassin's character, however, was seen in the treatment of Titian Peale. Peale had been engaged as mammologist and ornithologist on the United States Exploring Expedition of the Pacific, under Captain Charles Wilkes. This expedition left in August 1838, and was away for about four years, visiting Madeira, Brazil, Chile, the South Sea Islands, New Zealand, Hawaii, the west coast of North America, Japan, the Philippines, Singapore and the Cape of Good Hope, returning to New York in June 1842. Peale wrote the official report of the mammals and birds of the expedition, and this was published in 1848, but without the volume of plates which he had prepared. Peale described 109 birds, but no more than a third of them proved to be new. Only about 90 copies of this report were issued, the rest being destroyed by fire, and so this book, Peale's only ornithological publication, is of extreme rarity. What actually happened has never been clear. Was Peale dismissed, or did he withdraw in disgust owing to the failure to publish his illustrations? After a stormy correspondence with Wilkes, Cassin was engaged to write a new version of the report, which appeared in 1858, together with a volume of drawings, mostly by Peale. Peale considered himself very badly treated in the matter, and it seems that Cassin wrote his amended version of the report in such a way as to give himself as much, and Peale as little, credit as possible. Owing to the demise of his father's museum, Peale was appointed to an office job, and later became examiner in fine arts and photography. He died on 13 March 1885, having outlived Cassin by some 16 years. His library and collections were dispersed, and most is now lost, but his manuscript journals of the Exploring Expedition were subsequently found in a junk shop, and deposited in the Library of Congress. Cassin has been described as the only 19th-century American ornithologist who was as familiar with the birds of the Old World as of the New, but this knowledge was gained very greatly through the generosity of Thomas Wilson. Unfortunately, he did not think to endow the collection sufficiently to ensure the appointment of a permanent curator. In 1856, the Philadelphia collection was the largest in the world, numbering some 29,000 specimens, but after Cassin's death in 1869, active work ceased, and for 20 years the collection remained almost untouched by ornithologists.

So Washington, not Philadelphia, became the centre of American ornithology. In 1850, Spencer Fullerton Baird had founded the United States National Museum. His great plan was a comprehensive exploration of the mammals and birds of North America, comparable to the survey Pallas made of the Russian Empire. It was Baird who first noticed that in wide-ranging species, individuals tend to be larger in colder regions and smaller in warmer ones. He also noticed that birds from the Pacific coast were darker in colour than those from the interior. Because these trends could only be studied with the aid of large numbers of specimens, Baird arranged for the collection of a long series of North American species. At the time of his death, the collection contained over two and a half million specimens. Baird (1823–1887) was born at Reading, Pennsylvania, on 3 February 1823 and was educated at Dickinson College at Carlisle. In his spare time, he spent hours hiking through the woods searching for and collecting birds. At an early age he had been offered a place on one of Audubon's expeditions, but was obliged to decline because of a heart condition. After graduating, he went to New York to study medicine, but soon abandoned this discipline. After a number of years of self-education, during which he met and corresponded with many eminent naturalists, he purchased as many books as he could afford, and, as many of these were in foreign languages, he set himself to learn these languages. About 1844, he learned that the newly established Smithsonian Institution was looking for an assistant secretary and, with the aid of recommendations from a number of his friends, was offered the post, though as the building was not ready, confirmation was delayed for nearly three years. His own collections formed the nucleus of the new museum. His job as assistant secretary interfered with his fieldwork; but he was able to advise many younger men, and so build up a body of collectors in the remoter parts of the country, who would send

Spencer Fullerton Baird

specimens to him. Overseeing his collectors took up a great deal of his time, but he was a tremendous organiser with a remarkable memory, and a great letter writer. He was said to be without equal in his ability to inspire others with his own enthusiasm, and to flatter and cajole the best work out of them. His constant plea to his many collectors was to collect the commonest species. All too often these were scorned in favour of rarities, which may have been quite common elsewhere, whereas the commonest species often differed in subtle ways from place to place. It was an era of expansion over North America; government exploration, boundary surveys, geological surveys and Pacific railroad surveys meant that there were many places to which Baird could send collectors, and recruit those on other official duties to collect for him. Baird was the right man at the right time.

In a paper written in 1871, 'On the Mammals and Winter Birds of East Florida' (Allen 1871), J. A. Allen (1838–1921) had investigated Baird's 'climatic laws', found them to be justified, and tried to explain their causes. He pointed out that the large collection of birds at the Museum of Comparative Zoology was the more valuable from their having been collected at basically the same locality, and thus an examination of the material revealed "a hitherto unsuspected range of purely individual differentiation in every species thus far studied". Thus many 'species' already described were based on no more than individual variations. In discussing individual variation he pointed out that it almost exclusively involves intensity in colour, whereas in closely related species there are also differences in the style of coloration. He observed that birds are darkest in wet areas, palest in dry. In other words, he was observing what is now termed clinal variation, in which colour varies continuously across a wide area, and quite sensibly came to the conclusion that if names were to be applied to every single minor variant the number of species would become endless. He proposed that variants should simply be indicated in the description of the species. He had reached the same conclusion that had already been reached by Gloger when he wrote *Change in Birds produced by Climatic Influence* (Gloger 1833) though this book had been largely ignored. In 1877, Allen published his work *The Influence of Physical Conditions in the Genesis of Species* (Allen 1877). In this he attacked Darwin's theory of natural selection, and suggested that the direct influence of climatic and other environmental factors were more likely causes, thus returning more or less to the theories of Lamarck. Climatic variation he considered to be too imperfectly understood to be

Joel Asaph Allen

fully explained. The increase in colour to the south, he decided, coincided with increase in intensity of solar rays, and also humidity. This increase in depth of colour applied particularly to brilliancy, the maximum being reached in the tropics. Longitudinal variation – increase of colour to the west – seemed to be linked to the increase in humidity to the west. He found that in the USA, the darkest individuals invariably occurred in areas of wettest climate, and the palest in the driest. Humidity was, he decided, more important in influence on climatic variation in colour than solar intensity.

Joel Asaph Allen was the eldest son of a farmer and carpenter. His father was unsympathetic to his son's ambition to become a naturalist and expected him to do heavy farm work, as a result of which Allen suffered severe bouts of illness, and for a time was unable to do any kind of work at all. He could not begin his university course till he was 24, and although he spent three years at Harvard, his shyness was so excessive that he did not attempt the oral examination which was necessary to gain a degree. He went to Brazil in 1865, but was ill for a year as a result; then returned to Harvard to work at the Museum of Comparative Zoology where in 1871 he was promoted to assistant in ornithology and until 1885 acting curator of birds and mammals. All his other major expeditions were to the west of the US, and had been made before he was 36, thereafter ill health tied him to a desk for the rest of his active life. In 1885 he became curator of birds and mammals at the American Museum of Natural History, and then in 1908, curator of mammals, remaining there till his death at the age of 83. He was the first president of the American Ornithologists' Union, a position he held for eight years, and a member of Council for 38 years. His major writings on birds and mammals were in the areas of systematics and biogeography. At first he had objected to the use of trinomials, but later admitted that with more material and better collecting methods, more refined subdivisions were needed. However, he was a Lamarckian, and although this view was popular with American ornithologists for a time, Elliot Coues and others led the mainstream of American ornithological thought away from it. Allen disagreed with Wallace's conclusions in *The Geographic Distribution of Animals* (Wallace 1876). He stressed that faunal boundaries must be based primarily on climatic zones, since this is the case in North America (but not necessarily so in the tropics). It was because of this disagreement that Allen's views never really gained recognition in Europe. Today, however, his views have gained wider acceptance.

Meanwhile, another solution to the question of the nomenclature of geographical variation was proposed by Elliott Coues (1842–1899) who, in his *Key to North American Birds* (Coues 1872), proposed that all the variations between the two extremes in any species should be given extra names preceded by 'var.' for variation. A decade later, Ridgway in *Nomenclature of North American Birds* (Ridgway 1881) dropped the 'var.' and thus took the final step to the use of trinomials. As we have seen, Schlegel had already used them in his work many years earlier, but his use of them was quite different. Elliot Coues was born at Portsmouth, New Hampshire, and came under the influence of Baird at an early age. He attended Columbian College (now George Washington University) where he obtained three degrees, including an MD. He gained a place on an expedition to Labrador organised by Baird, and it became obvious from the notes made and specimens collected, that he was going to make a mark in ornithology. During the Civil War he served as an army doctor, while continuing to collect birds under difficult circumstances. He retired from the army in 1881, and spent the rest of his life writing – he was one of the most prolific of American ornithological writers, his most famous work being the *Key to North American Birds*, which achieved three further editions in his lifetime and two after his death. He was one of the principal founders of the American Ornithologists' Union, of which he was the first vice-president, under Allen, and was chairman of the Union's Committee on Nomenclature. However he was outspoken and many found him antagonising. This is probably why he never obtained from Baird the official position at the Smithsonian which he had coveted.

Elliott Coues

Baird's greatest protegé, however, was Robert Ridgway (1850–1929), a giant of American systematics; his magnum opus, *The Birds of North and Middle America* (Ridgway 1901–19), is often considered one of the finest single contributions to ornithology ever made by one man. He was born at Mount Carmel, Illinois, and was encouraged by his parents to take an interest in nature. From the age of ten he was given a shotgun by his father and encouraged to collect birds. It was in 1864 that he first came under the influence of Baird. He found a finch that he was unable to identify. So he was encouraged to send a letter and picture of it to the commissioner of patents in Washington, and the packet was directed to Baird. The bird (a Purple Finch) resulted in the commencement of a correspondence between the two. In

Robert Ridgway

1867, Baird offered Ridgway an appointment as zoologist on a survey of the fortieth parallel, though he was no more than 16. The expedition lasted two and a half years. At the conclusion of it, Baird was anxious to retain Ridgway's services, but, as there was no permanent appointment available, he gave the young man the task of preparing drawings and descriptions for *A History of North American Birds* (Baird, Brewer and Ridgway 1874), which Baird was writing with Thomas Brewer. In 1880, Ridgway became curator of birds at the Smithsonian, an appointment which infuriated Coues who had expected to obtain this post himself. Later in life, Ridgway took part in a number of other field expeditions, to Alaska and Costa Rica among other places. After his retirement he returned to Illinois near the place where he had been born and continued to work on *The Birds of North and Middle America*. The work was never finished, although after his death, Herbert Friedman produced three more volumes. Ridgway was extraordinarily loyal, foolishly so by today's standards. In 1869, the Director of the American Museum of Natural History offered him a position at an initial salary (with generous annual increments) nearly treble that which he was receiving at the Smithsonian with little hope of promotion. He declined the post, and recommended Allen for it. This was the sort of loyalty and affection Baird commanded from his employees.

Ridgway's other great contribution to ornithology was his publication *Color Standards and Color Nomenclature* (Ridgway 1912), which for many years was a bible on the subject, though it is now largely superseded. It was published privately, and in the necessary paperwork Ridgway was assisted by his wife, who, it is said, personally wrapped and posted nearly every copy sent out, and checked each page of each copy to ensure that there were no defects, duplication of plates or transposition of colours. It was a complex and work-intensive task. As Ridgway later explained in a letter to a friend, 5,000 copies of the plates were made, but only 1,000 bound as a first 'printing'. It was all handwork. Each separate colour pigment was mixed in one batch, then large sheets of paper were coated by hand, these then cut into small pieces and pasted on to the appropriate places on the plates. Only in this way, he explained, could absolute uniformity be obtained, and the possibility of variation in different copies be eliminated.

For a long time the question of species and subspecies continued to raise headaches. In a paper in *Science* on 14 May 1897, C. Hart Merriam discussed the ways of distinguishing between species and subspecies. He pointed out that under current ruling, the rule stood as:

> Forms known to intergrade, no matter how different, must be treated as subspecies and bear trinomial names; forms not known to intergrade, no matter how closely related, must be treated as full species and bear binomial names (Merriam 1897).

But, as Merriam pointed out, only in a small percentage of cases was enough material (in the form of specimens from a sufficient number of definite localities) available to the worker to enable him to determine whether related forms did or did not intergrade. The result of this was that authors usually exercised personal judgement as to the probable existence or non-existence of intergradation. This inevitably resulted in inconsistency. After further studies it often turned out that forms treated as species turned out to be only subspecies, and vice versa, resulting in the constant alternation between binomials and trinomials. It would be many years, he complained, before it would be possible to be certain of the status of even North American forms, and until then, nomenclatural stability would be impossible. He concluded:

> It would seem, therefore, since it is impossible for our nomenclature to tell everything we wish to know about a species, that it would serve a more useful purpose if the terms species and subspecies were so used as to indicate degree of difference, rather than the author's opinion as to the existence or non-existence of intergrades.

The American taxonomic principles were not immediately followed in Europe, where binomials were still adhered to. Some, like Finsch, Gadow and Radde refused to name subspecies at all, others like Sharpe, Cabanis and Salvadori gave binomials to every constant variation discovered, even the smallest. Needless to say, the result was chaos. Curiously enough, it was a steel manufacturer, Henry Seebohm, who managed to bring a breath of sanity to the warring bird men. With wealth behind him, he devoted himself to his obsession, paleaearctic ornithology. Seebohm was born in Bradford in 1832 and died in London on 26 November 1895. From an early age he had been fascinated by natural history, and devoted to it every hour he could spare from business. He had travelled widely to Greece, Scandinavia, Turkey, Siberia, South Africa and other places, and acquired a large collection of bird skins and eggs, both by purchase and by his own collecting. These were eventually donated to the Natural History Museum in London. In his two books, *Siberia in Europe* (Seebohm 1880), *Siberia in Asia* (Seebohm 1882) and the revised combined (posthumous) edition *The Birds of Siberia* (Seebohm 1901), Seebohm was the first ornithologist to notice the importance of isolation and secondary contact zones in producing variation. In 1881, in the 5th volume of *Catalogue of the Birds in the British Museum*, he argued strongly against the prevailing opinion that structural characteristics, such as the shape of beak, tail and wings, were older in origin than plumage patterns. Seebohm maintained that the opposite is true. His views were violently attacked at first, but later vindicated. His most important contribution to ornithology may have been his forceful arguments in favour of the American trinomial system made in *The Geographical Distribution of the Family Charadriidae* (Seebohm 1887).

Three years previously, Elliot Coues had travelled to London to address a meeting in the lecture hall of the Natural History Museum on trinomialism. Alone among the audience, Seebohm had been convinced. This momentous meeting took place in early July 1884, and was reported in *Nature* (Anon 1884). The chair was taken by Professor Flower, the Director of the Natural History Museum. R. Bowdler Sharpe spoke first. He pointed out that he had admitted the existence of subspecies in the first volume of the *Catalogue of the Birds in the British Museum* (Sharpe *et al.* 1874), and again freely in the third, and said that it would be quite possible to render these subspecies by the use of trinomials. However, he pointed out that American ornithologists had a great advantage over their European friends, in that they had a clear idea of the natural geographical divisions of their continent and were therefore in a better position

Henry Seebohm

R. Bowdler Sharpe

to determine boundaries of species and subspecies. Furthermore, he went on, trinomial nomenclature would: "open the door to a multiplication of species, or races, founded on insufficient materials, and bestowed by authors who have not sufficient experience of the difficulties of the subject." Seebohm spoke next. Initially he was in favour of a more complicated system, and considered trinomial nomenclature to be illogical. Much better, he declared, to indicate intergrades as follows:

Sitta caesia (Southern Nuthatch, a race of *Sitta europa*)
Sitta caesia-uralensis
Sitta uralensis
Sitta uralensis-sinensis
Sitta sinensis (Oriental Nuthatch, another race of *S.europa*)

After Coues had spoken, W. T. Blandford retorted that an equation containing three variables was much more complicated than one containing only two, and when one had three names, any one of which could be altered to suit personal opinions, the stability of nomenclature would be a long way off. Sharpe added that subspecies could be indicated quite easily by the old-fashioned system of 'var. a', 'var. b', and so on.

However, in Germany, Hartlaub had read the report of the event in *Nature*, and as a result several German writers ventured on a modest trial of the novelty. Of these the most important was a man who was to emerge as the most significant ornithological figure of the late 19th and early 20th centuries, Ernst Hartert. At the outset, the German acceptance of trinomialism was tentative. A five-man committee chaired by Count von Berlepsch decided that trinomials should be used only sparingly, in cases where binomials were obviously inadequate. This was adopted at Frankfurt in 1890 by a full meeting of the Deutsche Ornithologische Gesellschaft. Influenced by Seebohm's writings, however, Hartert, the secretary of the meeting, had published a few months earlier his own definition of a subspecies. At the meeting he found one other kindred spirit, an Englishman who had stopped off in Frankfurt en route to Budapest for the Second International Ornithological Congress. He was Richard Bowdler Sharpe (1847–1909). Sharpe has been described as one of the most tireless workers ever known in the development of ornithology. He was the son of a newspaper publisher, and in his spare time frequented the British Museum to study birds. He came to the attention of P. L. Sclater who helped him to a job as librarian with the Zoological Society. He was able to spend some time in Leyden studying with Schlegel, and, in 1872, was appointed head of the bird room on the death of G. R. Gray. Sharpe moved the collections from a fusty basement, to which no sane ornithologist would wish to go, into a temple where it was, if not exactly a pleasure, at least bearable to work. Donations of collections at once began to flood in, including those of A. R. Wallace in 1873, and in 1885 the vast Indian collections of Allan Hume, numbering 60,000 skins and 16,000 eggs.

As late as the 1820s almost nothing was known of the ornithology of India except for some collections of native drawings made by Indian army officers. One of the earliest and best known of these was the collection of General Hardwicke, a selection of which were engraved and published in 1830, edited by J. E. Gray of the British Museum. It was about 1830 that the beginnings of scientific investigation were laid by the researches of Major Franklin and Colonel Sykes, both of whom wrote papers in the *Proceedings of the Zoological Society* for 1831–32. In 1832, the *Journal of the Asiatic Society of Bengal* was commenced and published in Calcutta, which, along with a number of other journals, contained valuable papers by Hodgson, Hutton, Pearson, Tickell, McClelland and others. Hodgson, who resided in Nepal, was the pioneer in that country.

However, the dominant figure in early 19th-century Indian ornithology was Edward Blyth, who for many years was curator of the Asiatic Society's Museum at Calcutta. He was the first trained zoologist of his time to go to India. Blyth was born in London on 23 December 1810, of a Norfolk family. His father died in 1820, and his mother originally intended him for a career in the Church, but, on advice from his headmaster, sent him to study chemistry. His passion for natural history disinclined him for any other pursuits, and, on coming of age, he opened a chemist's shop in Tooting, London. But he gave this business so little attention that it inevitably failed. He devoted all his time to ornithology and entomology, and contributed many papers on zoology to the *Proceedings of the Zoological Society*. It was largely on the basis of his editing of Griffith's edition of Cuvier's *Le Regne Animal* that he was recruited by Horace Wilson in 1841 as curator of the Museum of the Asiatic Society of Bengal at Calcutta. For over 20 years he rendered devoted service to the Society, submitting detailed monthly reports (which often filled 15–20 pages), publishing many descriptions of new species. During all these years, he subsisted on a pittance of a salary, and his periodic applications for an increase in remuneration were cast aside. In spite of the disgraceful treatment he received at the hands of the Society, Blyth never made any complaint, and accepted the capricious and often unreasonable criticisms of his excellent work with equanimity. After his retirement, he wrote: "I had always a presentiment that my successor in the Museum would be more adequately remunerated, beginning with just double what I had after more than 20 years work ..." Fuel for a court case today, but Blyth dismissed it philosophically, as the great-hearted and underrated man he was. Grote, in writing Blyth's obituary in 1875, remarked: "Few men who have written so much have left in their writings so little that is bitter." Blyth died after a long and exceptionally productive career, as one of the pioneer zoologists in the continent of Asia. Sadly, he is little remembered today. He was the mentor of T. C. Jerdon, William Blandford, and particularly A. O. Hume. There is no rational explanation of why Blyth has been ignored by the history of zoology. He aroused the malice of J. E. Gray of the British Museum, as a result of a complaint made against his brother G. R. Gray. This does not, however, explain his neglect in the 20th century.

Allan Octavian Hume was born in 1829, the son of a Scottish MP. He served as a midshipman in the Royal Navy, studied medicine at University College Hospital and at the age of 20 was posted to the Bengal Civil Service. From 1849 to 1867, he served as district officer at Etawah in the NW Provinces,

Allan Octavian Hume

from 1867 to 1870 as commissioner to a centralised department, and from 1870 to 1879 as Secretary to the Government of India. He showed great gallantry during the Indian Mutiny in 1857, risking his life on several occasions. He worked tirelessly in promoting education, reforming the local police, and in 1859 founded a popular paper *The People's Friend*, published at a very low price so that it would be accessible to the poorest of village youths. As Commissioner, he introduced agricultural reforms, suitable to the particular needs of the people. He was always sensitive to local and traditional practice and never sought to impose 'foreign' measures. Besides matters directly pertaining to agriculture and horticulture, Hume had to deal with forestry, including the conservation of forests, the restoration of denuded areas, and the supply of firewood to the public. He was also required to attend to fisheries, emigration, meteorological observations, museums, and exhibitions of art and industry, shipping, harbours, lighthouses and customs. In 1879, after 30 years of devoted service, he was summarily dismissed from his post in the secretariat of the Government of India, because he expressed his views freely, without regard to the opinions or intentions of his superiors. Hume retired from public service in 1882 and became founder of the Indian National Congress. For this purpose, he travelled to England in 1883 to seek the support of influential friends there. The First Session of the Congress was held from 25 to 30 December 1885. He seems to have been one of the few Englishmen completely trusted by the people of India, and he had received many forewarnings of the Mutiny, in the hope that he could do something to avert the tragedy. He is a figure unjustly forgotten in Indian political history. On his return to Britain in 1894, he settled in Dulwich and threw himself into local politics, serving on many committees, often as Chairman or President. This was the man, who, in his 'spare time', studied the ornithology of the Indian sub-continent, and spent about £20,000 of his own money (a huge sum in those days) in accumulating an ornithological museum and library, which was the largest in Asia at that time. (It consisted of 63,000 bird skins and 19,000 eggs). He had planned to write a vast book on the ornithology of India and had made voluminous notes, but in 1884, while absent from home, a servant sold the manuscript in the market for waste paper. As a result, in 1885 he presented his entire collection to the Natural History Museum in London. In 1872, he had started, at his own expense, the journal *Stray Feathers*, edited it, and wrote many of the articles himself. It was published until 1899 when it had to be discontinued, as by then Hume was living in England.

It was not simply workaholism which drove Sharpe, he had ten daughters to provide for! In order to earn money, he completed monograph after monograph, and embarked on one of the most ambitious projects of all time, the 27-volume *Catalogue of the Birds in the British Museum*. Most of this work was done in the evenings at home. By this time the British Museum bird collections were by far the largest in the world. Still they remained housed in a basement in Bloomsbury until 1883, when they were moved to the new Natural History Museum in South Kensington. But not all the work on the *Catalogue of Birds* was done by Sharpe. Many of the volumes were farmed out to specialists in particular groups such as Edward Hargitt, who wrote the volume on woodpeckers. He was born on 3 May 1835 in Edinburgh, where his father, a well-known professor and composer of music, then lived. Hargitt's father was also a great lover of art, and the son was encouraged to become a painter, a calling which he pursued for many years. He showed a great interest in the down and fledgling plumages of birds – a topic which until then had received scant attention, and this resulted in his interest in ornithology and the collection of specimens. When it became too large to house it was disposed of, his eggs being acquired by Seebohm. Hargitt began to specialise in woodpeckers, in which he became the world expert of his day. He was a shy and retiring man, and generous to a fault. He died on 19 March 1895. Other authors included Seebohm, Sclater, Salvin, Ogilvie-Grant, Shelley, Saunders and Salvadori. Osbert Salvin (1835–1898) travelled in North Africa and Central America, and with his friend Frederic duCane Godman, built up a huge collection of bird specimens which were eventually deposited in the Natural History Museum. Salvin is believed to have died of arsenic poisoning as a result of preparing so many specimens. He wrote the volumes on petrels and hummingbirds, as well as many papers in the *Ibis* and other journals. William Robert Ogilvie-Grant (1863–1924) worked in the museum as second in command to Sharpe. Captain George Ernest Shelley (1840–1910) was a nephew of the poet. He travelled in South Africa and Ethiopia and for some years had an office in the museum, though he was not on the staff. Howard Saunders (1835–1907) was a merchant banker who devoted himself to the ornithology of Spain, publishing several papers in the *Ibis*. He was an expert on gulls and terns of the world, and was therefore appointed to write up these groups for the catalogue. Fittingly, both a gull and a tern are named after him.

Count Tomaso Salvadori

Count Tomaso Salvadori was born on 30 September 1835 at Porto S. Giorgio in Umbria, Italy, the son of Count Luigi Salvadori and his wife, the former Miss Ethel Welby of Lincoln. He studied medicine in Rome and Pisa, and in 1860 followed Garibaldi to Sicily, taking part in several campaigns as medical officer. He had already begun to study the birds of his country, and to make collections. In 1862, he visited his friend Antorini in Sardinia, and as a result of this visit published a catalogue of the birds of that country in the Proceedings of the Italian Society of Natural Science, at Milan. In 1863, he settled in Turin, where he was appointed assistant at the museum, becoming Vice-Director in 1879, a post he held for the rest of his life. The museum rapidly increased in importance through donations from the King of Italy, and many private individuals and travellers. At the invitation of Sharpe, Salvadori spent the years 1890 and 1891 at the Natural History Museum preparing three volumes on parrots, pigeons, and ducks and ratites, for the *Catalogue of Birds*.

A forerunner of the *Catalogue of Birds* had been G. R. Gray's *The Genera of Birds* (Gray 1844–49) with illustrations by D. W. Mitchell, an enormous piece of work in three folio volumes. Although Gray was not an ornithologist, save for the chance which placed this conscientious clerk in charge of the ornithological collection of the British Museum, this publication, and indeed the mere concept of it, was a remarkable achievement. It was closely followed by the same author's *A Catalogue of the Genera and Subgenera of Birds* (Gray 1855), and *Hand List of Genera and Species* (Gray 1869–71) in three octavo volumes. The main drawback of the latter work is that the specific names are identified in its index not with the accompanying generic name, but only by the author's name and date, which is a serious inconvenience to the user. Nonetheless, the *Hand List* was a most important tool of the working ornithologist at that time, and is still of considerable use today to anyone searching out synonyms and obsolete names.

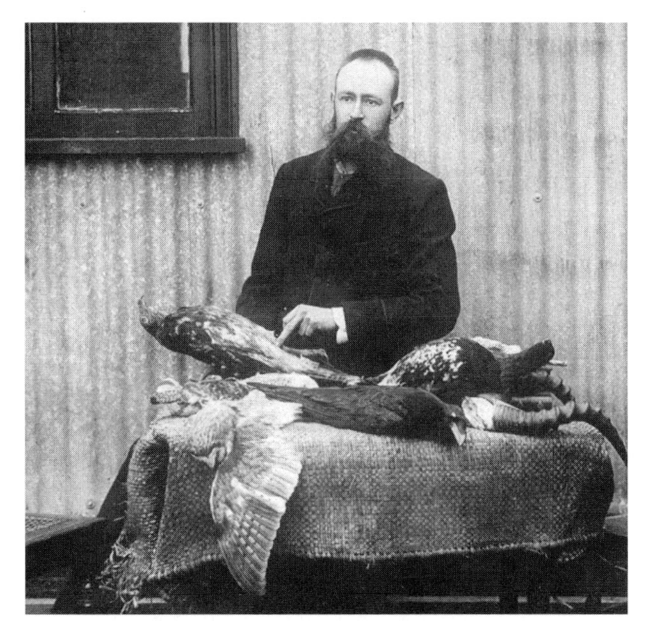

Ernst Hartert

Hartert was 32 and too poor to think of getting married, when he met Sharpe in Frankfurt. He had spent all his formative years studying birds. Born in Hamburg on 29 October 1859, he was the son of an army officer. At the age of 24 he had joined the Flegel expedition to collect in Nigeria, barely a year later he was in south Asia to earn some money by collecting specimens. He then spent two years wandering the forests of Assam, Sumatra and Perak. Faced now with the prospect of earning a living (his father had retired from the army and could no longer support him) he managed to obtain a post at Frankfurt. But his life totally changed after his meeting with Sharpe and they remained close friends until Sharpe's untimely death in 1909 at the age of 62. Five weeks after the Frankfurt meeting Hartert received a letter from the British Museum inviting him to work on the swifts and nightjars for the 16th volume of the *Catalogue of Birds*. A few days later he packed his bags, got married, and left for England, little realising that it would be 38 years before he returned. A meeting with H. E. Dresser resulted in an introduction to Walter Rothschild, and after a short collecting trip to Venezuela, he was appointed director of Tring Museum. Under Rothschild and Hartert, the collections at Tring became larger than any other in the world, except for the British Museum. Rothschild's money enabled him to send collectors to all the 'unexplored' regions of the world, particularly the Indo-Australian islands, and New Guinea. Hartert had the job of working on these collections, arranging them, and describing the many new species.

H. E. Dresser

After Seebohm's death in 1895, Hartert was the only ornithologist in Europe who favoured trinomial nomenclature. In 1897, he addressed a meeting of German ornithologists at Dresden, where Berlepsch was to deliver a lecture on the concept of subspecies. Count Hans von Berlepsch (1850–1915) had long been a Corresponding Member of the American Ornithologists' Union. He was a leading authority in his day on the birds of South America, had been the mentor of von Ihering, the Brazilian ornithologist, had himself published many papers on that subject, and amassed a considerable collection of South American birds. His father had invented the wooden four-sided frames used in beehives for collecting honey. (Although an Honorary Member of the British Ornithologists' Union, his death elicited only a brief notice in the *Ibis*, owing to the "sad state of our relations with Germany" at the time.) There were three speakers against Hartert – Berlepsch, Wiglesworth and Kleinschmidt. Berlepsch took the somewhat obscure view that Hartert's proposal would burden nomenclature with hypotheses, and that his lumping of previous species and their demotion to subspecies was contrary to nature. Lionel Wiglesworth objected to the concept of subspecies because their recognition required positive knowledge, which is rarely available, and that there was no way of knowing when a minor variation in a species was important enough for it to be recognised as a subspecies. Otto Kleinschmidt rejected evolution altogether. Hartert gained one convert at that meeting, A. B. Meyer, director of the Dresden Museum. The cause was won slowly. Among his many achievements, Hartert's most important major work was *Die Vögel der Paläarktischen Fauna*, which commenced publication in 1903. In this he employed the trinomial system, whereupon he was immediately attacked by P. L. Sclater, editor of the *Ibis* since 1859, and still as powerful an influence as ever. "How dare anyone," he thundered, "call upon us to give up the binomial system, in use since Linnaeus?" Sharpe agreed with him, and in his *Hand List of the Genera and Species of Birds* (Sharpe 1899–1909), he listed all taxa as full species, a total of 18,939. The influence of Sharpe and Sclater was so strong that Rothschild and Hartert were regarded as outsiders. However, Harry Forbes Witherby (1873–1943) decided to advocate the use of Hartert's system, and in 1907 founded the periodical *British Birds*, in which he employed the trinomial system. This system was used in the *Hand List of British Birds* (Witherby, F. C. R. Jourdain and N. F. Ticehurst 1912) and began to infiltrate the pages of the *Ibis* itself. As Stresemann put it: "The stoutest bulwark of binomialism had fallen after a twenty-year siege" (Stresemann 1975).

Events were now moving towards the acceptance of a system of classification which would serve ornithology through most of the 20th century. Carl Gegenbauer was born in 1826 at Würzburg. He studied medicine there, eventually becoming a professor specialising in biological research and comparative anatomy. He died in 1903. In 1861 he published an essay on the evolution of the egg. In this he showed for the first time that eggs of birds are single cells; hitherto it had been supposed that they were multicellular. His disciple, Max Fürbringer, born in 1846, followed Gegenbauer to Heidelberg where he eventually succeeded him. Fürbringer carried out anatomical investigations in many fields, but is mainly remembered for his work on birds, culminating in his massive monograph *Untersuchungen zur Morphologie und Systematik der Vögel* (Fürbringer 1888). The first half consists of a comparative study of the breast, shoulder and wing throughout the entire class of birds. This is followed by a bird classification based on comparative studies of representatives of all the orders and families. His arrangement assumed that birds began with toothed forms of small or medium size, with long tails, four lizard-like feet and bodies clothed in down. Next came birds with the down transformed into feathers for warmth, not flight. In this state they became bipedal, the forelimbs developed for prehension, and the hind legs and pelvis developed into a similar form to that of the bipedal dinosaurs. Some became climbers, and from them developed true flying birds. In the course of evolution there were many groups which aborted or degraded, such as the Ratites, which he considered were not a natural group, but represented a series of examples of evolution in the same direction. His system was complicated; he divided birds into sub-classes, orders, sub-orders, gens and families. The table (Appendix 29) is simplified, omitting, among other things, the 'gens'. Another unusual feature of Fürbringer's arrangement is that he listed a number of 'intermediate sub-orders' which he did not, or could not, actually include within a particular order.

It was Hans Gadow, a disciple of Alfred Newton, who in 1892 laid the foundation for modern classification, which he derived from Fürbringer. In his preface, Gadow pointed out that in recent years a number of classifications had appeared, in most cases the authors had given no reason for their arrangements, which were presumably based on 'personal convictions'. Gadow averred that the proposer of a classification ought to give his reasons. He must not merely list the characters used, but explain how they are used. To take all the characters would inevitably result in an artificial system; the systematist must decide which characters are important, and which are trivial. Gadow selected about 40 characters which he considered to be of taxonomic value. These were weighed family by family. Thus, for example, the parrots agreed with the cuckoos in 31 points, with the woodpeckers in 29, the rollers in 25, the falcons in 25, the owls in 22, the hornbills in 22, the game birds in 21, as against 19 points of difference. Gadow's results have stood the test of time. He realised that the owls were not closely related to the raptors, but were "a lateral branch of the lower Coracine stock". For Gadow's arrangement see Appendix 30.

The study of migration was boosted by the introduction of ringing or banding. This is a phenomenon of the 20th century, although sporadic attempts to mark birds had been made for nearly 200 years. One of the earliest instances known of its use was a copper ring placed on a swallow by an aristocrat hiding from the mob during or just after the French Revolution. He discovered that the same bird returned in three successive years to the same nest. A less reliable record (which occurred in either 1710 or 1770 according to different sources) was of a heron caught in Germany which carried rings placed on it in Turkey several years earlier. In 1884, Borggreve recommended cutting off the middle toe of birds to establish migration routes. Mercifully, this suggestion was never adopted, and would in any case, probably have been of very doubtful value. Around 1830 a Dutchman, Baron van der Heyden, had marked the offspring of ducks and geese with brass neck collars. This was commented on in the journal *Rhea* in 1849 (Stresemann 1975). Some years later a Danish teacher, Hans Christian Cornelius Mortensen, decided to try a similar experiment with starlings. He wrapped zinc strips round their legs with the locality and date 'Viborg 1890' written on them in ink. But he discovered that the birds objected strongly to the rings, and abandoned the experiment. Soon after this, aluminium became available at a reasonable price and so Mortensen repeated the experiment. In 1899, he ringed 162 starlings, but had no recoveries. In 1900 he tried again, and received two, one bird shot in Holland and one in Norway. When Johannes Thienemann repeated the experiment in 1903 at the Rossitten Bird Station, using Mortensen's methods a storm of protest resulted from the newspaper publicity. Animal protectionists demanded that the government forbid ringing, because it encouraged people to kill large numbers of birds in order to recover one from Rossitten. It was some time before the practice became respectable. Interest increased greatly after the First World War so that by 1927 there were ringing stations in 17 European countries. Now ringing or banding takes place throughout the world.

Ringing greatly increased knowledge of when and where birds migrate, but it did not explain why. In 1905 for the International Ornithological Congress, the Hungarian, Otto Herman, drew up a list of some of the differences of opinion as to the basis of migration. He posed these as alternatives, such as:

> Naumann: There are definite routes of migration.
> Homeyer: There are no definite routes of migration.
> Weismann: Birds learn how to migrate.
> Gätke: Birds migrate by instinct.
> Palmén: Orientation is traditional.
> Weismann: Orientation is congenital.
> Gätke: There is no leadership.
> Weismann: There is leadership.
> Wallace: The weather has no essential influence.
> Homeyer: The weather has an essential influence.
> Naumann: Temperature plays a very important part.
> Angot: Temperature is not an incentive.
> Lucanus: Flight occurs at 3000 ft.
> Gätke: Flight occurs at 35000 ft.
> Braun: The original home of the birds is the tropics.
> Deichler: The original home is not the tropics. (Herman 1907)

Even now, the questions posed by this list are unanswered, although for many species patterns of migration, flight paths and other factual information have been copiously documented. The mechanisms of migration and the nature of the urge to migrate remain elusive.

❧

One of Strickland's contributions to ornithology had been to speak out for the rule of priority in nomenclature. Until his time it had been customary for authors to replace specific names thought to be unsuitable or imperfectly formed by others more suitable. Strickland's 'Code' insisted that the first-used name must have priority over all later ones. One of the reasons for the strong opposition in England to the use of trinomials was because their use required the abandonment of the Strickland Code, which had been adhered to since 1842. It had been published as 'Report of a Committee Appointed "to consider the Rules by which the Nomenclature of Zoology May Be Established on a Uniform and Permanent Basis".' Prior to this, there had been no agreement on the species names to be used. In 1751, Linnaeus in *Philosophia botanica* had set forth aesthetic and philological rules to be followed in nomenclature. In ornithology, Illiger (1811) had felt obliged to replace all names not classically formed with 'purer' ones. Wagler in his *Systema Avium* (Wagler 1827) required that all scientific names should be suitable and dignified, and criticised those who preferred to stick to familiar names even when these were badly chosen, rather than invent new ones. Swainson in *Natural History and Classification of Birds* (Swainson 1836) considered that he had the right to 'clean up' nomenclature periodically, and choose between good and bad names, without any regard for priority. As late as 1879, Coues had criticised the new names introduced in 1863 by Cabanis and Heine in *Museum Heineanum*: "Most of the genera are useless and burdensome synonyms, resulting from the authors' purism in rejecting prior names not 'classical' in form – a practice totally indefensible, unless in exceptional cases." It was against this arrogance of classicism that Strickland had reacted, and introduced a strict rule of priority, founded on the twelfth edition of Linnaeus (1766). The new American system tightened this up and priority became even more rigid. Strickland had allowed the use of a new name when the original one was manifestly wrong, such as *Muscicapa atra* for an olive-green bird, and the invention of a new specific name if this were already used as a generic name, to avoid a tautology. (Thus if *Turdus merula*, the Blackbird, were made the type of a new genus *Merula*, the specific name would have to be altered to, say, *Merula vulgaris*). Coues and Allen swept all this aside. Even when the formation of a name was wrong, it must be kept, and could only be altered or 'corrected' in the case of a typographical error. The starting point for nomenclature was moved back from the twelfth edition of Linnaeus, to the tenth (1758). When the last part of *Die Vögel der paläarktischen Fauna* (Hartert 1903–22) appeared, Hartert had won over all the European ornithologists to the use of the new system. In 1930, he retired from Tring, to return to his native Germany. Only two years later, the Tring collections (280,000 bird skins), largely built up by Hartert, were sold to the American Museum of Natural History because Rothschild, blackmailed by an actress with whom he was having a clandestine affair, required the money to pay her off. Hartert was heartbroken and died on 11 November 1933. Otto Kleinschmidt (1870–1954), who had spoken against Hartert in 1897, was still an opponent of the conventional theory of evolution. There were many cases where it was difficult to decide whether a particular form was a species or a subspecies, and workers faced with this problem found much to attract them in Kleinschmidt's theory of the *Formenkreis*. He proposed that species were largely immutable, that a subspecies is always a subspecies and can never develop into a species. He believed, therefore, that the systematist's task was to bring together races which formed a 'natural' species or *Formenkreis*. This idea was for a time held in some regard, but it resulted in some absurdities, such as lumping the large 'black-backed' gulls into one group as a polytypic species. The *Formenkreis* theory was eventually, and mercifully, abandoned.

Ornithology and Ornithologists in the Twentieth Century

by

John Coulson

The twentieth century witnessed remarkable changes in higher education and scientific research, and ornithology, as a branch of zoology, has been well represented in this expansion. At the end of the 19th century, ornithology (and much of zoology) was still mainly descriptive, with taxonomy, anatomy and geographical distribution as the main fields. There were only a very small number of people professionally employed, mostly in museums, and field ornithologists were still few. Without a gun, little could be achieved. Binoculars (then called field-glasses) were little more than opera glasses, with a small field and low magnification. Telescopes were long, with several draw-tubes, and were difficult to keep steady in the field. Prismatic optical aids did not exist. With the instruments available, field identification of birds was difficult. Guides to assist the field identification of birds had not been published and the first would not appear for several decades. In Britain, clergymen were in the forefront in the study of natural history. The system of church patronage allowed appreciable time for the incumbent to follow hobbies and interests, such as ornithology and botany.

The administration of the then large British Empire enabled civil servants and army officers to visit many parts of the world. They shot animals and a few collected bird skins. The navies of the western powers visited every part of the world. They carried out exploration and mapping expeditions, with some of their vessels carrying a naturalist on board. Visits by naturalists to oceanic islands on Royal Navy and US vessels became frequent. For example, the many visits to the Hawaiian Islands during the 19th century are well documented by Olson and James (1993 [1994 in refs]). Major contributions to ornithology came from people who were wealthy and could employ others while on expeditions to make collections of birds. For example, Lord Rothschild's extensive personal collection of birds was housed in his museum at Tring, England. It later developed into a centre for taxonomists employed by the Natural History section of the British Museum and now houses extensive national and international bird collections. Private collections, similarly, formed the nuclei of major museum collections in the United States and elsewhere. At the beginning of the 20th century, the pattern of ornithology continued as in the second half of the previous century. Collecting skins and eggs was the basis of most ornithological study. "Birds recorded from ... " and "A new species of ... " were frequent titles in scientific communication. Egg collecting was a major hobby and interest. However, the reliability of taxonomic and distributional studies even in those days was not without problems. For example, the 'Hastings rarities' – bird species said to have been shot in southern England in the early years of the 20th century – had probably been obtained abroad and imported dead and refrigerated. Although in essence a fraud, it was only made possible by the advent of refrigeration. In all, some 30 species were struck off the British List in 1962 after the scam had been exposed in *British Birds*. Curiously, all but one, the Black Lark *Melanocorypha yeltoniensis*, have subsequently been re-instated, but based on other records. It has taken most of the 20th century to detect that some bird specimens purported to have been collected in various locations throughout the world by Col. Richard Meinertzhagen, a pillar of the British ornithological establishment, had been given incorrect locality labels. Some, indeed, were re-labelled specimens 'borrowed' from other collections and museums. At the beginning of the 20th century hoaxing was rife in natural history and its extent is still coming to light. Most large museums have some material whose locality labels are suspect and some have faked specimens of birds and other animals. Perhaps the best known of these is 'Piltdown Man' – a human skull fraudulently assembled from bones claimed to be those of an early human ancestor, the 'missing link'. Some have suggested that the same group of upper class hooligans were responsible for both the Hastings rarities and Piltdown Man hoaxes. Hastings and Piltdown are only some 20 miles distant from each other in the county of Sussex. Successfully tricking colleagues in a museum was a challenge to many taxonomists and taxidermists!

At the beginning of the 20th century, there was no indication of how the study of ornithology would expand. One measure of its increase can be obtained by examining the number of scientific papers listed each year in abstracting publications. A case in point is the Aves section of the Zoological Record, published by the Zoological Society of London, which shows a dramatic change. From 1900 to1939, there was a gradual increase from about 500 to 1,700 scientific ornithological articles on birds published each year. This was followed by a 20% dip during the Second World War, which continued for about ten years. Then came a dramatic increase uninterrupted for the next 40 years. By 1990, the total reached some 14,000 scientific articles published in that year alone, i.e. a 28-fold increase over the course of the century. At the same time, the opportunity to publish increased. Journals increased in size, many new ones were produced and new fields of scientific study involving birds developed. In the last ten years of the 20th century, the increase in publications levelled off, which may well be a relief to some! The cause of the publication eruption was a huge increase in the numbers both of professional and amateur ornithologists. Since 1950, many individuals have made contributions to ornithology of equal merit to those of earlier centuries. To analyse and evaluate their work here would involve several volumes.

In 1950, of the few professional ornithologists throughout the world, most worked in museum posts. Ornithology was not considered by universities to be an important aspect of zoology and so few posts for its teaching existed. For example, in the 1950s, W. H. Thorpe explored bird behaviour while holding the post of Reader in Entomology at Cambridge University. At the same time, but at another English university, ornithology in an undergraduate course in zoology extended to three lectures only, which were presented by another very able entomologist. At that time, zoology courses did not on the whole include ornithology. One very disparaging zoologist, a Fellow of the Royal Society, commented, "Birds and butterflies are part of botany." He despaired of the greater attraction of these subjects to the layperson. This bias against birds was not so pronounced in North America, but, again, a student of zoology was fortunate to be at a university where he/she received a course on birds and bird biology from a person researching in this area.

Following the First World War, there was a dramatic increase throughout the world in both the number and size of universities. Posts in ornithology multiplied and many new fields of zoology developed that used birds as a key group in their studies. The study of animal behaviour, stimulated by the works of Konrad Lorenz (1952) and Niko Tinbergen (1951, 1953), has depended extensively on the use of birds as study subjects. Biogeography took the distribution of birds as a topic because of the (relative) ease of identifying species, birds' predominantly diurnal activity, the ease of finding them and the extensive collecting that had already taken place. Following the introduction of the study of animal ecology in the late 1920s by Charles Elton (1927), avian ecology blossomed and contributed much to the principles of ecology. The study of free-living birds in their natural habitats had arrived. Ornithology flourished and its growth increased with the establishment of new universities and the expansion of old ones throughout the world.

It was recognised that some birds were major pests. Studies were initiated into the control of pest species such as Red-billed Quelea *Quelea quelea* in Africa, Wood Pigeon *Columba palumbus* in Britain and the Emu *Dromaius novaehollandiae* in Australia. An early study during the Second World War in Britain was that of the Rook *Corvus frugilegus*, considered as a possible major pest of agricultural crops. It was followed by detailed work on Wood Pigeons. In 1967, the potential involvement of birds, particularly Common Starlings *Sturnus vulgaris*, in the spread of Foot and Mouth disease in Britain gave rise to another investigation. The impact of wintering geese on grassland and of birds, particularly cormorants, on freshwater fish stocks has been studied.

Interest in game animals was another impetus to avian research. Protection of geese and ducks was the stimulus to signing by Canada and the United States of the Migratory Birds Convention in 1916. Studies of game birds were instituted in the UK by the Imperial Chemical Company, prompted by their interest in shooting and the manufacture of cartridges. The study of waterfowl gave rise to the employment of many ornithologists. In 1937, Ducks Unlimited was established in the USA and a year later, Duck Unlimited Canada was formed. In 1947, the Wildfowl Trust (now the Wildfowl and Wetlands Trust) was set up in the UK by Peter Scott and Government-based conservation and wildlife management developed. In the same year, the Canadian Wildlife Service, originally known as the Dominion Wildlife Service, was also established, while in the USA, state-run conservation bodies arose to supplement the work of the federal Fish and Wildlife Service. In 1950, the British Government established the Nature Conservancy, subsequently becoming the Nature Conservancy Council after losing its research arm (which separated as the Natural Environmental Research Council). In 1991, it was split regionally to form English Nature, while its Scottish and Welsh components were each amalgamated with the Countryside Commission. At each change, more biologists and wardens, many of them ornithologists, were appointed to deal with the increasing responsibility and the developing conservation legislation. More recently, legislation arising out of the European Community has greatly increased the workload in conservation, requiring the employment of yet more professional ornithologists.

Non-government organisations (NGOs) involved in conservation and supported by public membership and subscriptions have become major employers. In the USA, the Audubon Society and, more recently, their Nature Conservancy have flourished. In Britain, the Royal Society for the Protection of Birds (RSPB), founded in 1892, has grown from a membership of about 30,000 in 1950 to over one million in 2002. There has been a similar explosion in the numbers of members of bird societies throughout the world. These NGOs have become involved in the management and conservation of birds, particularly in the development and implementation of bird protection legislation. In England for instance, the British Trust for Ornithology (BTO) and RSPB are responsible for handling a number of important government contracts concerned with conservation and ecology.

Because the Antarctic held considerable reserves of minerals and other resources, political interest in that continent was sharpened. Several nations, the USA, UK, France, New Zealand, Australia and Russia in the vanguard, developed research programmes that included work on seabirds. The hostile environment that demands special vessels and bases, makes this research very expensive, but is politically driven. In the 1970s, the Natural Environmental Research Council (NERC) in the UK had insufficient funds from government sources to replace their main support vessel for Antarctic research and major cutbacks of staff and projects were contemplated. However, the dispute with Argentina over ownership of the Falkland Islands and South Georgia saw these financial problems resolved. Since then, there has been a greater British investment of money and staff into Antarctica and consequently into ornithological research.

Global warming has become the current cause for concern, and money, effort and research are being focused on the topic. Birds have probably the most active and controlled means of movement and dispersion of all animals. If global warming reaches the magnitude predicted, then the distribution and abundance of many birds and their habitats will be affected. No doubt many of the existing nature reserves will be maintained in the future, although key habitats therein will have changed. There will be controversy over how much should be invested into retaining species in unsuitable, let alone optimal, conditions. Even though so far only minimal changes in mean temperatures have been recognised, in ornithology as in other fields, many changes in the performance of animals have already been blindly attributed to global warming. To sort out the real influences from the others claimed will involve much research. In the near future, more attempts will be made to mitigate population and distributional changes; birds will be introduced to or from areas, either because they have been lost to that region or because conditions have now become right for them to occur and breed there. Little is known about how these climate changes will impact on birds. A new ornithological field of habitat management will develop – for years to come a source of employment for ornithologists.

At the end of the 20th century, there were at least a hundred times more professional ornithologists in post in the world for every one in 1960. Add to this number the legions of amateurs whose field identification skills are outstanding and it is not surprising that relatively few of the eminent ornithologists of the 20th century carried out their studies before 1950. Of those who did, the names of Margaret Nice, Robert Cushman Murphy, Ernst Mayr and Arthur Cleveland Bent in North America; David Lack, P. Palmgren, Finn Salomonsen, Julian Huxley and Erwin Stresemann in Europe; Reginald E. Moreau in Africa; and Lancelot Eric Richdale in New Zealand stand out. Konrad Lorenz and Niko Tinbergen also reached the peak of their scientific productivity in the middle of the century and, in 1973, together with Karl von Frisch were awarded the Nobel Prize for their contribution to the understanding of animal behaviour. Roger Tory Peterson, whose first field guide was published in 1932 and became a model for most subsequent bird guides, greatly facilitated and extended the identification of birds in the field. The ease with which birds may be identified has changed dramatically. In 1900 if an unusual bird was seen, it was shot and identified from texts that described it feather by feather. The advent of field glasses at the beginning of the century, and then prismatic binoculars, greatly enhanced the accuracy of field identification. Now, vast improvements in spotting scopes on tripods and in cameras and lenses have made identification in the field even safer. Indeed, the improvement in the accuracy of field identification of bird species, sexes and age classes is a major advance of the 20th century. We live in the era of the 'birder' and the 'twitcher'. The consequences of these advances in identification techniques have not usually been recorded in the scientific literature, but in field guides and in the many magazines and journals aimed at an amateur birdwatching readership.

But the popularity of birding has bred its own problems. The validity of field identifications is often challenged. In some regions, more than one observer is necessary before a record is accepted. Within the last 20 years, identifications supported by photographs and accepted by records committees have been challenged and in some cases overturned. Doubts are expressed about the quality of some field identification and confirmation in various forms is increasingly required. There was a time when things were simpler. For many years, the record of the first cuckoo of the year in the UK was published as a letter to *The* [London] *Times*. One year, the record was exceptionally early (in February), which two days later produced a highly sceptical reply from an eminent ornithologist, giving several reasons why the record was suspect. Two days later, the same ornithologist received a parcel by post. In it was the body of the cuckoo! Will the day arrive when records are not accepted without a DNA sample or a photograph that includes irrefutable evidence of the date and place?

In 1933, the British Trust for Ornithology (BTO) was established in the UK. Its aim was to provide a forum for amateur ornithologists spread throughout Britain. Regional representatives were appointed and there was active support from at first one, and then a few, professionals. The countrywide nature of the network prompted national surveys of many species, culminating in the production of atlases and distribution maps and through the nest record card scheme and censuses, the collection of data on breeding success. Together with the journal *British Birds*, the BTO developed an extensive ringing (banding) scheme. In many cases, national surveys were planned, co-ordinated and published by amateurs.

Over time, the role of the BTO has developed and now, with a large staff of professional ornithologists, it primarily serves as an organisation collecting national data for professional and government organisations. Some regret that this unique organisation gives much less scope for the able amateur to initiate, plan and execute research studies than was its founders' intention. Now, professional ornithologists analyse data required by government bodies, although it is still collected mainly by amateurs. Similarly, the Wildfowl and Wetlands Trust (formerly the Wildfowl Trust) uses a network of amateur observers and counters to record monthly wildfowl numbers nationwide. Another amateur group, the Seabird Group, carries out national census work in the face of inadequate financial support from statutory government organisations. The UK Government has obtained data necessary to meet its legal obligations in Britain and Europe at a bargain price, although as a result there are probably fewer professional ornithologists! The co-operation between British amateur and professional ornithologists is remarkable. In North America, most of the comparable work is carried out by professionals.

The 12 plenary sessions of the XXI International Ornithological Congress held in Vienna in 1994 involved major presentations on molecular genetics, physiology, bird song, neurobiology, conservation, endocrinology, immunology, evolutionary ecology, population ecology, physiological ecology and social behaviour. At the start of the century most of these fields of study did not exist! In addition, another ten fields of study, including several aspects of applied ornithology, can be found within the 52 symposia sessions, 60 round-table discussions and in some 800 poster displays of research work contributed by a thousand ornithologists. The main feature of 20th century ornithology has been the dynamic expansion in the fields of study. As examples of these changes, two fields have been selected: molecular studies in avian taxonomy and the value of marking birds as individuals. Molecular studies start as late as 1985 with the development of DNA fingerprinting first used in human forensic and legal medicine by Alec J. Jefferies and his co-workers (1985) and then two years later in birds by Burke and Bruford (1987). In contrast, marking studies on free-living individuals started at the beginning of the 20th century and developed in many ways throughout the following hundred years. Both these developments are considered in greater detail in the following chapters.

MARKING BIRDS TO STUDY INDIVIDUALS

In the 20th century the capture and marking of birds became a fine art. Its development is linked to changes in the approach to the study of ornithology, particularly the growth of field studies, i.e. work carried out in the field where wild birds live and breed.

Several centuries ago, seamen on whaling ships visiting the Antarctic captured albatrosses and inscribed names or messages on their bills; but in the 19th century the first systematic attempt to ring birds was made by Lord William Percy in Northumberland. Annually from 1891, he marked young Woodcock *Scolopax rusticola* with copper rings inscribed with the letter 'N' and the year attached to the leg. Records of local movements and longevity were obtained. No long-distance recoveries were made and hardly could have been expected, since no address was given. In the 1890s, Christian Mortensen in Denmark instituted the first properly recognised scheme which used rings that gave an adequate address and used unique numbers to identify individuals and to record the date and place of marking. Larger scale schemes in Germany, Hungary and in Britain were then developed. The rings (bands) were made of aluminium.

In 1909, the first ringing group was established in the USA. In the same year, two schemes started in Britain, one at Aberdeen University and the other by H. F. Witherby through the journal *British Birds*. The latter scheme continues to this day under the auspices of the British Trust for Ornithology. Further schemes followed in Sweden and the Netherlands and, outside Europe, in Egypt, India and Japan. In North America, banding or ringing, as it is known in Europe, on a national scale did not start until the 1920s with the support of the US Biological Survey, now the US Fish and Wildlife Service. Ringing has been carried out throughout the world for most of the 20th century and several million birds have been marked. In Europe, amateurs have carried much of the ringing with a relatively small input from professionals, such as those working on wildfowl and in universities. In North America, professional researchers form the core of those contributing to banding schemes.

Ringing brought to light the extent and nature of avian migration. Rings, returned from Africa which had been taken from Barn Swallows *Hirundo rustica* and Common Cuckoos *Cuculus canorus* ringed in Europe, put to rest the notions which had persisted into the 20th century, that swallows did not leave Europe in winter but lay dormant in the mud at the bottom of lakes and ponds or that cuckoos turned into hawks in late summer and remained as such until the next spring. Ringing has not disproved the belief that some birds journeyed to the moon, although the first ringing recovery from that natural satellite is still awaited!

Soon, ringing produced more information. In the northern hemisphere winter, Barn Swallows from Britain were mainly reported from South Africa, whereas those from Germany were mainly reported from equatorial Africa. Evidence that some seabirds crossed the Atlantic Ocean came to light. The speed with which migrating birds could fly was estimated from recaptures of birds at a considerable distance from where they had been marked only a few days earlier. More (and different) species marked in Europe were recovered in Africa and a Northern Lapwing *Vanellus vanellus* from England crossed the Atlantic Ocean and was recovered in Newfoundland.

The main limitation of inscribed metal leg-rings is that so few individuals are reported again. The return rate may approach 25% for geese and cormorants, but for small birds it can be less than one individual in a thousand marked. Typically, only 1–2% are recovered. Even in species such as House Sparrow *Passer domesticus*, which are closely associated with human habitation, only about 1% of rings are recovered. Where do they die? They are rarely found. However, despite the low return rate, our knowledge of bird biology has increased. Individuals of species that migrate, despite journeys of thousands of kilometres to different continents, were found to return to the area in which they had previously bred or hatched. Small passerines in the wild have been shown to live more than ten years and gulls, terns, waders and albatrosses may live over 30 years in the wild. Some petrels and shearwaters, indeed, may be proved soon to double this lifespan under natural conditions. However, these are the extremes, not the average life expectancies, and most individuals live much shorter periods.

Males of some species winter in different areas from the females (e.g. Common Chaffinch *Fringilla coelebs*) and young birds tend to move further than older birds (e.g. Northern Gannet *Morus bassanus*). During the course of the 20th century, some birds have changed their migration routes and wintering areas. Such conclusions are possible because their routes could be mapped from ringing recoveries. White Storks *Ciconia ciconia* migrating from Europe to Africa have apparently learnt to avoid crossing the central Mediterranean where they are liable to be shot, and now move either through Spain or Turkey and Israel. Differences in migration routes used by closely related species became evident. While most warblers breeding in Europe move south into Africa, the Lesser Whitethroats *Sylvia curruca* from western Europe migrate southeast through Italy to Israel and only then move due south into Africa, following the River Nile.

Rings are less than 1% of the bird's normal body mass. While little or no evidence has come to light that ringing has had an adverse effect on the behaviour of marked individuals, well into the second half of the century, there was reluctance to approve and even opposition to ringing from some conservation bodies. By the end of the century, this concern had dissipated as the important contribution of ringing to aspects of the management of wild birds, such as conservation, protection and pest control, became evident. Ringing is a key component in the suite of tools used by conservation and research bodies.

The first extensively used bird rings were made of aluminium that was light and malleable enough to allow numbers and addresses to be impressed into the rings. The rings were made by hand, with inscriptions and numbers laboriously punched one by one into the metal. It took many years before it was detected that in some species the rings wore out long before the lifespan of the bird was reached. For instance, rings placed on Black-legged Kittiwakes *Rissa tridactyla* became so worn that they lost their inscriptions within four years and fell off. Scarcely a Herring Gull *Larus argentatus* ringed as a chick was recovered over four years old, the earliest age of first breeding. It was not until 1960 that the first rings made of more durable metals were used. Now monel, incoloy and stainless steel rings are used on species where ring wear was heavy, to ensure that the rings survive their lifespan. As a result, we now know that many pelagic species live much longer than indicated by the use of aluminium rings.

Survival and mortality rates

Margaret Nice, in her remarkable study of colour-ringed Song Sparrows *Melospiza melodia*, made the first estimate of natural mortality rates in birds. David Lack was first to recognise that ringing recoveries could predict the survival rates of wild birds. While the length of life of most of the birds that were ringed was not known, because they were not found again, those recovered that had recently died could be regarded as a representative sample. Lack showed that the adult European Robin *Erithacus rubecula* had an annual mortality rate of close to 62% and that the average individual lived for only 1.1 years. He discovered that the mortality rate in the first year of life was appreciably higher than that of adult birds. These results chimed with the conclusions drawn by Margaret Nice from her work with Song Sparrows. But since many birds lived much longer in captivity, such theories were, at first, treated with scepticism. Now low annual survival rates in many passerines are accepted without reservation. If it were not so, given the number of birds that fledge, we would be knee-deep in warblers, finches and buntings!

As the number of birds ringed and recovered increased, calculations could be made of year-by-year mortality rates for individual species. The Common Starling *Sturnus vulgaris* was the first subject. Many other species have since been analysed. Awareness of the annual variation in mortality rates has given an important insight into conservation. It has highlighted the effect on wild bird populations of severe winters and of degradation of farmland habitat through intensive farming practices involving the use of agrochemicals, including insecticides such as DDT and dieldrin. Statisticians have produced mathematical models to facilitate these calculations, and their methods have become so sophisticated that they have probably outstripped the reliability of the actual ringing data. Even so, some assumptions still require critical examination, particularly those relating to the age at which young birds are ringed and the influence of this on the return and report rates.

One of Lack's deductions was that, after the first year of life, the mortality rate of birds remained constant (Lack 1943a,b). Botkin and Miller (1974) pointed out that the theoretical implication of this was that some individuals would be immortal. At some stage in old age, the mortality rate has to increase. More recent studies have done little to shed new light on the problem. Samples of recoveries of long-lived individuals are relatively few and therefore it is difficult to show at what point there is a significant change in the survival rate of old birds. There are still many questions remaining to be answered.

Colour-ringing

While the success of the metal ringing of birds primarily relies on the general public finding and reporting ring numbers, colour-ringing requires the marker to collect the information relating to the bird himself or herself. Normally, we cannot recognise birds as individuals in the way we do our own species. Colour-rings allow each bird to be recognised and studied as an individual. More recently, laminated coloured leg-rings placed on larger species have enabled numbers and letters to be added to the rings. Neck-collars and wing-tags have been attached to species such as swans and geese where the legs are often concealed in water or by vegetation.

The first colour-ringing study of consequence was that of Margaret Nice, who in Ohio during the 1930s, individually marked and then observed a group of Song Sparrows. She found that each year about 47% of her marked birds disappeared and this led to the first field estimate of natural mortality. Since that time, many colour-ringing studies have been made. Individuals can be followed and analyses of accumulated data can be made to reveal the effects of ageing and nest location. What difference in breeding success exists between birds nesting on the periphery and in the centre of colonies and in poor and good habitats?

Marking birds has made a major contribution to measuring the lifetime reproductive success of individuals within a species. In 1960, no such studies existed, but within thirty years, it became possible to produce a book devoted to such information (Newton 1989). The key outcome of such analyses is the awareness that a small proportion of individuals are highly productive, producing most of the next generation; the great majority, even of those reaching maturity, produced few or no offspring.

There is considerable variation in the age of first breeding, particularly in those species with high adult survival rates and which typically delay breeding until two or more years after hatching. In some species, the first breeding attempt may range over several years, depending upon the quality of the individual, the availability of food and of suitable nesting sites. Similarly, adult birds may miss one or more breeding seasons. Because of adverse environmental conditions or poor body condition caused by disease or malnutrition, it is probably a quite frequent occurrence.

Philopatry as a term in biology and ornithology is often misused. Literally it means a return to the place of birth to breed and its study provides a comparison of where an individual hatched and where it bred. It is therefore an inverse measure of gene flow. It should not be used as a synonym for nest-site tenacity, which is the extent to which an adult returns to the same nesting site irrespective of where it was hatched and is thus an incomplete measure of gene flow. It is now clear that philopatry varies considerably between species. In many albatrosses, for example, it is high, reaching 100% in some species with few colonies. However, in most species of birds it is variable and less complete, with a proportion of individuals showing a high tendency to return to where they have been hatched, while others move and breed at a considerable distance from the natal area. Often there is a marked difference between sexes – males showing a higher degree of philopatry than females (in mammals the reverse is true). In several species, there is a clear distinction between philopatric birds and those which are not. About 80% of Black-legged Kittiwakes return to breed either in the natal colony or within 20 km of it; 5% breed 20 km to 800 km away; and the remaining 15% at a distance of up to 1,500 km. Similar patterns have been found for Northern Lapwing *Vanellus vanellus*, Woodcock *Scolopax rusticola* and Mallard *Anas platyrhynchos*, but how the philopatric birds differ from the rest has yet to be investigated.

Philopatry is less frequent in species which nest in temporary habitats, such as sand spits or areas where vigorous plant growth takes place. It seems also to be less frequent in species whose food supply fluctuates from year to year such as Snowy Owl *Nyctea scandiaca*, Short-eared Owl *Asio flammeus* and Long-tailed Skua (jaegers) *Stercorarius longicaudus*.

Nest-site tenacity is frequent in birds and is often age related. An extreme example is the Common Eider *Somateria mollissima*. Some young birds hatched and fledged in Finland wintered and bred for one year in the Netherlands, but in subsequent years returned to breed in Finland. The European Shag *Phalacrocorax aristotelis* has been found to behave in the same way. In general, birds that are successful breeders are more likely to return to the same nest site than unsuccessful birds. Philopatry and site tenancity are difficult subjects to investigate quantitatively. Because of the large area that has to be searched as the distance from the natal area or previous breeding site increases, it is much easier to find a marked bird that has returned than one that has moved. No doubt DNA studies will give a new perspective on this topic in years to come.

Behaviour

Considerable advances have been made in determining the sex of study birds and in following their behaviour under natural conditions. Birds, whose sex in the field is not obvious, can be sexed by their behaviour at copulation; or when the bird is in the hand, from measurements or DNA samples. Once individuals have been sexed, their behaviour can be investigated. Lancelot Richdale (1949) was able to identify the sex of over a hundred individually marked Yellow-eyed Penguins *Megadyptes antipodes* as a result of dissecting and sexing one individual that was found dead.

Divorce, that is the establishment of a new pair, although both of the previous partners are still alive, has been reported in the Black-legged Kittiwake and more recently in several other seabirds.

Divorce rate decreased with age and was more prevalent on the edge of the colony. Further, it was more likely to occur when birds had failed to breed successfully in the previous year. The Eurasian Oystercatcher *Haemotopus ostralegus* appears to behave in the same way.

Telemetry

Many birds are difficult to observe all of the time, either because they become lost in the vegetation or move beyond the vision of the observer. The attachment of a simple transmitting radio to a bird allows that bird to be located at any time. The radio signal is received by a directional aerial (antenna) attached to a receiver. This indicates where and how far distant the bird is. More precise positions can be obtained by triangulation, identifying the direction of the signal from two or more positions. The nature of the terrain affects the reception of the signal. It can be distorted by woodland or waves at sea. On low-lying, open ground, signals can be received over several kilometres. Elevating the receiver high above the ground on a mast or on a cliff top or carrying it on an aircraft increases the distance at which the signal can be received. Naturally, the power of the transmitter also affects the distance at which signals are received.

The problem with telemetry is the effect the transmitter has on the bird carrying it. Early transmitters were large and were only suitable for use on large birds; but in the last few years, transmitters have been reduced in size and weight. However, the smaller the batteries, the shorter the active life of the transmitter, although solar powered batteries have been used with advantage.

The method of attachment of the transmitter to the bird is also important. On some birds, it is glued to the tail feathers; on others fitted with a harness that holds it as a backpack between the wings. In some studies the transmitter has been implanted under the skin, but this practice is not generally approved as it involves invasive action. Caution is also needed in attaching transmitters to diving birds as their buoyancy and resistance underwater can be affected. Telemetry has revealed unexpected aspects of bird behaviour. The recently fledged young of Common Buzzards *Buteo buteo* remain close to the nest site, but some of them fly tens of kilometres to forage in areas new to them. They stay away a day or so and then return to their natal area, such exploratory activity being rarely repeated.

The problem of locating birds carrying a transmitter has been greatly reduced by the use of satellites to receive the signal. High quality positioning can be achieved by this means, although the movement of the receiving satellites around the world determines the frequency and accuracy of the signals. At present, the good and bad quality signals are evenly mixed; but quality will be improved when more satellites are in orbit. Transmitters used in satellite telemetry are larger than conventional ones and are still only suitable for large or moderately sized birds. Such studies have told us a great deal about the movements of adult albatrosses as they feed prior to returning to their colony to feed their chicks. These studies pioneered in the 20th century will be developed in years to come.

Transponders

Transponders are small electronic devices that do not require their own power supply and depend on 'power' produced by the receiver, which records the coded number built into the transponder. As a result, they are much smaller than radio transmitters and fitted subcutaneously, remaining effective for many years. Their main disadvantage is that they need a powered sensor close to them to read their coded number. Such devices are used to mark domestic animals, so that ownership can be established if they are lost. Transponders have been used on penguins, where the birds have been forced to walk past the sensor by restriction of the routes available back to the colony. Sensors were placed near the nests of Common Terns *Sterna hirundo* in Germany, enabling the birds to be identified while incubating. A fascinating outcome of the use of transponders in conjunction with standard leg-rings has been the demonstration that a number of terns lost their leg-rings, despite their being made of a hard metal, presumably taken off by people capturing them in their wintering area in Africa. Thus a further limitation of leg-rings has been highlighted.

MOLECULAR STUDIES IN AVIAN TAXONOMY, SYSTEMATICS AND BIOLOGY

The acronym DNA (deoxyribonucleic acid) is now in common usage. DNA is the key to genetic inheritance. It is used in paternity claims because it can identify the parents with a high degree of certainty. Similarly, it is widely used to identify criminals as the DNA can be extracted from extremely small samples of body tissue or fluid left at the scene of a crime. In humans, DNA establishes the identity of an individual with greater accuracy than fingerprints or other traditional forensic evidence. In animals, generally, it reflects the unique genetic make-up of each individual and points to its evolution from ancestral animals.

DNA is extracted from chromosomes within the nuclei of cells or from the mitochondria in the cytoplasm. The red blood cells of mammals lack nuclei, but those of birds are nucleated, so very small samples of avian blood are rich sources of DNA. In any case, a recently developed method of increasing a small sample of DNA (amplification) has meant that even minute amounts can now be used in research studies.

Although less than 30 years old, the study and use of DNA has already developed a vocabulary of its own, which makes it difficult for the layman to understand the methodology and interpretations presented in scientific papers and reports. Nevertheless, the methodology is straightforward and well within the facilities of many laboratories. Most genetic studies are simpler in design than in execution because the instruments required to manipulate very small, fragile and sensitive biological samples are highly complex. In effect, to analyse DNA and to make comparisons between pairs of samples, the user must follow standard protocols, well within the capabilities of graduates in biochemistry. The interpretation is, however, often more complex and it is this that requires careful consideration and statistical examination.

Paternity and pair fidelity

DNA fingerprinting, comparable to that first used in human paternity studies in 1985 by Jefferies *et al.* (1985a,b,c), was soon taken up by ornithologists. The Common Cuckoo *Cuculus canorus* has been shown to be monogamous, the House Sparrow *Passer domesticus* extensively polygamous, and polygyny's effect on the productivity of male Red-winged Blackbirds *Agelaius phoeniceus* has been revealed. Extensive data have been produced on extra-pair mating which has led to the concept of 'sperm competition', the striving by males to leave a maximum number of surviving offspring. Much received wisdom about pair faithfulness in birds has had to be dramatically revised. For example, the first brood of young House Sparrows whose DNA samples were examined revealed that five young in the same nest had three different fathers. For birds it would appear that the old saw 'it's a wise youngster that knows its own father' is particularly relevant. Clutches containing eggs from more than one female have been recorded, for example, in ducks, the Ostrich *Struthio camelus* and in the Eastern Bluebird *Sialia sialis*. DNA studies have led to new ideas about social organisation and the role of monogamy. The great majority of birds breed as monogamous pairs but the extent of pair fidelity is being increasingly questioned.

Taxonomy and systematics

DNA analysis has been used in two main areas of avian systematics: the study of evolutionary relationships between the major taxa of birds, and evaluation of species and subspecies where allopatric populations exist.

Questions have been asked in the past about the inter-relationships of birds. For example, there has long been speculation about the affinities of birds classified as ratites – the flightless birds that occur in Africa (Ostrich), South America (Rhea), New Zealand (Kiwi), New Guinea (Cassowary) and Australia (Emu). Evidence has been obtained from comparative morphology, behaviour, breeding biology and plumage, but some of this has been contradictory. Could flightless birds have evolved as a homogeneous group derived from a common ancestor, or were they the result of convergent evolution? Auks and penguins and Old World and New World warblers are outwardly similar but unrelated. Again, Australian passerines are similar in structure and behaviour to northern hemisphere robins, wrens and nuthatches, but is this similarity the result of convergent evolution, or are the Australian species descended from common ancestral stock which repeatedly invaded that continent from the north?

At the species level, certain groups of birds, such as the chiffchaffs (a group within the Old World warblers in the genus *Phylloscopus*), differ relatively little in plumage and biometrics, but are renowned for the geographical variation of their songs and calls. Are the marked differences variations within a single species, or are several species involved?

Comparisons of the DNA show that the ratites are more similar to each other than to any group of flying birds and may therefore be descended from a common ancestor; but that Australian passerines are a homogeneous group and almost certainly radiated within Australia from a common stock. As a result of studies on their DNA and songs by Helbig *et al.* (1996), the chiffchaffs have been split into four distinct species and a further six subspecies. Two new species have been recognised (*P. canariensis* and *P. brehmii*) and the status of one taxon, the Siberian Chiffchaff (*P. tristis*), still remains uncertain. Of course, these answers are not infallible, but the evidence presented in most cases is convincing.

Thus DNA analysis is more precise and quantitative than tools previously available, and adds an important dimension to the classification of birds. Nevertheless, it is essentially a sampling technique. Only a small sample of the whole genome is compared, although the range of molecular techniques will increase.

The molecular examination of taxonomic status in birds was first applied using a DNA hybridisation technique which separated the DNA strands and formed artificially produced DNA molecules, where one of the two strands came from each of the species being compared. This hybrid material is then compared quantitatively with the same DNA strands obtained from each species separately. The difference is measured by determining how much heat is needed to separate the two hybrid strands. Differences in the structure of the two strands weaken their adhesion compared with those of the two parent species, where, in each case, the two strands are similar. The greater the genetic difference between the two species, the greater the dissimilarity will be between the hybrid and parent DNA and the strands will separate at a lower temperature. The difference between the hybrid material and that from the two species being investigated is called the 'numerical distance'. The greater the difference, the less closely related are the two species.

This method has been applied more extensively in avian than in other animal research. Charles Sibley, working first with Jon Ahlquist (1986, 1990) and later with Burt Monroe (1990), used it to revise the whole classification of the birds. A vast amount of repetitive work was involved. Many of their conclusions agreed with views previously held based on older, conventional taxonomic methods; but a quarter of all classifications have been challenged. But where does the truth lie? Some of the new classifications were unexpected. Old and New World quail; Old and New World vultures; divers (loons) and grebes; and swifts and swallows are said to be only distantly related; on the other hand birds of paradise may be more closely related to the crow family than to bowerbirds, and the Hoatzin *Opisthocomus hoazin* may belong to the cuckoos. With hindsight, many of Sibley's conclusions are probably correct; nevertheless they need confirmation using other methods, both molecular and non-molecular. Ernst Mayr and Walter Bock have pointed out (*Ibis*136: 12–18) that, because of its sheer volume, much of the data generated by Sibley and others have not been published. This makes reanalysis of aspects of the work more difficult. Some of these new conclusions have already been tested using an independent molecular test involving mitochondrial DNA (see below). Many of Sibley and Ahlquist's conclusions mentioned above were confirmed, although the relationship suggested between the Hoatzin and cuckoos was challenged.

A more recent molecular method of analysis uses mitochondrial DNA (mtDNA). This DNA comes from the mitochondrion, a structure which lies separate from the nucleus within each cell. It is present only in the ovum and so is passed from generation to generation entirely through the female line. The system is different from the DNA hybrid technique used in Sibley's studies. It depends upon identifying the sequence of cytochrome *b* nucleotide. Since there are 1,143 nucleotides in bird cytochrome *b*, there are plenty of opportunities for differences to occur. Some of these differences (mutations) affect the phenotype; the others are 'silent' mutations and do not appear to influence the characteristics of the individual. The figures obtained from the examination of two samples of material taken from different birds can be expressed as percentages of cytochrome *b* divergence. The value increases as the degree of separation between the two organisms becomes greater. This divergence, unlike that in hybrid DNA analysis, depends on mutations only. It was thought that the historical mutation rate in all birds was constant and so the degree of divergence was a measure of the time since the two populations separated during evolution, although this assumption is now challenged.

At the species level, Eric Pasquet (1998) used mtDNA analysis to demonstrate that of the ten nuthatch species (*Sittidae*), five comprise a monophyletic group of species spread around the world, but the remaining five are much less closely related and, presumably separated from the main line many millions of years ago. In another study, the taxonomic relationship of Berthelot's Pipit *Anthus berthelotii*, endemic to three groups of North Atlantic Islands, to other pipits was examined. MtDNA studies showed that it was closely related to the Tawny Pipit *A. campestris*, although the percentage of divergence was large enough to consider it a distinct species. Other empirical evidence suggests it should be lumped with Tawny Pipit.

One of the weaknesses of the method is that some species that are considered distinct by other criteria overlap in the extent of their divergence with those that are clearly subspecies. That the divergence ranges from as low as 1.1% (and in one case 0.1%!) for a distinct species to as high as 6.4% for a subspecies is a cause for concern. New data are being accumulated rapidly and the overlap range can only grow. For example, disagreement exists about the status of the Citril Finch *Serinus citrinella* populations on mainland Europe and those on the islands of Corsica and Sardinia in the Mediterranean. In 1997, two authors considered that a 2.7% mtDNA difference did not justify specific status, but three years later, Sangster (2000) proposed specific status for the Corsican and Sardinian populations, although accepting the same 2.7% difference.

As problems are solved, so new ones arise. Molecular studies help in making decisions in some specific cases, but they are not definitive or infallible in the way that was envisaged only 15 years ago. MtDNA analysis is superior to DNA hybrid methods in several respects, but it is too early to assess the potential of either or both of the methods. Clearly both raise new problems of avian classification. But the systems involved go way beyond the comprehension of previous generations of taxonomists. Factors that modify the DNA and mtDNA results, particularly the degree of difference, need to be identified and researched. The 21st century will undoubtedly see new advances, which will give clearer and more precise insight into evolution and speciation.

Appendix I

Walter Charleton

(a number of these names are obscure and not easily identifiable today)

1. LAND BIRDS

1. Carnivorous birds

Aquila - Eagles & the Osprey
Vultur - Vultures
Accipiter (including Circus, Buteo, Collurio, Lanius & Milvus)
Falco - Falcons
Cuculus - Cuckoos
Psittacus - Parrots
Corvus - Crows
Pica (including Mimus, the Oystercatcher, the Wax-wing, "Pica Brasiliensis" (the great-beaked Pie of Brasil) [? Toucan], Loxia curvirostra, "The Horned Pie of Ethiopia" [? = Hornbill]
Bubo (Owls, inc. Caprimulgus "Goat-sucking Owl" and Nycticorax, "Night raven" = Night Heron)
Struthio - Ostrich
Emeu (Cassowary)
Vespertilio (the bat)

2. Seed eaters

a. Dust-bathers

Pavo - Peafowl
Gallopavo - Turkeycock
Phasianus - Pheasants
Urogallus - Grouse
Grygallus - Capercaillie
Gallina corylorum "The Birch cock"
Attagen the Heathcock - Black Grouse
Otis - Bustards
Anas (also under Water birds) Ducks
Stella ?? based on Jonston's description
Oedicnemus - Stone Curlews
Perdix (inc. Lagopus) Partridges
Gallina - Guinea Fowl
Ortygometra - Rails
Coturnix - Quail
Cynchramus ?
Gallus - domestic fowl

b. Dust & water-bathing

Columba (including Sandgrouse)
Passeres

c. Singing

Carduelis - Goldfinch
Spinus - Siskin
Fringilla - Chaffinch
Montifringilla - Brambling
Chloris - Greenfinch
Linaria - Linnet
Citrinella - Yellowhammer
Alauda - Larks
Ficedula - Flycatchers

3. Berry-eaters

Turdus - Thrushes
Merula (not clear what this group includes, The Rosy Starling is there, and several thrushes).
Sturnus - Starling
Coccothraustes vulgaris - Hawfinch
Coccothraustes virginianus - Cardinal
Caryocatactes - Nutcracker

4. Insect-eaters

a. Non-singing

Picus (including Jynx, Certhia, Merops. ? Sitta
Caeruleo ("the Clot-bird, Smatch or Stone-check" - ??)

b. Singing

Regulus "The Wren" 1. Regulus cristatus, "The Copped Wren", 2. Regulus indicus
Asilus ??
Curruca - The Hedge Sparrow
Hirundo (including Martins, Swifts, and *Hirundo marina*, the "Sea Swallow" [? = Storm Petrel]
Parus - Titmice
Motacilla - Wagtails
Rubetra - Blackcap
Rubecula - Robin Redbreast
Ruticilla - Redstart
Anthus - Pipits
Cannevarola - ?
Oenanthe - Wheatear
Luscinia - Nightingale
Upupa - Hoopoe

2. WATER BIRDS

1. Palmipedes

Onocrotalus - Pelican
Avis Diomedea - ?
Larus (including Auks, Skuas and Terns)
Anser Bassanus - Gannet
Phalacrocorax - Cormorant & Shag
Graculus palmipos the "Sea Crow"
Mergus (including some sea ducks, divers & grebes)
Avosetta - Avocet
Trochilus "the fin-footed runner" - ?
Cygnus - Swans
Anser - Geese
Anas (also under seed eaters!) - Ducks
Boscas - Mallard
Querquedula - Teal
Puffinus
Fulica - Coots & Gallinules

2. Fissipeds

a. Fish-eaters

Ciconia - Storks
Ibis - Ibises
Phoenicopterus - Flamingoes
Ardea (including Platalea (Spoonbill), Bittern, Falcinellus ["Scyth-billed Heron" ?= Glossy Ibis])
Pugnax - Ruff
Porphyrion
Horion
Helorius
Limosa - Godwit
Barge
Himantopus "the Redshank"
Alcion (Alcedo)
Ispida - Kingfisher

b. Insect-eaters

Morinellus - the Dotterel
Arquata the Curlew (including the Stone Curlew)
Crex - Corncrake
Totanus the Pool Snipe
Calidris (the Godwit)
Gallinula
Scolopax - Woodcock
Gallinago - Snipe
Trynga "Water Thrush" - Sandpiper
Junco
Cinclus "Water Swallow" - Dipper
Merula aquatica
Vannellus - Lapwing
Pluvialis (Grey Plover, Green Plover)
Charadrius - Plovers
Cercio

c. Plant eaters

Grus - Crane

It will be seen that there are a number of apparent duplications.

APPENDIX 2

MÖHRING

1. Hymenopodes:

1. Picae: Collyrio, Paradisaea, Bucco, Tragopan, Coracias, Corvus, Sturnus, Cuculus, Picus, Merops, Ispida, Upupa, Troglodytes, etc.
2. Passeres:
   a. Crassirostrae: Fringilla, Ampelis etc.
   b. Tenuirostrae: Alauda, Motacilla, Erithacus, Ficedula, Luscinia, Parus, Hirundo, etc.

2. Dermatopodes:

1. Accipitres: Strix, Caprimulgus, Psittacus, Falco, Aquila, Vultur.
2. Gallinae: Pavo, Crax, Meleagris, Gallus, Tetrao, Coturnix, Columba.

3. Brachyptera: Struthio, Rhea, Casuarius, Didus, Otis.

4. Hydrophilae:

1. Odontorhynchae: Phoenicopterus, Platalea, Anas, Mergus, Plotus.

2. Platyrhynchae: Spheniscus.
3. Stenorhynchae: Pelecanus, Phalacrocorax, Phaeton, Alca, Larus, Procellaria, Uria, Colymbus.
4. Urinatrices: Colymbus, Fulica.
5. Scolopaces: Grus, Ibis, Ardea, Ciconia, Limicolae and others including crakes, guineafowl, the plains wanderer, hummingbirds, waders, plovers, dippers.

Appendix 3

J.C. Schaeffer

Class 1 Nudipedes [feet (i.e. legs) naked]
  Ordo 1. Fissipedes Didactyli [Split feet, 2 toes]
    Genus 1. Struthio (Ostrich)
  Ordo 2. Fissipedes Tridactyli [ditto, 3 toes]
    Genus 2. Rhea (Rhea)
    Genus 3. Casuarius (Cassowary [? and Emu])
    Genus 4. Otis (Bustard)
    Genus 5. Pluvialis (Stone Curlews)
    Genus 6. Ostralega (Oystercatcher)
    Genus 7. Himantopus (Stilt)
  Ordo 3. Fissipedes Tetradactyli [ditto, 4 toes]
    Genus 8. Vanellus (Plovers)
    Genus 9. Jacana (Jacana)
    Genus 10. Balearica (Crowned Cranes)
    Genus 11. Glareola (Pratincole)
    Genus 12. Platea (= Platalea, Spoonbills)
    Genus 13. Arenaria (? Turnstone)
    Genus 14. Rallus (Rails)
    Genus 15. Porphyrio (Swamphen)
    Genus 16. Ardea (Herons)
    Genus 17. Ciconia (Storks)
    Genus 18. Scopus (Hammerkop)
    Genus 19. Cochlearius (Boatbill)
    Genus 20. Tringa
    Genus 21. Limosa
    Genus 22. Scolopax (Waders)
    Genus 23. Cariama (Cariama)
    Genus 24. Anhima (Screamer)
    Genus 25. Raphus (Dodo)
    Genus 26. Numenius (Wader)
  Ordo 4. Pinnipedes Tetradactyli [Lobed (literally winged or feathered) feet, 4 toes]
    Genus 27. Gallinula (Gallinules)
    Genus 28. Fulica (Coots)
    Genus 29. Phalaropus (Phalaropes)
    Genus 30. Colymbus (Grebes)
  Ordo 5. Palmipedes Tridactyli [Palmed (webbed) feet, 3 toes]
    Genus 31. Uria (Auks)

Genus 32. Alca
Genus 33. Fratercula
Genus 34. Albatrus (?Albatross)
Ordo 6. Palmipedes Tetradactyli Postico Solvto [hind toe separated]
Genus 35. Mergus (Divers)
Genus 36. Spheniscus (Penguins)
Genus 37. Catarractes (?Penguin)
Genus 38. Corira ??
Genus 39. Puffinus (Shearwater)
Genus 40. Procellaria (Shearwater)
Genus 41. Stercorarius (Skuas)
Genus 42. Larus (Gulls)
Genus 43. Sterna (Terns)
Genus 44. Rynchopfalia (Skimmers)
Genus 45. Avocetta (Avocet)
Genus 46. Merganser (Merganser)
Genus 47. Anser (Goose)
Genus 48. Anas (Duck)
Genus 49. Phoenicopterus (Flamingo)
Ordo 7. Palmipedes Tetradactyli nullo solvto [hind toe not separated]
Genus 50. Sula (Gannets)
Genus 51. Anhinga (Anhinga)
Genus 52. Lepturus (Tropic Bird)
Genus 53. Phalacrocorax (Cormorant)
Genus 54. Onocrotalus (Pelecan)
Class 2. Plumipedes [feathered feet (i.e. legs)]
Ordo 1. Fissipedes Tetraanisodactyli
Genus 55. Picus (Woodpeckers)
Genus 56. Torquilla (Wryneck)
Genus 57. Goibula [pres. = Galbula] (Jacamars)
Genus 58. Bucco (Barbets)
Genus 59. Cuculus (Cuckoos)
Genus 60. Tucana (Toucans)
Genus 61. Trogon (Trogons)
Genus 62. Crotophagus (Anis)
Genus 63. Psittacus (Parrots)
Ordo 2. Fissipedes Tetraanisodactyli Aduncirostris [bill hooked]
Genus 64. Vultur
Genus 65. Aquila
Genus 66. Accipiter
Genus 67. Asio
Genus 68. Strix
Ordo 3. Fissipedes Tetraanisodactyli Conicoincurvirostres [bill conical, incurved]
Genus 69. Gallopavo
Genus 70. Gallus
Genus 71. Meleagris
Genus 72. Phasianus
Genus 73. Perdix
Genus 74. Lagopus

Ordo 4. Fissipedes Tetraanisodactyli Conicotenuirostres [bill slender]
Genus 75. Pyrrhula (Bullfinches)
Genus 76. Colius (Colies)
Genus 77. Passer (Sparrows)
Genus 78. Coccothraustes (Hawfinch)
Genus 79. Carduelis (Finches)
Genus 80. Emberiza (Buntings)
Genus 81. Tangara (Tanagers)
Genus 82. Loxia (Crossbills)
Ordo 5. Fissipedes Tetraanisodactyli Conicoprotensirostres [bill with a "car" or prominent cere]
Genus 83. Columba (Pigeons)
Genus 84. Manucodiata (Birds of Paradise)
Genus 85. Icterus (American Orioles & Blackbirds)
Genus 86. Galgulus (Rollers)
Genus 87. Corvus (Crows)
Genus 88. Pica (Magpies)
Genus 89. Nucifraga (Nutcrackers)
Genus 90. Garrulus (Jays)
Ordo 6. Fissipedes Tetraanisodactyli Convexirostres [bill convex]
Genus 91. Buphagus (Oxpeckers)
Genus 92. Sturnus (Starlings)
Genus 93. Cotinga (Cotingas)
Genus 94. Turdus (Thrushes)
Genus 95. Lanius (Shrikes)
Ordo 7. Fissipedes Tetraanisodactyli Subulirostres [bill like a shoemaker's awl]
Genus 96. Alauda (Larks)
Genus 97. Ficedula (Flycatcher)
Genus 98. Parus (Titmice)
Genus 99. Caprimulgus (Nightjars)
Genus 100. Hirundo (Swallows & Swifts)
Ordo 8. Fissipedes Tetraanisodactyli Cuneirostres [bill wedge-shaped]
Genus 101. Sitta (Nuthatches)
Ordo 9. Fissipedes Tetraanisodactyli Filirostres [bill thread-like]
Genus 102. Mellisuga (Hummingbirds, part)
Ordo 10. Fissipedes Tetraanisodactyli Falcirostres [bill sickle-shaped]
Genus 103. Upupa (Hoopoes)
Genus 104. Promerops (Promerops)
Genus 105. Certhia (Treecreepers)
Genus 106. Polytmus (Hummingbirds, part)
Ordo 11. Fissi-Anomolipedes, Tetradactyli
Genus 107. Rupicola (Cocks-of-the-Rock)
Genus 108. Manacus (Manakins)
Genus 109. Todus (Todies)
Genus 110. Ispida (Kingfishers)
Genus 111. Apiaster (Bee-eaters)
Genus 112. Momotus (Motmots)
Genus 113. Hydrocorax [unidentifiable ?= an African Hornbill]

Appendix 4
Brisson

Order 1		Genus 1 Columbinum - Pigeons
Order 2	Section 1	Genus 2-4 Gallopavo, Gallus, Meleagris-Turkeys & chickens
	Section 2	Genus 5-7 Lagopus, Perdix, Phasianus- Grouse, Pheasants & Partridges
Order 3	Section 1	Genus 8-10 Accipiter, Aquila, Vultur - Birds of prey
		Genus 11-12 Asio, Strix -Owls
Order 4	Section 1	Genus 13-17 Coracia, Corvus, Pica, Garrulus, Nucifraga - Corvids & Rollers
	Section 2	Genus 18-20 Galgulus (Coracias L), Icterus, Manucodiata (Paradisea L) - Icterids & Birds of Paradise
Order 5	Section 1	Genus 21-23 Lanius, Turdus, Cotinga - Shrikes, Thrushes, Cotingas
	Section 2	Genus 24 Muscicapa - Flycatchers
Order 6	Section 1	Genus 25 Buphagus - Oxpeckers
	Section 2	Genus 26 Sturnus - Starlings
Order 7	Section 1	Genus 27 Upupa - Hoopoe
	Section 2	Genus 28 Promerops -
Order 8		Genus 29 Caprimulgus - Nightjars
		Genus 30 Hirundo - Swallows & Swifts
Order 9	Section 1-3	Genus 31-38 Tangara, Carduelis, Passer, Coccothraustes, Emberiza, Colius, Pyrrhula, Loxia
Order 10	Section 1-2	Genus 39-41 Alauda, Ficedula (Motacilla L), Parus
Order 11		Genus 42 Sitta
Order 12	Section 1-2	Genus 43-45 Certhia, Polytmus, Mellisuga (Trochilus L)
Order 13	Section 1-5	Genus 46-54 Torquilla, Picus, Galbula, Bucco, Cuculus, Trogon, Crotophagus, Psittacus, Tucana
Order 14	Section 1-5	Genus 55-61 Rupicola, Manacus, (Pipra L), Momotus, Ispida, Todus, Apiaster, Hydrocorax (Buceros L)
Order 15	Section 1-3	Genus 62-65 Struthio, Rhea, Casuarius, Raphus
Order 16	Section 1-3	Genus 66-69 Otis, Himantopus, Ostralega, Pluvialis
Order 17	Section 1-12	Genus 70-87 Vanellus, Jacana, Arenaria, Glareola, Rallus, Tringa, Scolopax, Numenius, Platea, Ciconia, Ardea, Scopus, Cochlearius, Balearica, Cariama, Anhima, Porphyrio
Order 18	Section 1-2	Genus 88-90 Gallinula, Phalaropus, Fulica
Order 19		Genus 91 Colymbus (Podiceps L)
Order 20	Section 1-2	Genus 92-94 Uria, Fratercula, Alca
Order 21	Section 1-2	Genus 95-97 Spheniscus, Catarractes, Mergus (Colymbus L)
Order 22		Genus 98 Albatrus
Order 23	Section 1-2	Genus 99-104 Puffinus, Procellaria, Stercorarius, Larus, Sterna, Rynchopsalia
Order 24	Section 1-2	Genus 105-7 Merganser, Anser, Anas
Order 25	Section 1-2	Genus 108-112 Anhinga, Lepturus, Sula, Phalacrocorax, Onocrotalus
Order 26	Section 1-2	Genus 113-5 Phoenicopterus, Avocetta, Corrira

Appendix 5

Linnaeus 1758 (10th edition)

Order 1. Accipitres

1. Vultur. Vultures. [inc. Harpy Eagle]
2. Falco. Eagles and Hawks and Falcons.
3. Strix. Owls.
4. Lanius. Shrikes. [inc. some Tyrannids]

Order 2. Picae

5. Psittacus. Parrots.
6. Ramphastos. Toucans
7. Buceros. Hornbills.
9. Crotophaga. Anis.
10. Corvus. Crows, Drongos and some Rollers.
11. Coracias. Rollers. [inc. Orioles]
13. Gracula. Grackles. [some of these are Icterids, others Starlings]
14. Paradisaea. Birds of Paradise. [2 species. *apoda* and *regia*]
17. Cuculus. Cuckoos.
18. Jynx. Wryneck.
19. Picus. Woodpeckers.
20. Sitta. Nuthatches.
22. Alcedo. Kingfishers. [inc. Todies]
23. Merops. Bee-eaters. [inc. Promerops]
24. Upupa. Hoopoe. [inc. one Bird of Paradise, the Chough and the Waldrapp]
25. Certhia. Creepers. [inc. the Bananaquit]
26. Trochilus. Hummingbirds.

Order 3 Anseres

27. Anas. Swans, Geese & Ducks. ["beak lamellated"]
28. Mergus. Mergansers. ["beak serrated and hooked"]
29. Alca. Auks.
30. Procellaria. Petrels.
31. Diomedea. Albatrosses. [2 species, the Wandering Albatross & the Blackfoot Penguin]
32. Pelecanus. Pelicans. [inc. Gannets & Frigatebirds]
34. Phaeton. Tropicbirds. [inc. Blackfoot Penguin again]
35. Colymbus. Divers and Grebes.
36. Larus. Gulls. [inc. Skuas]
37. Sterna. Terns.
38. Rhynchops. Skimmers.

Order 4. Grallae: Waders.

39. Phoenicopterus. Flamingo.
40. Platalea. Spoonbills.
42. Mycteria. The Jabiru.
44. Ardea. Cranes, Herons & Storks. [and Wood Storks (later genus *Ibis*)]
45. Tantalus. Ibis. [1 sp.]

46. Scolopax. Snipes & Curlews. [inc. Scarlet & White Ibises (later returned to *Tantalus)*]
47. Tringa. Sandpipers. [inc. Phalaropes]
48. Charadrius. Plovers. [inc. Stone Curlews and Stilts]
49. Recurvirostra. Avocet.
50. Haematopus. Oystercatcher.
51. Fulica. Coots, Moorhens and Gallinules. [inc. Jacana]
53. Rallus. Rails.
54. Psophia. Trumpeters.
55. Otis. Bustards.
56. Struthio. Ostrich, Rhea & Cassowary. [and Dodo]

Order 5. Gallinae: Gallinaceous Birds.

58. Pavo. Peafowl. [inc. Peacock Pheasants]
59. Meleagris. Turkey. [inc. one Curassow and the Satyr Tragopan]
60. Crax. Curassows.
61. Phasianus. Pheasants. [inc. Guineafowl]
63. Tetrao. Grouse and Partridges. [inc. Quails]

Order 6. Passeres.

64. Columba. Pigeons.
65. Alauda. Larks.
66. Sturnus. Starlings. [inc. Dippers]
67. Turdus. Thrushes. [inc. some Sylvid warblers]
69. Loxia. Grosbeaks. - a selection of seedeaters + the Mousebirds.
70. Emberiza. Buntings. [inc. Weavers]
72. Fringilla. Finches.
74. Motacilla. Wagtails & Warblers. [inc. Flycatchers, Nightingale, Chats, the Guira Cuckoo, Wrens, Goldcrests and some Parulids]
76. Parus. Tits. [inc. Manakins]
77. Hirundo. Swallows. [inc. Swifts]
78. Caprimulgus. Goatsuckers.

Linnaeus 1766 (12th edition)

Order 1. Accipitres

1. Vultur. Vultures [inc Harpy Eagle]
2. Falco. Eagles and Hawks and Falcons.
3. Strix. Owls.
4. Lanius. Shrikes. [inc. some Tyrannids]

Order 2. Picae

5. Psittacus. Parrots.
6. Ramphastos. Toucans.
7. Buceros. Hornbills.
8. Buphaga. Oxpeckers.
9. Crotophaga. Anis.
10. Corvus. Crows, Drongos and the Waldrapp!
11. Coracias. Rollers.

12. Oriolus. Orioles. [inc. some Icterids]
13. Gracula. Grackles. [some of these are Icterids, others Starlings.
14. Paradisaea. Birds of Paradise. [3 species. *apoda*, *regia* and *tristis* which = a Starling]
15. Trogons.
16. Bucco. Barbets.
17. Cuculus. Cuckoos.
18. Yunx. Wryneck.
19. Picus. Woodpeckers.
20. Sitta. Nuthatches.
21. Todus. Todies.
22. Alcedo. Kingfishers.
23. Merops. Bee-eaters.
24. Upupa. Hoopoe. [inc. Promerops]
25. Certhia. Creepers. [a very general grouping including a number of sunbirds and the Bananaquit]
26. Trochilus. Hummingbirds.

Order 3 Anseres

27. Anas. Swans, Geese & Ducks. ["beak lamellated"]
28. Mergus. Mergansers. ["beak serrated and hooked"]
29. Alca. Auks.
30. Procellaria. Petrels.
31. Diomedea. Albatrosses. [2 species, the Wandering Albatross & the Blackfoot Penguin]
32. Pelecanus. Pelicans. [inc. Gannets & Frigatebirds]
33. Plotus. Darter.
34. Phaeton. Tropicbirds. [inc. Blackfoot Penguin again]
35. Colymbus. Divers and Grebes. [inc. Guillemots & Diving Petrels]
36. Larus. Gulls. [inc. Skuas]
37. Sterna. Terns.
38. Rhynchops. Skimmers.

Order 4. Grallae: Waders.

39. Phoenicopterus. Flamingo.
40. Platalea. Spoonbills.
41. Palamedea. Screamers.
42. Mycteria. The Jabiru.
43. Cancroma. The Boatbill.
44. Ardea. Cranes, Herons & Storks.
45. Tantalus. Ibises and Wood Storks. (later genus *Ibis)*
46. Scolopax. Snipes & Curlews. [inc. the Limpkin and larger Sandpipers]
47. Tringa. Sandpipers. [inc. Phalaropes]
48. Charadrius. Plovers. [inc. Stone Curlews and Stilts]
49. Recurvirostra. Avocet.
50. Haematopus. Oystercatcher.
51. Fulica. Coots, Moorhens and Gallinules.
52. Parra. Jacana.
53. Rallus. Rails.
54. Psophia. Trumpeters.

55. Otis. Bustards.
56. Struthio. Ostrich, Rhea & Cassowary.

Order 5. Gallinae: Gallinaceous Birds.
57. Didus. Dodo.
58. Pavo. Peafowl. [inc. Peacock Pheasants]
59. Meleagris. Turkey. [inc. one Curassow and the Satyr Tragopan]
60. Crax. Curassows.
61. Phasianus. Pheasants. [inc. Motmots]
62. Numida. Guineafowl.
63. Tetrao. Grouse and Partridges. [inc. Quails and Sandgrouse]

Order 6. Passeres.
64. Columba. Pigeons.
65. Alauda. Larks.
66. Sturnus. Starlings. [inc. Dippers]
67. Turdus. Thrushes. [inc. some Sylvid warblers]
68. Ampelis. Chatterer (i.e. Cotingas). [inc. Waxwings]
69. Loxia. Grosbeaks. - a selection of seedeaters + the Mousebirds
70. Emberiza. Buntings. [inc. Weavers]
71. Tanagra. Tanagers.
72. Fringilla. Finches.
73. Muscicapa. Flycatchers. [inc. some Tyrannids]
74. Motacilla. Wagtails & Warblers. [inc. Nightingale, Chats, the Guira Cuckoo, Wrens, Goldcrests and some Parulids]
75. Pipra. Manakins. [inc some Parulids]
76. Parus. Tits.
77. Hirundo. Swallows. [inc. Swifts, Pratincoles and the Striped Cuckoo (*Tapera*)]
78. Caprimulgus. Goatsuckers.

Appendix 6

Brünnich

FALCO Birds of prey
lanarius: This name appears in many old lists. It is thought to refer to a young Peregrine, certainly not a Lanner Falcon, which does not occur in northern Europe.
milvus = Red Kite
tinnunculus = Kestrel
gentilis = Goshawk
islandus = rusticolus = The Gyrfalcon
subbuteo = Hobby
albicilla = White-tailed Eagle
ossifraga = albicilla = White-tailed Eagle
torquatus = Apparently taken from Brisson. Probably a Harrier
lagopus = Rough-legged Buzzard
aeruginosus = Marsh Harrier
apivorus = Honey Buzzard
nisus = Sparrowhawk
haliaetus = Pandion = Osprey
buteo = Common Buzzard

STRIX Owls [Curiously, Aegiolus funereus of Linnaeus (Tengmalm's Owl) is not listed]
otus = Asio otus = Long-eared Owl
aluco = Tawny Owl
stridula = aluco = Tawny Owl
ulula = Surnia = Hawk Owl
passerina = Pygmy Owl
bubo = Eagle Owl

nyctea = Snowy Owl
LANIUS Shrikes, including the Waxwing and the Bearded Tit!
excubitor = Great Grey Shrike
collurio = Red-backed Shrike
garrulus = Waxwing
biarmicus = Bearded Tit
CORVUS Corvids [The Rook is not included. It occurs very locally in Denmark and the southern parts of Norway and Sweden.]
corax = Raven
varius = Faeroe Raven
corone = Carrion Crow
cornix = Hooded Crow
monedula = Jackdaw
pica = Magpie
glandarius = Jay
caryocatactes = Nutcracker
infaustus = Siberian Jay
CORACIAS Roller
garrulus = Roller
CUCULUS Cuckoo
canorus = Cuckoo
JYNX Wryneck
torquilla = Wryneck
PICUS Woodpeckers
martius = Black Woodpecker
viridis = Green Woodpecker
major = Great Spotted Woodpecker
medius = Medium Spotted Woodpecker
minor = Lesser Spotted Woodpecker
SITTA Nuthatch
europea = Nuthatch
UPUPA Hoopoe
epops = Hoopoe
CERTHIA Treecreeper
familiaris = Treecreeper
ANAS All Anatidae except Sea Ducks (Mergus)
cygnus = Whooper Swan
tadorna = Shelduck
fusca = Velvet Scoter
marila = Scaup
bernicla = Brent Goose
anser = Grey Lag
anser domesticus = tame goose
erythropus = Pink-footed Goose
mollissima = Eider
platyrhynchos = ?Mallard
clangula = Bucephala clangula = Goldeneye
penelops = penelope = Widgeon
hiemalis = Clangula hyemalis = Long-tailed Duck
ferina = Pochard
querquedula = Garganey
crecca = Teal
circia = Anas querquedula = Garganey
histrionica = Harlequin Duck
minuta = female of Harlequin
boschas = platyrhynchos = Mallard
glaucion = nyroca = Ferruginous Duck
fuligula = Tufted Duck
latirostra. Still unidentified, but clearly not mythical, for Brunnich lists a specimen from Christiansoe and describes it carefully.
nigra = Scoter
strepera = Gadwall
clypeata = Shoveler
acuta = Pintail
MERGUS
merganser = Goosander
rubricapilla = merganser Goosander
cristatus ?= Hooded Merganser
serratus = serrator (L) Red-breasted Merganser
albellus = Smew
glacialis = albellus Smew
ALCA
torda = Razorbill
balthica = torda = Razorbill
unisulcata = torda = Razorbill
arctica = Fratercula = Puffin
deleta = Fratercula = Puffin
impennis = Great Auk
alle = Little Auk
candida = Alle alle = Little Auk
URIA
lomvia = Brunnich's Guillemot
troille = U. aalge = Common Guillemot
svarbag = U. lomvia = Brunnich's Guillemot
ringvia = "Bridled Guillemot"
alga ?= aalge
grylle = Cepphus grylle = Black Guillemot
grylloides = Cepphus grylle
balthica = Cepphus grylle
PROCELLARIA
pelagica = Storm Petrel
glacialis = Fulmar
puffinus = Manx Shearwater (first description of this bird)
PELECANUS Cormorants & Gannet
carbo = Phalacrocorax carbo = Cormorant
graculus = aristotelis = Shag
phalacrocorax. From the description, this seems to be P. carbo also.

cristatus = aristotelis
bassanus = Gannet
CATHARACTA
skua = Great Skua
cepphus = parasiticus (pale morph) Arctic Skua
parasitica = longicaudus = Buffon's Skua
coprotheres = parasiticus (dark morph) Arctic Skua
[The Pomarine Skua had not been described at this time. It was first described by Temminck]
COLYMBUS Divers & Grebes
immer = Great Northern Diver
stellatus = Red-throated Diver
borealis = stellatus
lumme = stellatus
arcticus = Black-throated Diver
torquatus = immer
cristatus = Great crested Grebe
auritus = Slavonian or Horned Grebe
LARUS Gulls
rissa = Kittiwake
canus = Common Gull
fuscus = Lesser Black-backed Gull
marinus = Great Black-backed Gull
maculatus ? young bird
glaucus = hyperboreus = Glaucous Gull
argentatus = Herring Gull
varius = argentatus
STERNA Terns
hirundo = Common Tern
paradisea = Arctic Tern
nigra = Black Tern
PLATALEA Spoonbill
leucorodia = Spoonbill
ARDEA Herons, Storks & Crane
ciconia = White Stork
nigra = Black Stork
stellaris = Bittern
cinerea = Heron
grus = Crane
SCOLOPAX Larger Scolopacids
totanus = Common Redshank
arquata = Curlew
phaeopus = Whimbrel
gallinago = Snipe
minima = Jack Snipe
rusticola = Woodcock
lapponica = Bar-tailed Godwit
falcinellus = Limicola = Broad-billed Sandpiper (ex Pontoppidan)
glottis = nebularia = Greenshank
TRINGA Smaller Scolopacids + Lapwing & Grey Plover (but not Golden!)
pugnax = Ruff
vanellus = Lapwing
lobata = Red-necked Phalarope
fulicaria = Grey Phalarope
alpina = Dunlin
hypoleucos = Common Sandpiper
interpres = Turnstone
squatarola = Grey Plover
littorea = pugnax = Ruff
cinerea = canutus = Knot
ferruginea = Curlew Sandpiper
variegata. Unidentified.
maritima = Purple Sandpiper
undata. Unidentified.
ocropus = Green Sandpiper
CHARADRIUS Plovers
hiaticula = Ringed Plover
morinellus = Dotterel
apricarius = Golden Plover
pluvialis = apricarius Golden Plover
alexandrinus = Kentish Plover
RECURVIROSTRA Avocet
avosetta = Avocet
HAEMATOPUS Oystercatcher
ostralegus = Oystercatcher
FULICA Coot & Moorhen
atra = Coot
chloropus = Moorhen
RALLUS Crakes
crex = Corncrake
aquaticus = Water Rail
TETRAO Grouse, Partridges & Quail
urogallus = Capercaillie
tetrix = Black Grouse
lagopus = Willow Grouse
bonasia = Hazel Grouse
perdix = Partridge
coturnix = Quail
COLUMBA Doves
oenas = Stock Dove
palumbus = Wood Pigeon
[then a number of domestic varieties]
risoria = Barbary Dove
ALAUDA Larks & Pipits
arvensis = Skylark
pratensis = Meadow Pipit
arborea = Woodlark
campestris = Tawny Pipit
trivialis = Tree Pipit
cristata = Crested Lark

STURNUS Starling & Dipper
vulgaris = Starling
cinclus = Dipper
TURDUS Thrushes (sensu stricto)
viscivorus = Mistle Thrush
pilaris = Fieldfare
merula = Blackbird
musicus = Song Thrush or Redwing
torquatus = Ring Ousel
LOXIA Greenfinch, Bullfinch, Crossbill, etc.
curvirostra = Crossbill
enucleator = Pinicola = Pine Grosbeak
pyrrhula = Bullfinch
chloris = Greenfinch
atra ? unidentified
EMBERIZA Buntings
nivalis = Snow Bunting
calandra = Corn Bunting
citrinella = Yellow Bunting
schoeniclus = Reed Bunting
FRINGILLA Chaffinch, Brambling, Goldfinch, Canary, Linnet, Redpoll, Sparrows
coelebs = Chaffinch
montifringilla = Brambling
carduelis = Goldfinch
canaria = Canary
cannabina = Linnet
domestica = House Sparrow
montana = Tree Sparrow
linaria = flammea = Redpoll
MOTACILLA Dunnock, Wagtails, Chats, Goldcrest, Warblers
modularis = Dunnock
philomela = Nightingale
alba = White Wagtail
flava = Yellow Wagtail
sylvia = communis = Whitethroat
oenanthe = Common Wheatear
rubetra = Whinchat
atricapilla = Blackcap
phoenicurus = Redstart
ficedula = Pied Flycatcher
rubecula = Robin
troglodytes = Wren
regulus = Goldcrest
trochilus = Chiffchaff/Willow Warbler
curruca = Lesser Whitethroat
hippolais = either hortensis (Garden Warbler) or colybita (Chiffchaff)
luscinia = Nightingale
PARUS Titmice
major = Great Tit
coeruleus = Blue Tit
palustris = Marsh Tit
ignotus = Great Tit
HIRUNDO Swallows & Swifts
rustica = Barn Swallow
urbica = House Martin
riparia = Sand Martin
apus = Swift
CAPRIMULGUS Nightjars
europaeus = Nightjar

Appendix 7
Latham

Division 1. Land Birds

Order 1. Accipitrine

[Genera. 1. Vulture (inc. Secretary Bird) 2. Falcon (all other raptors) and 3. Owl].

Order 2. Pies

a. with legs for walking.

b. with climbing feet.

c. feet made for leaping.

[4. Shrike (includes some Tyrannids & others), 5. Parrot, 6. Toucans, 7. Motmots, 8. Hornbills, 9. Beef-eater (Oxpecker), 10. Ani, 11. Wattlebird, 12. Crows, 13. Rollers, 14 Orioles (Orioles and Icterids and a number of other birds which clearly do not belong), 15. Grackles (inc. some Icterids & Mynas), 16. Birds of Paradise, 17. Trogons, 18. Barbets, 19. Cuckows, 20. Wryneck, 21. Woodpeckers, 22. Jacamars, 23. Kingfishers, 24. Nuthatches (includes a few sub-oscines), 25. Todies (incs some Barbets and Manakins), 26. Bee-eaters, 27. Hoopoes (includes Promerops, Sicklebills & others), 28. Creeper (contains a wide range of items, inc. some Drepanids), 29. Humming birds.

Order 3. Passerine.

a. with thick bills.

b. with curved bills, upper mandible bent at tip.

c. with bills leaving upper mandible emarginated near the tip.

d. simple-billed - bill straight, integral attenuated.

[30. "Stares" (Starlings & some Icterids), 31. Thrushes (a very amorphous assemblage), 32. Chatterer (Waxwings & Cotingas), 33. Colies, 34. Grosbeaks (large billed seedeaters), 35. Buntings, 36. Tanagers, 37. Finches, 38. Flycatcher (included a number of Tyrannids), 39. Larks, 40. Wagtails, 41. Warblers (contains a great deal more than the Sylvidae!), 42. Manakin (inc. the Cocks-of-the-Rock), 43. Titmice, 44. Swallows (inc. Swifts), 45. Goatsuckers].

Order 4. Columbine. Bill sharpish on the edge, nostrils gibbous covered with an obsolete membrane.

[46. Pigeons]

Order 5. Gallinaceous.

a. with 4 toes.

b. with 3 toes.

[47. Peacocks, 48. Turkeys (inc. Guans), 49. "Pintado" (Guinea Fowl), 50. Curassows, 51. Pheasants (inc the Hoatzin), 52. Tinamous, 53. Grouse, 54. Partridges, 55. Trumpeters, 56. Bustards].

Order 6. Struthious

a. with 4 toes.

b. with 3 toes, placed forwards.

c. with 2 toes placed forwards.

[57. Dodo, 58. Ostrich, 59. Cassowary]

Division 2. Water Birds

Order 7. With cloven feet.

a. with 4 toes.

b. with 3 toes placed forwards.

[60. Spoonbills, 61. Screamers, 62. Jabiru, 63. Boatbill, 64. "Umbre" (Hammerkop), 65. Heron, 66. Ibis (inc. some Storks), 67. Curlews, 68. Snipe (inc. Woodcocks and some sandpipers), 69. Sandpipers, 70. Plovers, 71. Oister-Catcher, 72. Pratincoles, 73. Rails, 74. Jacanas, 75. Gallinules (inc. some Porzanas), 76. Sheathbill].

Order 8. with Pinnated (lobed) feet

[77. Phalaropes, 78. Coots, 79. Grebes]

Order 9. Web-footed

a. with long legs.

b. with short legs.

[80. Avocets, 81. "Courier" (Correira - unidentified), 82. Flamingos, 83. Albatrosses, 84. Auks (broadbilled - Great Auk, Razorbills, Puffins, etc), 85. Guillemot (narrow-billed Auks), 86. Divers, 87. Skimmers, 88. Terns, 89. Gulls, 90, Petrels (inc. Shearwaters, Diving Petrels & Storm Petrels), 91. Merganser, 92. Ducks (inc. Geese and Swans), 93. Penguins, 94. Pelecans (inc. Frigate-birds, Cormorants & Gannets), 95. Tropic-birds, 96. Darters].

Appendix 8
Pallas

Note: Trinomials are given only where Pallas's name is a valid geographical race.

ORDO 1. PRAEPETES

1. Stryges

1. Stryx Bubo = Bubo bubo = Eagle Owl
2. Stryx otus? = Long-eared Owl
3. Stryx aegiolus = Aegiolus funereus = Tengmalm's Owl
4. Stryx scops = Otus scops = Scops Owl
5. Stryx nyctea = Nyctea scandiaca - Snowy Owl
6. Stryx doliata (ulula?) = Hawk Owl
7. Stryx barbata = Strix nebulosa
8. Stryx uralensis = Strix uralensis = Ural Owl
9. Stryx aluco = Strix aluco = Tawny Owl
10. Stryx ulula = Surnia ulula = Hawk Owl
11. Stryx passerina = Pygmy Owl

2. Falcones

12. Falco gyrfalco = Falco rusticolus = Gyrfalcon
13. Falco peregrinus = Peregrine
14. Falco laniarius - of old authors, probably a young Peregrine, certainly not a Lanner Falcon.
15. Falco subbuteo = Hobby
16. Falco tinnunculus = Kestrel
17. Falco vespertinus = Red-footed Falcon
18. Falco aesalon = Falco columbarius = Merlin

3. Aquilae

19. Aquila nobilis = Aquila chrysaetos = Golden Eagle
20. Aquila chrysaetos = Golden Eagle
21. Aquila pelagica = Haliaeetus pelagicus = Steller's Eagle
22. Aquila albicilla = Haliaeetus albicilla = White-tailed Eagle
23. Aquila leucocephala = Bald Eagle
24. Aquila ossifraga =? immature H. albicilla
25. Aquila clanga = Spotted Eagle
26. Aquila leucorypha = Haliaeetus leucoryphus - Pallas' Eagle

4. Accipitres

27. Accipiter hypoleucos = Circaetus gallicus = Short-toed Eagle
28. Accipiter haliaetus = Pandion haliaetus = Osprey
29. Accipiter milvus = Milvus milvus = Red Kite
30. Accipiter regalis = Milvus migrans = Black Kite
31. Accipiter lacertiarius = Pernis apivorus = Honey Buzzard
32. Accipiter lagopus = Buteo lagopus = Rough-legged Buzzard
33. Accipiter buteo = Buteo buteo = Common Buzzard
34. Accipiter circus = Circus aeruginosus = Marsh Harrier
35. Accipiter variabilis ?= Circus cyaneus = Hen Harrier

36. Accipiter astur = A. gentilis = Goshawk
37. Accipiter nisus = Sparrowhawk

5. Vultures

38. Vultur barbatus = Gypaetus barbatus - Lammergeier
39. Vultur percnopterus = Neophron percnopterus - Egyptian Vulture
40. Vultur persicus "new species" ?= Gyps fulvus - Griffon Vulture
41. Vultur meleagris = Neophron percnopterus - Egyptian Vulture

ORDO 2. OSCINES

6. Corvi

42. Corvus corax = Raven
43. Corvus corone = Carrion Crow
44. Corvus cornix = Hooded Crow
45. Corvus frugilegus = Rook
46. Corvus monedula = Jackdaw
47. Corvus dauuricus = Daurian Jackdaw
48. Corvus pica = Pica pica = Magpie
49. Corvus cyanus = Cyanopica cyana = Azure-winged Magpie
50. Corvus stelleri = Cyanocitta stelleri = Steller's Jay
51. Corvus glandarius = Garrulus glandarius = Common Jay
52. Corvus mimus (?= infaustus) ?Siberian Jay
53. Corvus caryocatactes = Nucifraga caryocatactes = Nutcracker
54. Corvus graculus = Chough, but unclear which one, the name "graculus" has been used for both species.

7. Lanii

55. Lanius major = Lanius excubitor = Great Grey Shrike
56. Lanius excubitor = Lanius excubitor = Great Grey Shrike
57. Lanius vigil = Lanius minor = Lesser Grey Shrike
58. Lanius collurio = Red-backed Shrike
59. Lanius phoenicurus = Lanius cristatus = Brown Shrike
60. Lanius brachyurus = ?Unidentified

8. Pici

61. Picus martius = Dryocopus martius = Great Black Woodpecker
62. Picus viridis = Green Woodpecker
63. Picus chlorio = Picus canus = Ashy or Grey-headed Woodpecker
64. Picus cirris = an east Siberian race of Dendrocopos leucoto
65. Picus cissa = Picus major = Great Spotted Woodpecker
66. Picus cynaedus = Dendrocopos medius = Medium Spotted Woodpecker
67. Picus pipra = Dendrocopos minor = Lesser Spotted Woodpecker
68. Picus tridactylus = Picoides tridactylus = Three-toed Woodpecker
69. Picus jynx = Jynx torquilla = Wryneck

9. Sturni

70. Sturnus vulgaris = Common Starling
71. Sturnus roseus = Rose-coloured Starling
72. Sturnus dauuricus ?= Sturnus sturninus = Daurian Starling
73. Sturnus cinclus = Cinclus cinclus = Common Dipper

10. Xanthorni

74. Xanthornus caucasicus = Emberiza melanocephala = Black-headed Bunting

75. Xanthornus pendulinus = Remiz pendulinus = Penduline Tit
11. Certhiae
76. Certhia muraria = Tichodroma muraria = Wallcreeper
77. Certhia scandulaca = Certhia familiaris = Treecreeper
12. Upupa
78. Upupa vulgaris = Upupa epops = Hoopoe
13. Alcedines
79. Alcedo ispida = Alcedo atthis = Little Blue Kingfisher
80. Alcedo alcyon = Ceryle alcyon = Belted Kingfisher
14. Meropes
81. Merops apiaster = European Bee-eater
82. Merops persicus = Blue-cheeked Bee-eater
15. Coracias
83. Coracias garrula = Roller
16. Cuculus
84. Cuculus borealis = Cuculus canorus = Cuckoo
17. Turdi
85. Turdus oriolus = Oriolus oriolus = Golden Oriole
86. Turdus saxatilis = Monticola saxatilis
87. Turdus auroreus ?= female of Zoothera sibiricus
88. Turdus varius = Zoothera dauma
89. Turdus merula = Blackbird
90. Turdus leucocillus = male of Zoothera sibiricus
91. Turdus torquatus = Ring Ousel
92. Turdus fuscatus = Turdus naumanni
93. Turdus ruficollis = Turdus ruficollis
94. Turdus viscivorus = Turdus viscivorus
95. Turdus musicus = Turdus philomelos
96. Turdus pilaris = Turdus pilaris
97. Turdus illas = iliacus
98. Turdus pallens = Turdus obscurus = Eyebrowed or Dark Thrush
99. Turdus junco = Acrocephalus arundinaceus
100.Turdus aedon = Phragmaticola aedon
18. Muscicapa
101. Muscicapa grisola = Muscicapa striata - Spotted Flycatcher
102. Muscicapa albicilla = Ficedula parva albicilla - Red-breasted Flycatcher
103. Muscicapa fuscedula = Muscicapa sibirica
104. Muscicapa atricapilla = Muscicapa hypoleuca - Pied Flycatcher
105. Muscicapa eridea - based on S.G. Gmelin's MS - ?Unidentified
106. Muscicapa guttata - based on Latham's Oonalaschka Thrush ?Unidentified
19. Motacillae
1. Oenanthe
107. Motacilla rubetra = Saxicola rubetra = Whinchat
108. Motacilla rubicola (? not of Linnaeus) Unidentified.
109. Motacilla luteola = Ficedula mugimaki
110. Motacilla montanella = Prunella montanella
111. Motacilla cyane = Larvivora (Erithacus) cyane = Siberian Blue Chat

112. Motacilla vitiflora = Oenanthe oenanthe = Common Wheatear
113. Motacilla strapazina = Oenanthe isabellina
114. Motacilla erithacus = Phoenicurus ochruros
115. Motacilla phoenicurus = Phoenicurus ochruros
116. Motacilla aurorae = Phoenicurus aurorae
117. Motacilla leucomela (pleschanka)
118. Motacilla coerulecula (?= svecica)

2. Philomelae Lusciniae Kleinii

119. Motacilla calliope = Luscinia calliope
120. Motacilla philomela. Unidentified
121. Motacilla luscinia = Luscinia megarhynchos
122. Motacilla aedon = Luscinia luscinia
123. Motacilla curruca = Sylvia curruca
124. Motacilla sylvia = Sylvia curruca affinis
125. Motacilla cyanura = Tarsiger cyanurus
126. Motacilla rubecula = Erithacus rubecula
127. Motacilla salicaria = ?Acrocephalus streperus
128. Motacilla trochilus = Phylloscopus trochilus
129. Motacilla rubiginosa = Dendroica petechia rubiginosa
130. Motacilla acredula = Phylloscopus collybita tristis
131. Motacilla pileolata = Wilsonia pusilla pileolata
132. Motacilla regulus = Regulus regulus
133. Motacilla proregulus = Phylloscopus proregulus
134. Motacilla troglodytes = Troglodytes troglodytes

3. Pallenurae Motacillae Scopoli

135. Motacilla melanope = Motacilla cinerea
136. Motacilla flaveola = Motacilla flava
137. Motacilla citrinella = Motacilla citreola
138. Motacilla campestris = Motacilla flava lutea
139. Motacilla albeola = Motacilla alba

4. Corydales; Alaudis

140. Motacilla locustella - not identified, probably a Lark
141. Motacilla certhiola = Locustella certhiola
142. Motacilla cervina = Anthus cervinus
143. Motacilla spipola = Anthus trivialis

21. Alaudae

144. Alauda tatarica (yeltoniensis) = Alauda yeltonensis
145. Alauda mongolica = Alauda mongolica
146. Alauda calandra = Alauda calandra
147. Alauda leucoptera = Alauda leucoptera
148. Alauda nivalis = Eremophila alpestris
149. Alauda cristata = Galerida cristata
150. Alauda galerita = Galerida cristata
151. Alauda coelipeta = Alauda arvensis
152. Alauda grandior - ?not identified
153. Alauda testacea probably Calandrella brachydactyla, Gmelin, but not certainly identified
154. Alauda pispoletta = Calandrella rufescens

21.bis Hirundines

1. Chelidones

155. Hirundo domestica = Hirundo rustica
156. Hirundo lagopoda = Delichon urbica lagopoda
157. Hirundo alpestris = Hirundo daurica
158. Hirundo riparia = Riparia riparia
159. Hirundo melba = Apus melba

2. Apodes vel Falculae

160. Hirundo apus = Apus apus

160 bis. Hirundo ciris

3. Caprimulgi

161. Hirundo caprimulgus = Caprimulgus europaeus

22. Sitta

162. Sitta europaea = Sitta europaea

23. Pari

163. Parus bombycilla = Bombycilla garrulus
164. Parus barbatus = Panurus biarmicus
165. Parus caudatus = Aegithalos caudatus
166. Parus cyanus = Parus cyanus
167. Parus coeruleus = Parus coeruleus
168. Parus cristatus = Parus cristatus
169. Parus fringillago = Parus major
170. Parus carbonarius = Parus ater
171. Parus palustris = Parus palustris

24. Columbae

172. Columba oenas = Columba livia
173. Columba palumbes = Columba palumbus
174. Columba turtur = Streptopelia turtur
175. Columba rupicola = Streptopelia orientalis
176. Columba fusca - not certainly identified, possibly Columba eversmanni

25. Loxiae

177. Loxia curvirostra = Loxia curvirostra
178. Loxia psittacea = Loxia psittacea

26. Pyrrhula

179. Pyrrhula rubicilla = Pyrrhula pyrrhula
180. Pyrrhula erythrina = Carpodacus erythrinus
181. Pyrrhula caudata = Uragus sibiricus

27. Coccothraustes

182. Coccothraustes vulgaris = Coccothraustes c. japonicus
183. Coccothraustes caucasicus = Carpodacus rubicilla
184. Coccothraustes chloris = Carduelis chloris

28. Passeres

185. Passer carduelis = Carduelis carduelis
186. Passer spinus - probably Carduelis spinus, but not certainly identified
187. Passer spiza = Fringilla coelebs
188. Passer montifringilla = Fringilla montifringilla
189. Passer calcaratus = Calcarius lapponicus

190. Passer alpicola = Montifringilla nivalis alpicola
191. Passer arctous = Leucosticte arctoa
192. Passer roseus = Carpodacus roseus
193. Passer linaria = Carduelis flammea
194. Passer cannabina = Carduelis cannabina
195. Passer papaverina = Carduelis cannabina
196. Passer pusillus = Serinus pusillus
197. Passer domesticus = Passer domesticus
198. Passer montanina = Passer montanus

29. Emberizae

199. Emberiza nivalis = Plectrophenax nivalis
200. Emberiza hyperborea - not identified, but probably a Plectrophenax. It came from the Chutschi Peninsula, so possibly the same bird later called (quite coincidentally) P. hyperboreus, by Ridgway.
201. Emberiza miliaria = Emberiza calandra
202. Emberiza citrinella = Emberiza citrinella
203. Emberiza pithyornus = Emberiza leucocephala
204. Emberiza cia = Emberiza cia
205. Emberiza fucata = Emberiza fucata
206. Emberiza pusilla = Emberiza pusilla
207. Emberiza rustica = Emberiza rustica
208. Emberiza coronata = Zonotrichia atricapilla
209. Emberiza chrysops = Ammodramus sandwichensis
210. Emberiza chrysophrys = Emberiza chrysophrys
211. Emberiza schoeniclus = Emberiza schoeniclus
212. Emberiza pyrrhuloides = Emberiza schoeniclus pyrrhuloides
213. Emberiza passerina = Emberiza schoeniclus passerina
214. Emberiza hortulanus = Emberiza hortulana
215. Emberiza spodocephala = Emberiza spodocephala
216. Emberiza aureola = Emberiza aureola
217. Emberiza rutila = Emberiza rutila

ORDO 4. PULVERATRICES

30. Tetraones

218. Tetrao urogallus = Tetrao urogallus
219. Tetrao tetrix = Lyrurus tetrix
220. Tetrao lagopus = Lagopus lagopus
221. Tetrao bonasia = Tetrastes bonasia
222. Tetrao chata (= alchata) Pterocles alchata
223. Tetrao arenaria = Pterocles sp.
224. Tetrao paradoxa = Syrrhaptes paradoxus
225. Tetrao caucasica = Tetraogallus caucasicus
226. Tetrao perdix = Perdix perdix
227. Tetrao rufa = Alectoris rufa
228. Tetrao coturnix = Coturnix coturnix

31. Phasiani

229. Phasianus colchicus = Phasianus colchicus
230. Phasianus pictus = Chrysolophus pictus

231. Phasianus auritus = Crossoptilon auritum = Blue Eared-Pheasant
232. Phasianus gallinaceus = Gallus gallus - Domestic Fowl
32. Meleagris
233. Meleagris gallopavo = Meleagris gallopavo = Turkey
33. Pavo
234. Pavo cristatus = Pavo cristatus = Peafowl
34. Numida
235. Numida meleagris = Numida meleagris - Guineafowl

ORDO 5. GRALLAE
35. Otides
236. Otis tarda = Otis tarda
237. Otis tetrix = Otis tetrax
36. Grues
238. Grus antigone = Grus antigone
239. Grus leucogeranus = Grus leucogeranus
240. Grus vulgaris = Grus grus
241. Grus virgo = Anthropoides virgo
242. Grus vipio = Grus vipio
37. Ciconiae
243. Ciconia alba = Ciconia alba
244. Ciconia nigra = Ciconia nigra
38. Ardeae
245. Ardea cinerea = Ardea cinerea
246. Ardea purpurea = Ardea purpurea
247. Ardea alba = Egretta alba
248. Ardea garzetta = Egretta garzetta
249. Ardea comata = Ardeola ralloides
250. Ardea stellaris = Botaurus stellaris
251. Ardea nycticorax = Nycticorax nycticorax
252. Ardea minuta = Ixobrychus minutus
39. Haematopus
253. Haematopus hypoleuca = Haematopus ostralegus
254. Haematopus niger = Haematopus bachmani
40. Charadrii
255. Charadrius vanellus = Vanellus vanellus
256. Charadrius gregarius = Vanellus gregarius
257. Charadrius morinellus = Eudromius morinellus
258. Charadrius caspius ?= Eupoda asiatica = Caspian Plover
259. Charadrius mongolicus = Charadrius mongolus
260. Charadrius hypomelanus = Pluvialis squatarola
261. Charadrius apricarius = Pluvialis apricaria
262. Charadrius pluvialis = Pluvialis dominicus
263. Charadrius pardela = Pluvialis squatarola
264. Charadrius alexandrinus = Charadrius alexandrinus
265. Charadrius hiaticula = Charadrius hiaticula
266. Charadrius minutus = Charadrius dubius

267. Charadrius oedicnemus = Burhinus oedicnemus
268. Charadrius cinclus = Arenaria interpres

41. Glareola

269. Glareola pratincola = Glareola pratincola

42. Ralli

270. Rallus crex = Crex crex
271. Rallus aquaticus = Rallus aquaticus
272. Rallus porzana = Porzana porzana
273. Rallus minutus = Porzana pusilla

43. Fulicae

274. Fulica porphyrio = Porphyrio porphyrio
275. Fulica chloropus = Gallinula chloropus
276. Fulica atrata "Greater Coot" of Latham
277. Fulica pullata "Common Coot" of Latham

44. Recurvirostra

278. Recurvirostra avosetta = Recurvirostra avosetta

45. Platalea

279. Platalea leucorodia = Platalea leucorodia

46. Numenii

280. Numenius ibis ?= Threskiornis aethiopica
281. Numenius falcinellus = Plegadis falcinellus
282. Numenius arquata = Numenius arquata
283. Numenius phaeopus = Numenius phaeopus

47. Scolopaces

284. Scolopax rusticola = Scolopax rusticola
285. Scolopax palustris = Gallinago media
286. Scolopax gallinago = Gallinago gallinago
287. Scolopax gallinula = Lymnocryptes minima
288. Scolopax alpina = Calidris alpina

48. Limosae

289. Limosa aegocephala = Limosa lapponica
290. Limosa glottis = Tringa nebularia
291. Limosa ferruginea = Limosa lapponica
292. Limosa barge = Limosa sp.
293. Limosa recurvirostra = Xenus cinereus
294. Limosa totanus = Tringa totanus
295. Limosa calidris - ? Unidentified
296. Limosa himantopus = Himantopus himantopus
297. Limosa fusca = Tringa erythrypus

49. Tryngae

298. Trynga falcinella = Calidris testacea
299. Trynga arquatella = Tringa maritima
300. Trynga pugnax = Philomachus pugnax
301. Trynga naevia = Calidris canutus
302. Trynga ochropus = Tringa ochropus
303. Trynga glareola = Heteroscelus brevipes
304. Trynga guinetta = Tringa stagnatilis

305. Trynga littorea = Tringa glareola
306. Trynga leucoptera = Calidris (Actitis) hypoleucos
307. Trynga canutus = Calidris canutus
308. Trynga tridactyla = Calidris alba
309. Trynga salina = Calidris ruficollis
310. Trynga cinclus = Calidris temminckii
311. Trynga calidris = Calidris canutus

50. Phalaropi

312. Phalaropus ruficollis = Phalaropus lobatus
313. Phalaropus cinerascens = Phalaropus lobatus
314. Phalaropus rufus = Phalaropus fulicarius

51. Phoenicopterus

315. Phoenicopterus roseus = Phoenicopterus roseus

ORDO 7. HYDROPHILAE

52. Cygni

316. Cygnus olor = Cygnus cygnus
317. Cygnus sibilus = Cygnus olor

53. Anseres

318. Anser cygnoides = Anser cygnoides
319. Anser grandis - Unidentified
320. Anser vulgaris = Anser anser
321. Anser erythropus = Anser erythropus
322. Anser hyperboreus = Anser caerulescens
323. Anser brenta = Branta bernicla - Brent Goose
324. Anser bernicla = Branta leucopsis - Barnacle Goose
325. Anser canadensis = Branta canadensis
326. Anser ruficollis = Branta ruficollis
327. Anser pictus = Anser canagicus

54. Anates

328. Anas cutberti = Somateria molllissima
329. Anas spectabilis = Somateria spectabilis
330. Anas stelleri = Polysticta stelleri
331. Anas tadorna = Tadorna tadorna
332. Anas rutila = Tadorna ferruginea
333. Anas carbo = Melanitta fusca
334. Anas atra = Melanitta nigra
335. Anas marila = Aythya marila
336. Anas ferina = Aythya ferina
337. Anas penelope = Anas penelope
338. Anas rufina = Netta rufina
339. Anas strepera = Anas strepera
340. Anas boschas = Anas platyrhynchos
341. Anas adunca = Anas platyrhynchos
342. Anas moschata = Cairina moschata
343. Anas falcata = Anas falcata
344. Anas glocitans = Anas formosa
345. Anas crecca = Anas crecca

346. Anas querquedula = Anas querquedula
347. Anas fuligula = Aythya fuligula
348. Anas colymbis "Brown Duck" of Pennant = Aythya fuligula
349. Anas glaucion = Aythya nyroca
350. Anas hyemalis = Bucephala clangula
351. Anas clangula = Bucephala clangula
352. Anas histrionica = Histrionicus histrionicus
353. Anas glacialis = Clangula hyemalis
354. Anas caudacuta = Anas acuta
355. Anas clypeata = Anas clypeata
356. Anas mersa = Oxyura leucocephala

55. Mergi

357. Mergus merganser = Mergus merganser
358. Mergus serrator = Mergus serrator
359. Mergus albellus = Mergus albellus
360. Mergus cristatus - not certainly identified. Thought possibly to be M. cucullatus

56. Pelecanus

361. Pelecanus onocrotalus = Pelecanus onocrotalus

57. Phalacrocorax

362. Phalacrocorax carbo = Phalacrocorax carbo
363. Phalacrocorax graculus = Phalacrocorax aristotelis
364. Phalacrocorax pygmaeus = Phalacrocorax pygmaeus
365. Phalacrocorax bicristatus = Phalacrocorax urile
366. Phalacrocorax pelagicus = Phalacrocorax pelagicus
367. Phalacrocorax perspicillatus = Phalacrocorax perspicillatus
368. Phalacrocorax albidus - Unidentified, but possibly a now-extinct Gannet.

58. Sula

369. Sula bassana = Morus bassanus

59. Diomedea

370. Diomedea albatrus = Diomedea albatrus

60. Catarractae

371. Catarractes skua = Catharacta skua
372. Catarractes parasita = Stercorarius longicaudus

61. Procellariae

373. Procellaria glacialis = Fulmarus glacialis
374. Procellaria aequinoctialis = Procellaria aequinoctialis
375. Procellaria orientalis = Oceanodroma furcata
376. Procellaria pelagica = Hydrobates pelagicus

62. Lari

377. Larus cachinnans = Larus cachinnans
378. Larus niveus = Larus canus
379. Larus glaucus = Larus hyperboreus
380. Larus marinus = Larus marinus
381. Larus rissa = Rissa tridactyla
382. Larus ichthyaetus = Larus icthyaetus
383. Larus atricilla = Larus ridibundus
384. Larus cinerarius = Larus ridibundus

385. Larus naevius = Larus ridibundus
386. Larus torquatus = Rissa tridactyla
387. Larus gavia = Rissa tridactyla
388. Larus canus = Larus canus
389. Larus minutus = Larus minutus

63. Sternae

390. Sterna caspia = Sterna caspia
391. Sterna hirundo = Sterna hirundo
392. Sterna camtschatica - Unidentified
393. Sterna minuta = Sterna albifrons
394. Sterna naevia = Hydrochelidon nigra
395. Sterna hybrida = Hydrochelidon hybrida
396. Sterna fissipes = Hydrochelidon leucoptera

64. Cepphi

397. Cepphus torquatus = Gavia immer
398. Cepphus arcticus = Gavia arctica
399. Cepphus septentrionalis = Gavia stellata
400. Cepphus imber = Gavia immer
401. Cepphus stellatus = Gavia stellata
402. Cepphus lomvia = Uria lomvia
403. Cepphus arra = Uria lomvia arra
404. Cepphus columba = Cepphus columba
405. Cepphus carbo = Cepphus carbo
406. Cepphus perdix = Brachyramhus marmoratus perdix

65. Colymbi

407. Colymbus cornutus = Podiceps cristatus
408. Colymbus cucullatus = Podiceps griseigena
409. Colymbus naevius = Podiceps griseigena
410. Colymbus auritus = Podiceps auritus
411. Colymbus minutus = Podiceps ruficollis

66. Alcae

412. Alca torda = Alca torda
413. Alca pica = Alca torda
414. Alca monocerata = Cerorhinca monocerata

67. Lundae

415. Lunda cirrhata = Lunda cirrhata
416. Lunda arctica = Fratercula corniculata
417. Lunda psittacula = Cyclorrhynchus psittacula

68. Uriae

418. Uria senicula = Synthliborhamphus antiquus
419. Uria alle = Alle alle
420. Uria aleutica = Ptychoramphus aleuticus
421. Uria cristatella = Aethia cristatella
422. Uria dubia = Aethia cristatella
423. Uria tetracula = Aethia cristatella
424. Uria mystacea = Aethia pygmaea
425. Uria pusilla = Aethia pusilla

Appendix 9

Gloger

1. Aves rapaces:
Vultur [= Gyps]
Falco
Strix
2. Aves passerinae
Lanius
Corvus
Sturnus
Turdus
Saxicola
Sylvia
Cinclus
Anthus
Alauda
Accentor
Emberiza
Fringilla
Loxia
Parus
Sitta
Tichodroma
Certhia
Troglodytes
Bombycilla
Oriolus
Regulus
Muscicapa
Hirundo
3. Aves passerinae anomalae
Cypselus [= Apus]
Caprimulgus
Coracias
Merops
Alcedo
Cuculus
Picus
Iynx [=Jynx]
Upupa
4. Aves peristeroides
Columba
5. Aves gallinaceae
Phasianus
Tetrao
Perdix
Glareola
Tachydromus [= Cursorius]
Otis
Oedicnemus
Charadrius
Haematopus
Himantopus
Totanus
Tringa
Limosa
Scolopax (inc. Gallinago)
Ibis [= Plegadis]
Grus
Ciconia
Ardea (all Ardeidae)
Platalea
Recurvirostra
Crex
Rallus
Gallinula (inc. Porzana)
6. Aves natatoriae grallariae
Fulica
Phalaropus
Sterna
Larus
Lestris [= Stercorarius & Catharacta]
Procellaria [= Hydrobates]
Pelecanus
Anser
Cygnus
Anas
Mergus
Halieus [= Phalacrocorax]
Colymbus [= Podiceps]
Eudytes [= Gavia]

Appendix 10
Merrem

I. AVES CARINATAE
- 1. Aves aereae.
  - A. Rapaces (Accipitres, Owls)
  - B. Hymenopodes
    - a. Chelidones (Nightjars, Swallows and Swifts)
    - b. Oscines (hard-billed seedeaters; all other passerines, Todies and Rollers)
  - C. Mellisugae (Hummingbirds & Treecreepers)
  - D. Dendrocolaptidae (Woodpeckers & Wryneck)
  - E. Brevilingues (Hoopoes, Kingfishers)
  - F. Levirostres (Toucans, Parrots)
  - G. Coccyges (Cuckoo, Trogons, Puffbirds, Anis)
- 2. Aves terrestres (Pigeons; Game birds)
- 3. Aves aquaticae.
  - A. Odontorhynchi (Ducks, Geese & Swans; Flamingoes)
  - B. Platyrhynchi (Pelicans, Tropic-birds, Anhingas)
  - C. Aptenodytes (Penguins)
  - D. Urinatrices (Auks, Grebes, Divers)
  - E. Stenorhynchi (Petrels, Albatrosses, Gulls, Terns, Skimmers)
- 4. Aves palustres.
  - A. Rusticolae.
    - a. Phalarides (Rails, Gallinules & Jacanas)
    - b. Limosugae (Plovers, sandpipers, Avocets)
  - B. Grallae.
    - a. Erodii (Herons without crests)
    - b. Pelargi (Storks, Ibises, Spoonbills)
    - c. Gerani (Egrets [i.e. crested herons], Cranes, Trumpeters)
  - C. Bustards.

II. AVES RATITAE - Ostriches etc.

Appendix 11
Blainville

1. Pelvis medium
   - Anomalous
     - Prehensores - Psittacus
     - Raptatores - Serpentarius, Falco, Strix
     - Scansores
     - Heterodactyles - Caprimulgus, Crotophaga
     - Zygodactyles - Cuculus, Picus
     - Syndactyles - Alcedo

Normal

Saltatores - Trochilus, Corvus, Turdus, Motacilla, Lanius, Fringilla

Sponsores - Columba

Gradatores - Phasianus, Perdix

2. Pelvis long

Cursores - Struthio

Grallatores - Psophia, Otis, Tringa, Scolopax, Phoenicopterus, Ciconia, Rallus

3. Pelvis short

Natatores - Larus, Procellaria, Pelecanus, Anas, Mergus, Aptenodytes

Appendix 12

L'Herminier

A. Oiseaux normaux (Carinatae)

1. Birds of Prey
2. Secretary birds
3. Owls
4. Touracos
5. Parrots
6. Hummingbirds
7. Swifts
8. Nightjars
9. Cuckoos
10. Trogons
11. Rollers
12. Bee-eaters
13. Kingfishers
14. Hornbills
15. Toucans
16. Woodpeckers
17. Epopsides (An artificial assemblage consisting of Promerops, Hoopoes, Philemon Honeyeaters and Horneros, etc).
18. Passerines
19. Pigeons
20. Game birds
21. Tinamous
22. Coots
23. Cranes
24. Herons
25. Ibises and Spoonbills
26. Waders
27. Gulls
28. Petrels
29. Pelicans
30. Ducks
31. Grebes
32. Divers
33. Auks
34. Penguins

B. Oiseaux anomaux (Ratitae)

Appendix 13

Nitzsch

1. Aves Carinatae

A. Aves carinatae aereae
   1. Birds of Prey
   2. Passerines
   3. Hummingbirds & Swifts
   4. Nightjars, Cuckoos, Trogons, Rollers, Bee-eaters
   5. Toucans, Woodpeckers
   6. Parrots
   7. Kingfishers, Woodpeckers, Epopsides [see note in L'Herminier's system]
   8. Touracos

B. Aves carinatinae terrestres
   1. Pigeons
   2. Game birds

C. Aves carinatae aquaticus
   1. Alectorides
   2. Cranes
   3. Rails (Coots)
   4. Herons
   5. Herons (part), Ibises & Spoonbills
   6. Flamingoes
   7. Limicolae
   8. Gulls
   9. Petrels
   10. Ducks, Geese & Swans
   11. Pelicans
   12. Grebes, Divers, Auks, Penguins

2. Aves Ratitae

Appendix 14

Wagler

Order 1. Striges (Owls)
Order 2. Hirundines
   1. Nightjars, Swallows, Bee-eaters
   2. Pratincoles
   3. Gulls & Terns
Order 3. Accipitres (Birds of Prey)
Order 4. Gallinae
   1. Game birds
   2. Cranes
   3. Rails, Sheathbills
Order 5. Columbae
   1. Pigeons
   2. Sandgrouse
Order 6. Otides
   1. Bustards, Limicolae
   2. Oystercatchers, Avocets, Phalaropes, etc.
Order 7. Cuculi
   1. Cariamas
   2. Hoatzin
   3. "Schizorhis"
   4. Colies
   5. Toucans, Puffbirds, Jacamars, Trogons, Barbets, Honeyguides, Cuckoos

Order 8. Psittaci Parrots
Order 9. Passeres Passerines
Order 10. Corvi Crows, Starlings, Titmice, Todies, etc.,
Order 11. Spelecti
  1. Treecreepers, Nuthatches, Dendrocolaptidae, etc.
  2. Honeyeaters, Sunbirds.
Order 12. Trochili Hummingbirds
Order 13. Pici Woodpeckers & Wryneck
Order 14. Tantali
  1. Hoopoes, Hornbills, Kingfishers
  2. Storks
  3. Steganopodes & Tubinares
Order 15. Ardea
  1. Divers, Auks, Penguins, Kiwis, Dodos
  2. Herons, Crab Plover
Order 16. Anseres
  1. Flamingoes
  2. Geese, Swans & Ducks
Order 17. Struthiones
  1. Megapodes & Tinamous
  2. Ratites

Appendix 15

Lacépède

Sub-class 1. Lower leg feathered, toes divided.

Division 1. Thick, strong toes, zygodactyl.
  Order 1. Ara. Psittacus.
  Order 2. Rhamphastos, Trogon, Touraco, Musophaga.
  Order 3. Bucco.
  Order 4. Galbula, Picus.
  Order 5. Jynx.
  Order 6. Cuculus, Crotophaga.
Division 2. Three toes in front, one behind.
  Sub-division 1. Claws strong and crooked.
    Order 7. Vultur, Gypaetos (=? Gyps), Aquila, Astur, Nisus, Buteo, Circus, Milvus, Falco, Strix.
  Sub-division 2. Claws not very crooked, external toes mostly free.
    Order 8. Phytotoma.
    Order 9. Lanius, Tyrannus, Muscicapa, Muscivora, Turdus, Myrmecophaga, Oriolus, Ampelis, Tanagra.
    Order 10. Cacicus, Icterus, Xanthornis, Sturnus, Loxia, Pyrrhula, Fringilla, Emberyza (Emberiza).
    Order 11. Gracula, Corvus, Coracias, Paradisea, Sitta, Buphaga, Picioides.
    Order 12. Parus, Alauda, Sylvia, Motacilla.
    Order 13. Hirundo, Caprimulgus.
    Order 14. Glaucopsis, Upupa, Certhia, Trochilus.
    Order 15. Orthorhynchus (a hummingbird)
  Sub-division 3. External toes united.
    Order 16. Buceros, Momot.

Order 17. Alcedo, Ceyx.
Order 18. Todus.
Order 19. Pipra.
Order 20. Merops.
Sub-division 4. Gallinaces. Front toes united by membrane at base.
Order 21. Columba, Tetrao, Perdix, Tinamou, Tridactylus (?), Pavo, Phasianus, Numida, Meleagrix (Meleagris), Crax, Penelope, Gouan (? Guan).

Sub-class 2. Water Birds. Lower leg naked, or toes webbed.

Division 1. Three toes in front, one or none behind.
Sub-division 1. Front toes webbed.
Order 22. Phoenicopterus, Diomedea, Pelecanoides, Procellaria.
Order 23. Anas, Mergus, Prion.
Order 24. Rhynchops, Urinator (?= Gavia), Colymbus (= Podiceps "Grebe"), Uria, Alca, Pinguinus, Aptenodytes.
Order 25. Sterna.
Order 26. Recurvirostra.
Order 27. Larus.
Sub-division 2. All four toes webbed.
Order 28. Fregata, Carbo (= Phalacrocorax).
Order 29. Sula, Phaeton, Plotus (= Anhinga).
Order 30. Pelecanus.
Sub-division 3. Three toes before, one or none behind.
Order 31. Serpentarius, Palamedea (?), Glareola.
Order 32. Psophia, Vaginalis (?).
Order 33. Grus, Ciconia, Ardea, Hians (? = Open-bill Stork), Rallus, Scopus, Haematopus.
Order 34. Cancrona (?), Platalea.
Order 35. Scolopax.
Order 36. Mycteria, Ibis, Tantalus, Macrotarsus (?).
Order 37.Hydrogallina (?= Hydrophasianus), Fulica, Jacana, Parra, Phalaropus, Charadrius, Otis.
Division 2. Two, three or four very strong toes.
Order 38. Struthio, Touyou (?).
Order 39. Rhea. ("Casoar")
Order 40. Didus (Dodo).

Appendix 16

Illiger

Order 1 Scansores
1. Psittacini: Parrots
2. Serrati: Toucans, Puffbirds, Touracos.
3. Amphiboli: Anis, Scythrops, Barbets, Cuckoos, Coucals.
4. Sagittilingues: Woodpeckers & Wrynecks
5. Syndactyli: Jacamars

Order 2. Ambulatores
6. Angulirostres: Kingfishers and Bee Eaters

7. Suspensi: Hummingbirds
8. Tenuirostres: Sunbirds, Wallcreepers and Hoopoes
9. Pygarrhichi: Treecreepers, Dendrocolaptidae
10. Gregarii: Xenops, Nuthatches, Oxpeckers, Orioles, New world Orioles, Starlings.
11. Canori: Thrushes, Dippers, Accentors, Wagtails, Flycatchers, Myiothera, Shrikes, Todies, Manakins.
12. Passerini: Tits, Larks, Buntings, Tanagers, Finches, Crossbills, Mousebirds, Wattlebirds, Plantcutters.
13. Dentirostres: Motmots, Hornbills.
14. Coraces: Crows, Rollers, Birds of Paradise, Fruit-crows, Mynas.
15. Sericati: Cotingas
16. Hiantes: Swallows, Swifts, Nightjars.

Order 3 Rapatores

17. Strix: Owls
18. Falco, Gypogeranus (Secretary Bird), Gypaetus (Lammergeier).
19. Vultures, Cathartids.

Order 4 Rasores

20. Gallinacei: Guineafowl, Turkeys, Guans, Curassows, Hoatzin, Peafowl, Pheasants, Fowl, Lyrebirds, Grouse, Partridges.
21. Epollicati: Buttonquail, Sandgrouse.
22. Columbini: Pigeons
23. Crypturi: Tinamous
24. Inepti: Dodos.

Order 5 Cursores

25. Proceri: Cassowaries, Ostriches, Rheas.
26. Campestres: Bustards
27. Littorales: Plovers, Sandpipers (Calidris), Stilts, Oystercatchers, Coursers, Thicknees.

Order 6 Grallatores

28. Vaginati: Sheathbills.
29. Alectorides: Pratincoles, Cape Barren Goose, Screamers, Trumpeters.
30. Herodii: Cranes, Storks, Herons, Sunbittern, Hammerhead, Boatbill, Openbill Stork.
31. Falcati: Ibises, Woodstorks
32. Limicolae: Curlews, Woodcocks & Snipe, Sandpipers (Ereuntes, Actitis, Tringa), Turnstones.
33. Macrodactyli: Jacanas, Rails, Corncrakes.
34. Lobipedes: Coots, Finfoots, Phalaropes.
35. Hygrobatae: Correira, Avocets, Spoonbills, Flamingoes.

Order 7 Natatores

36. Longipennes: Skimmers, Terns, Gulls, Skuas.
37. Tubinares: Petrels, Diving Petrels, Prions, Albatrosses.
38. Lamellosodentati: Ducks, Geese, Mergansers.
39. Steganopodes: Pelecans, Cormorants, Gannets, Tropicbirds, Anhinga.
40. Pygopodes: Divers, Grebes, Guillemots, Puffins, Razorbills.
41. Impennes: Penguins.

Appendix 17
Vieillot

1. Order Accipitres:
    - Tribe 1. Diurnal (Birds of Prey)
    - Tribe 2. Nocturnal (Owls)
2. Order Sylvicolae:
    - Tribe 1. Zygodactyli
        - Family 1. Psittacini (Parrots)
        - Family 2. Macroglossi (Woodpeckers & Wryneck)
        - Family 3. Aureoli (Jacamars)
        - Family 4. Pteroglossi (Toucans)
        - Family 5. Barbati (Barbets, Trogons, Puffbirds)
        - Family 6. Imberbi (Cuckoos, Honeyguides, Cuckoo Roller)
        - Family 7. Frugivori (Touracos)
    - Tribe 2. Anisodactyli
        - Family 8. Granivori (Hard-billed seedeaters, Plantcutters, Colies)
        - Family 9. Aegithali (Titmice, some Tyrannids, Pardalotes, Manakins)
        - Family 10. Pericalles (Tanagers, Vireos)
        - Family 11. Textores (Icterids, Orioles)
        - Family 12. Leimonites (Starlings, Oxpeckers)
        - Family 13. Carunculatus (Mynas, Wattlebirds)
        - Family 14. Paradisei (Birds of Paradise, except Astrapias)
        - Family 15. Coraces (Corvids, Astrapias, some Icterids,some Rollers)
        - Family 16. Baccivori (some Rollers, Waxwings, Cotingas, Swallow Tanager)
        - Family 17. Chelidones (Swallows, Swifts, Nightjars)
        - Family 18. Myiotheres (Todies, Conopophagas, Cuckoo-Shrikes, Flycatchers, some Tyrannids)
        - Family 19. Colluriones (Shrikes, Vangas, Drongos, Wood Swallows)
        - Family 20. Canori (Thrushes & Chats, Palm Chat, Figbirds,Grallina, Pittas, Accentors, Larks, Pipits, Wagtails, Warblers, Kinglets, true Wrens)
        - Family 21. Anerpontes (Nuthatches, Treecreepers etc., Piculets).
        - Family 22. Anthomysi (Bananaquit, Hummingbirds)
        - Family 23. Epopsides (Promerops, Hoopoes, Philemon Honeyeaters, Horneros)
        - Family 24. Pelmatodes (Bee Eaters, Kingfishers)
        - Family 25. Antriades (Cocks of the Rock)
        - Family 26. Prionoti (Motmots, Hornbills)
        - Family 27. Lyriferi (Lyrebirds)
        - Family 28. Ophiophagae (Hoatzin)
        - Family 29. Columbini (Pigeons)
        - Family 30. Alectrides (Guans)
3. Order Gallinacei
    - Family 1. Nudipedes (Curassows, Turkeys, Pheasants, Partridges, Guineafowl, Tinamous, Buttonquails)
    - Family 2. Plumipedes (Grouse, Sandgrouse)
4. Order Grallatores
    - Tribe 1. Di-Tridactlyi
        - Family 1. Megistanes (Ostriches, Rheas, Emus, Cassowaries).

Family 2. Pedionomi (Plains Wanderer)
Family 3. Aegialites (Thick-knees, Stilts, Oystercatchers, Charadrine Plovers, smaller Sandpipers)
Tribe 2. Tetradactlyi
Family 4. Elonomi (Lapwings, larger Sandpipers, Painted Snipe)
Family 5. Falcirostres (Ibises)
Family 6. Latirostres (Spoonbills, "Cancroma" ?= *Shoebill)*
Family 7. Herodiones (Herons, Hammerkop, Storks)
Family 8. Aerophoni (Cranes)
Family 9. Coleoramphi (Sheathbills)
Family 10. Uncirostres (Cariamas, Secretary Bird, Cereopsis geese, Pratincoles, Screamers)
Family 11. Hilebatae (Trumpeters)
Family 12. Macronyches (Jacanas)
Family 13. Macrodactlyi (Rails, except Coots)
Family 14. Pinnatipedes (Coots, Phalaropes)
Family 15. Palmipedes (Avocets, Flamingoes)
5. Order Natatores
Tribe 1. Telepodes
Family 1. Syndactyli (Frigate Birds, Pelecans, Cormorants, Gannets, Tropic Birds, Anhingas)
Family 2. Urinatores (Sungrebe, Divers, Grebes)
Family 3. Dermorhynci (Geese, Swans & Ducks)
Family 4. Pelagii (Skuas, Gulls, Terns, Skimmers)
Tribe 2. Atelopodes
Family 5. Siphorini (Petrels, Albatrosses)
Family 6. Brachypteri (Auks)
Tribe 3. Ptilopteri
Family 7. Sphenisci (Penguins)

Appendix 18

Temminck

1. Order Rapaces (Birds of Prey, Owls)
2. Order Omnivores (Crows, Rollers, Orioles, Starlings)
3. Order Insectivores (Shrikes, Flycatchers, Thrushes, Dippers, Warblers, Chats, Accentors, Wagtails, Pipits)
4. Order Granivores (Larks, Titmice, Buntings, Finches)
5. Order Zygodactyli (Cuckoos, Woodpeckers, Parrots)
6. Order Anisodactyli (Nuthatches, Treecreepers, Wallcreeper, Hoopoes, Sunbirds, Hummingbirds)
7. Order Alcyones (Bee-eaters, Kingfishers)
8. Order Chelidones (Swallows, Swifts, Nightjars)
9. Order Columbae (Pigeons)
10. Order Gallinae (Pheasants, Grouse, Sandgrouse, Partridges)
11. Order Alectorides (Pratincoles, Trumpeters, Screamers, Cariamas)
12. Order Cursores (Ratites, Cursors, Bustards)
13. Order Grallatores

Division 1. "with three toes" (Thick-knees, smaller Sandpipers, Oystercatchers, Stilts, Charadrine Plovers)

Division 2. "with four toes" (Lapwings, Cranes, Storks, Herons, Flamingoes, Avocets, Spoonbills, Ibises, larger Sandpipers, Rails & Gallinules)

14. Order Pinnatipedes (Coots, Phalaropes, Grebes)

15. Order Palmipedes (Terns, Gulls, Skuas, Petrels, Ducks, Swans, Geese, Pelecans, Cormorants, Gannets, etc., Divers, Auks)

Appendix 19

Bonaparte

Bonaparte's 1851 arrangement:

Subclass 1. Insessores
- 1. Psittaci (Parrots)
- 2. Accipitres (Raptors)
- 3. Passeres (modern passerines and nightjars, kingfishers and picarian birds)
- 4. Columbae (Pigeons)

Subclass 2. Grallatores
- 5. Gallinae (Game birds)
- 6. Struthiones (Ratites)
- 7. Grallae (Waders, Herons, Ibises, etc).
- 8. Anseres (Duck family & seabirds)

By 1853, this had been amended as follows:

Subclass 1. Insessores
- 1. Psittaci (Parrots)
- 2. Accipitres (Raptors)
- 3. Passeres (modern passerines and nightjars, kingfishers and picarian birds)
- 4. Columbae (Pigeons)
- 5. Herodiones (including *Dromas*, the Crab Plover)
- 6. Gaviae (Procellarids, etc.)

Subclass 2. Grallatores
- 7. Struthiones (Ratites)
- 8. Gallinae (Game birds)
- 9. Grallae (Waders, Herons, Ibises, etc).
- 10. Anseres (Duck family, flamingoes, Auks, Divers, grebes, Penguins).

Appendix 20

Avium Conspectus of Tschudi

Order Accipitres
- Family Vulturidae Cathartids
- Family Falconidae Other raptors
- Family Strigidae Owls

Order Passeres

Family Caprimulgidae Nightjars etc.
Family Hirundinidae Swallows
Family Ampelidae Cotingas
Family Pipridae Manakins, Cocks-of-the-Rock
Family Muscicapidae Tyrannids
Family Laniadae sub-oscines
Family Myiotheridae sub-oscines, Dipper, Wrentit
Family Turdidae Thrushes, Mockingbirds,
Family Troglodytidae Wrens
Family Sylviadae sub-oscines, Pipits
Family Tanagridae Tanagers
Family Fringillidae Buntings, Plantcutters, Finches
Family Sturnidae Icterids
Family Corvidae Jays
Family Certhiadae sub-oscines, some Tanagers
Family Trochilidae Hummingbirds
Family Momotidae Motmots
Family Halcyonidae Kingfishers

Order Scansores
Family Galbulidae Jacamars
Family Cuculidae Cuckoos & Trogons
Family Bucconidae Puffbirds, Barbets
Family Ramphastidae Toucans
Family Picidae Woodpeckers
Family Psittacidae Parrots

Order Columbae
Family Columbidae Pigeons

Order Rasores
Family Chionidae Seedsnipe
Family Tetraonidae Partridges
Family Crypturidae Tinamous
Family Cracidae Curassows
Family Penelopidae Guans

Order Cursores
Family Struthionidae Rheas

Order Grallatores
Family Charadridae Plovers
Family Ardeadae Herons, Storks, Spoonbills
Family Tantalidae Ibises
Family Scolopacidae Sandpipers
Family Rallidae Rails
Family Recurvirostridae Stilts
Family Phoenicopteridae Flamingoes

Order Natatores
Family Laridae Gulls, Terns, Petrels, Diving Petrels
Family Anatidae Geese, Ducks
Family Pelecanidae Pelecans, Boobies, Cormorants, Frigates, Tropicbirds, Anhingas
Family Colymbidae Grebes
Family Alcadae Penguins

Appendix 21

Cuvier

1. Birds of prey & Owls. Beak crooked, with a cere, talons crooked, 3 toes forward, one behind.
    1. Diurnal.
        1. Vultur
            1. Vultur, proper (i.e. Great Old World Vultures)
            2. Sarcorhamphus (i.e. Condors & King Vulture)
            3. Percnopterus (Small Old World Vultures & Turkey Vultures)
        2. Griffon (i.e. Lamergeier).
        3. Falco
            1. Nobiles Falcons
            2. Hierofalco Gyrfalcon
            3. Ignobiles
                1. Aquila
                    1. Eagles proper
                    2. Haliaetus (Haliaetus, Pandion, Caracara)
                    3. Harpyia
                    4. Morphnus (Urubitinga, Urutaurana i.e. Crowned Eagles)
                    5. Cymides
                2. Astur
                3. Milvus (inc. Elanus & Milvus)
                4. Pernis
                5. Buteo
                6. Circus
                7. Serpentarius
    2. Nocturnal (i.e. Owls) Divided into several groups based on the presence or absence of ear-tufts, size of the facial discs, etc., etc.
2. Passeres. All such as cannot be classed in the other orders.
    1. Exterior toe united to the middle toe by one or two joints only.
        1. Dentirostres
            1. Lanius (inc. Vangas and Cuckoo-Shrikes)
            2. Tanagra Tanagers (inc. Cardinals & Grosbeaks)
            3. Muscicapa Flycatchers (inc. Tyrant Flycatchers, Picathartes and Umbrellabirds)
            4. Cotingas (except Umbrellabirds)
            5. Edolius Drongos
            6. Turdus Blackbirds & Thrushes
            7. Pyrrhocorax Choughs
            8. Oriolus
            9. Myrothera Ant thrushes
            10. Cinclus Water Ousels
            11. Philedon = Philemon
            12. Gracula Grackles (Starlings)
            13. Menura Lyrebirds
            14. Pipra Manakins (inc. Cock-of-the Rock)
            15. Motacilla

1. Saxicola
2. Sylvia
3. Curruca
4. Regulus Wrens
5. Troglodytes Wrens
6. Motacilla Wagtails
7. Anthus Pipits

2. Fissirostres
   1. Hirundo Swallows, Martins & Swifts
   2. Caprimulgus Nightjars
   3. Podarges Potoos etc.
3. Conirostres
   1. Alauda Larks
   2. Parus Titmice
   3. Emberiza Buntings
   4. Fringilla Finches, Weavers, Widow-birds, etc.
   5. Loxia Crossbills
   6. Corythrus
   7. Colius Colies
   8. Glaucopis Wattle-birds
   9. Buphaga Oxpeckers
   10. Cassicus Icterids
   11. Sturnus Starlings
   12. Sitta Nuthatches
   13. Corvus Crows, Magpies, Jays, Nutcrackers, Temia.
   14. Coracias Rollers
   15. Paradisea Birds of Paradise
4. Tenuirostres
   1. Upupa Hoopoes (inc. Promerops & Epimachus, Sicklebill)
   2. Certhia Creepers (inc. Dendrocolaptes, Tichodroma [Wallcreeper], Nectarinia, Dicaeum, Cinnyris).
   3. Trochilus Hummingbirds

2. Exterior toe almost as long as the middle toe and united with it as far as last joint but one.
   1. Merops Bee-eaters
   2. Prionites
   3. Alcedo Kingfishers (part)
   4. Ceyx Kingfishers (part)
   5. Todus Todies
   6. Buceros Hornbills
3. Climbers. (inc. parrots). Toes two forward, two back.
   1. Jacamar, Galbula
   2. Picus
   3. Yunx (Jynx)
   4. Cuculus (Cuckoos, Couas, Coucals, Indicators (Honeyguides) etc.
   5. Malcohas.
   6. Scythrops "Psittaceous Hoen-bills"
   7. Bucco (Barbets)
   8. Trogon
   9. Crotophaga (Anis)

10. Ramphastos (Toucans)
11. Psittacus (subdivided into Ara, Parrakeet Aras, Arrow-tailed Parrakeets, Parrakeets proper, Cockatoos, Parrakeets with trunks [*sic*], Pezoporus).
12. Touracos. Beak not ascending the forehead.
13. Musophaga ("Banana-eaters" = Plantain eaters) Beak forming a discal shield.

4. Gallinaceae. Upper mandible vaulted, nostrils covered by a cartilaginous scale, anterior toes united at base by a short membrane.
    1. Pavo (inc. Polyplectron - peacock-pheasants.)
    2. Meleagris Turkeys
    3. Alectoris Curassows (inc. Hoatzin)
    4. Satyrus Tragopans
    5. Phasianus (inc. Gallus, Phasianus proper, Crested Pheasants) etc.
    6. Numida Guineafowl
    7. Tetrao Grouse (inc. Pterocles - Sandgrouse [except Syrrhaptes], partridges, francolins, quails and "colins" [= American quails])
    8. Tridactylus Hemipodes, inc. Syrrhaptes.
    9. Tinamous
    10. Columbae
5. Grallae. (inc. ostriches, ibis) Thighs naked.
    1. Brevipennes
        1. Struthio Ostriches & Rheas.
        2. Casuarius Cassowaries & Emu.
    2. Pressirostres
        1. Otis Bustards
        2. Charadrius Plovers (inc. Oedicnemus, Stone Curlews)
        3. Tringa Lapwings & Squatarola
        4. Haematopus Oystercatchers
        5. Cursorius
        6. Cariama
    3. Cultirostres
        1. Grus Cranes (inc. Crowned Cranes and Psophia, Trumpeters)
        2. Cancroma Boatbills
        3. Ardea Herons (inc. Egrets, Bitterns & Night Herons considered sub-groups)
        4. Ciconia Storks
        5. Mycteria Jabiru
        6. Scopus
        7. Anastomus Open-billede Storks
        8. Tantalus Wood Storks
        9. Platalea Spoonbills
    4. Longirostres
        1. Scolopax - Ibises, Curlews, Whimbrels, Pygmy-curlews, Woodcocks, godwits, sandpipers, stints, ruffs, Turnstones, phalaropes, snipes, stilts.
        2. Recurvirostra - Avocets.
    5. Macrodactyla
        1. Jacana Jacanas
        2. Palamedea Screamers
        3. Rallus Rails, coots & gallinules.
    6. Glareola - Pratincoles & Sea Partridges

6. Palmipedes. The only birds whose neck excedes the length of their legs, palmated feet, set far back, plumage imbued with an oily juice.
    1. Brachypteres
        1. Colymbus
            1. Podiceps Grebes
            2. Colymbus proper Divers
            3. Uria Guillemots
            4. Cephus Pigeon Guillemots
        2. Alca
            1. Fratercula Auks
            2. Alca Penguins
        3. Aptenodytes
            1. Aptenodytes Patagonian Penguins
            2. Catarrhactes
            3. Spheniscus
    2. Longipennes
            1. Procellaria Petrels (inc. Prions, Shearwaters and Diving Petrels
            2. Diomedea Albatrosses
            3. Larus Gulls (inc. Skuas)
            4. Sterna Terns
            5. Rhynchops Skimmers
    3. Totipalmes
        1. Pelecanus
            1. Onocrotalus Pelecans
            2. Phalacrocorax Cormorants
            3. Frigate birds
            4. Sula Gannets, Boobies
        2. Plotus Darters
        3. Phaeton Tropicbirds
    4. Lamellirostres
        1. Anas
            1. Cygnus Swans
            2. Anser Geese
            3. Anas proper
            4. Mergus Sea ducks
7. Phoenicopterus Flamingoes

Appendix 22

J.J. Kaup

Order Zygodactyli
- Psittacidae Parrots
- Cuculidae Cuckoos
- Rhamphastidae Toucans
- Picidae Woodpeckers
- Musophagidae Trogons

Order Passeres
- Sub-order Conirostres
  - Fringillidae Finches
  - Artamidae Woodswallows
  - Sturnidae Starlings
  - Buphagidae Oxpeckers
  - Alaudidae Larks
- Sub-order Fissirostres
  - Muscicapidae Flycatchers
  - Hirundinidae Swallows
  - Eurylaimidae Broadbills
  - Coracidae Rollers
  - Ampelidae Cotingas
- Sub-order Syndactyli
  - Prionitidae
  - Meropidae Bee-eaters
  - Bucerotidae Hornbills
  - Alcedidae Kingfishers
  - Pipridae Manakins
- Sub-order Dentirostres
  - Sylvidae Warblers
  - Oriolidae Orioles
  - Corvidae Crows
  - Laniidae Shrikes
  - Paridae Titmice
- Sub-order Tenuirostres
  - Certhidae Treecreepers
  - Trochilidae Hummingbirds
  - Upupidae Hoopoe
  - Sittidae Nuthatches
  - Meliphagidae Honeyeaters

Order Grallae
- Sub-order Pressirostres
- Sub-order Longirostres
- Sub-order Brevipennes
- Sub-order Cultirostres
- Sub-order Macrodactyli

Order Icthyornithes
- Sub-order Rapaces
- Sub-order Longipennes
- Sub-order Brachypteri
- Sub-order Totipalmati
- Sub-order Lamellirostres

Order Gallinae
- Cracidae Curassows
- Columbidae Pigeons
- Crypturidae Tinamous
- Tetraonidae Grouse
- Gallinae Pheasants & Partridges

Appendix 23

Boie

1 Order Raptores
1. Family Gypogeranidae (Secretary Bird)
2. Family Vulturidae (Vultures & Cathartids)
3. Family Falconidae (all other raptors)
4. Family Strigidae (Owls)
5. Family Caprimulgidae (Nightjars & Frogmouths)

2. Order Insessores
1. Family Hirundinidae (Swallows & Swifts)
2. Family Pipridae (Waxwings, Cotingas & Manakins)
3. Family Meropidae (Bee-eaters)

4. Family Trochilidae (Hummingbirds)
5. Family Nectariniadae (Sunbirds)
6. Family Sylviadae (Flycatchers, Warblers, Flowerpeckers, Vireos, Accentors, some Chats, Minivets)
7. Family Merulidae (Thrushes, Mockingbirds, Pittas, Cock-of-the Rock)
8. Family Motacillidae (Wagtails, Forktails, some Chats)
9. Family Myiotheridae (Babblers, Cuckoo-shrikes, Bulbuls)
10. Family Muscicapiadae (some Flycatchers, Tyrannids, Drongos)
11. Family Laniadae (Shrikes, Vangas, Wood-swallows)
12. Family Tangaridae (Tanagers)
13. Family Fringillidae (Finches, Viduas, Buntings)
14. Family Loxiadae (some American Emberizids, Crossbills, Plantcutters, Colies, Parrotbills)
15. Family Alaudidae (Larks, Pipits, Snow Buntings!)
16. Family Paridae (Kinglets, Titmice, Long-tailed Tits, Ioras, Megalurus warblers, Blue Wrens, true Wrens, Pardalotes, some sub-oscines)
17. Family Certhiadae (Promerops, Hoopoes, Hawaiian Honeycreepers, Wallcreeper, Treecreepers, Nuthatches, Sharpbill, some sub-oscines)
18. Family Melliphagidae (nothing specified, but presumably Honeyeaters)
19. Family Garrulidae (Jays, Nutcrackers, Oxpeckers)
20. Family Sturnidae (Starlings)
21. Family Oriolidae (Orioles, Icterids, Broadbills, Mynas, Fairy Bluebirds, Rollers)
22. Family Paradisiadae (Birds of Paradise and ?Bower birds)
23. Family Corvidae (Crows except those in Garrulidae, Criniger Bulbuls)
24. Family Buceridae (Hornbills)
25. Family Rhamphastidae (Toucans)
26. Family Halcyonidae (Kingfishers, Jacamars)
27. Family Bucconidae (Puffbirds, Barbets, Trogons)
28. Family Cuculidae (Cuckoos, Cuckoo Roller, Honeyguides)
29. Family Picidae (Woodpeckers & Wrynecks)
30. Family Psittacidae (Parrots & Touracos)

3. Order Rasores
   1. Family Columbidae (Pigeons)
   2. Family Phasianidae (Pheasants, Guineafowl, Turkeys)
   3. Family Tetraonidae (Grouse, Quails, Partridges, Sand-grouse, Tinamous)
   4. Family Struthionidae (Ratites, Dodos, Kiwis, Bustards)
   5. Family Cracidae (Curassows, Guans, Lyrebirds, Megapodes, Hoatzin)

4 Order Grallatores
   1. Family Charadriadae (Coursers, Plovers, Thick-knees, Calidris sandpipers, Pratincoles, Oystercatchers, Stilts, Avocets)
   2. Family Scolopacidae (Larger Scolopacids, Phalaropes, some Ibises)
   3. Family Ardeidae (some Ibises, Spoonbills, Storks, Flamingoes, Herons, Bitterns, Limpkin, Sunbittern, Hammerkop)
   4. Family Gruidae (Cranes, Screamers, Trumpeters)
   5. Family Rallidae (Rails, Jacanas)

5. Order Natatores
   1. Family Colymbidae (Penguins, Auks, Divers, Grebes)
   2. Family Procellaridae (Petrels, Shearwaters)
   3. Family Laridae (Skuas & Jaegers, Albatrosses, Gulls, Terns)
   4. Family Pelecanidae (Frigate Birds, Cormorants, Finfoots, Gannets & Boobies, Anhinga, Tropic birds [Pelicans not included, presumably an error!])
   5. Family Anatidae (Ducks, Geese, Swans)

Appendix 24

Hogg

Subclass 1. Aves Constrictipedes
- Division 1. Terrestres
  - Order 1. Raptores
    - Tribe 1. Planicerirostres
      - Subtribe 1. Diurni
        - Family 1. Sarcoramphidae
          - Genus Neophron
        - Family 2. Vulturidae
          - Genus Gyps, Vultur
        - Family 3. Gypaeetidae
          - Genus Gypaeetus
        - Family 4. Aquilidae
          - Genus Haliaeetus, Aquila, Pandion, Circaeetus
        - Family 5. Falconidae
          - Genus Falco, Accipiter, Astur, Milvus, Nauclerus, Elanus
        - Family 6. Buteonidae
          - Genus Buteo, Pernis, Circus, Strigiceps
    - Tribe 2. Tecticerirostres.
      - Subtribe 2. Nocturni
        - Family 1. Strigidae
          - Genus Surnia, Nytea, Strix, Ulula, Syrnium, Athene
        - Family 2. Bubonidae
          - Genus Bubo, Otus, Scops
  - Order 2. Prehensores
    - Tribe Rotundirostres
      - Subtribe 1. Laevilingues
        - Family 1. Plyctolophidae
        - Family 2. Psittacidae
        - Family 3. Macrocercidae
        - Family 4. Pezoporidae
        - Family 5. Psittaculidae
      - Subtribe 2. Hirtilingues
        - Family 6. Loriadae
      - Subtribe 3. Tubilingues
        - Family 7. Microglossidae
  - Order 3. Insessores
    - Tribe 1. Curvirostres
      - Subtribe 1. Scansores
        - Family Cuculidae
          - Genus Cuculus, Oxylophus, Coccyzus
    - Tribe 2. Cuneirostres
        - Family 1. Picidae
          - Genus Dryotomus, Picus, Jynx
        - Family 2. Apternidae

Genus Apternus
Family 3. Sittidae
Genus Sitta
Tribe 3. Conirostres
Subtribe 2. Clamatores
Family 1. Coraciadidae
Genus Coracias
Family 2. Corvidae
Genus Garrulus, Pica, Nucifraga, Corvus, Pyrrhocorax, Fregilus
Subtribe 3. Cantatores
Family 3. Sturnidae
Genus Sturnus, Pastor, Agelius
Family 4. Loxiadae
Genus Loxia, Pyrrhula, Corythus, Erythrospiza, Coccothraustes
Family 5. Fringillidae
Genus Petronia, Passer, Linota, Serinus, Carduelis, Fringilla
Family 6. Emberizidae
Genus Emberiza, Plectrophanes
Family 7. Alaudidae
Genus Philemus, Alauda, Galerida
Tribe 4. Dentirostres
Family 1. Anthidae
Genus Anthus
Family 2. Motacillidae
Genus Budytes, Motacilla
Family 3. Paridae
Genus Aegithalus, Calamophilus, Mecistura, Parus
Family 4. Aedonidae
Genus Regulus, Melizophilus, Sylvia, Curruca, Aedon, Salicaria, Accentor, Calliope
Family 5. Saxicolidae
Genus Phoenicura, Erithacus, Saxicola, Vitiflora
Family 6. Ampelidae
Genus Bombycilla
Family 7. Merulidae
Genus Oriolus, Haematornis, Turdus, Petrocincla, Merula, Cinclus
Subtribe 4. Latrones
Family 8. Laniadae
Genus Lanius, Collurio
Family 9. Muscicapidae
Genus Muscicapa
Tribe 5. Teniurostres
Subtribe 5. Anisodactyli
Family 1. Certhiadae
Genus Troglodytes, Certhia, Tichodroma
Family 2. Upupidae
Genus Upupa
Tribe 6. Fissirostres
Subtribe 6. Syndactyli

Family 1. Halcyonidae
Genus Halcyon
Family 2. Meropidae
Genus Merops.
Subtribe 7. Allodactyli
Family 3. Hirundinidae
Genus Cypselus, Progne, Hirundo, Chelidon
Family 4. Caprimulgidae
Genus Caprimulgus, Scotornis
Tribe 7. Cutinarirostres
Subtribe 8. Gyratores
Family Columbidae
Genus Columba, Turtur, Ectopistes
Subclass 2. Aves Inconstrictipedes
Order 4. Rasores
Tribe Convexirostres
Subtribe 1. Podarcees
Family 1. Phasianidae
Genus Phasianus
Family 2. Tetraonidae
Genus Tetrao, Lagopus, Bonasia
Family 3. Pteroclidae
Genus Pterocles
Family 4. Perdicidae
Genus Francolinus, Perdix, Ortyx, Coturnix
Family 5. Hemipodiadae
Genus Hemipodius
Subtribe 2. Podenemi
Family 6. Otitidae
Genus Otis
Division 2. Aquaticae
Order 5. Grallatores
Tribe 1. Pressirostres
Subtribe 1. Cursores
Family 1. Charadriadae
Genus Oedicnemus, Cursorius, Charadrius, Hoplopterus
Family 2. Vanellidae
Genus Squatarola, Vanellus, Glareola, Strepsilas
Family 3. Haematopodidae
Genus Haematopus
Tribe 2. Cultrirostres
Subtribe 2. Ambulatores
Family 1. Gruidae
Genus Balearica, Anthropoides, Grus
Family 2. Ardeidae
Genus Ciconia, Ardea, Ardeola, Erogas, Nycticorax
Tribe 3. Pyxidirostres
Family Phaenicopteridae
Genus Phoenicopterus

- Tribe 4. Spathulirostres
    - Family Plataleidae
        - Genus Platalea
- Tribe 5. Longirostres
    - Family 1. Tantalidae
        - Genus Tantalus, Ibis
    - Family 2. Recurvirostridae
        - Genus Recurvirostra
    - Family 3. Numeniadae
        - Genus Terekia, Limosa, Numenius
    - Family 4. Scolopacidae
        - Genus Totanus, Machetes, Rusticola, Scolopax, Macrorhamphus, Erolia, Tringa
    - Family 5. Phalaropodidae
        - Genus Phalaropus, Lobipes
    - Family 6. Calidridae
        - Genus Himantopus, Calidris
- Tribe 6. Diversirostres
    - Subtribe 3. Macrodactlyi
        - Family Rallidae
            - Genus Rallus, Crex, Zapornia
- Tribe 7. Frontiscutirostres
    - Family Fulicidae
        - Genus Gallinula, Porphyrio, Fulica

Order 6. Natatores

- Tribe 1. Lamellirostres
    - Subtribe 1. Simplicipollices
        - Family 1. Anseridae
            - Genus Bernicla, Anser, Chen, Cygnus, Olor, Plectropterus, Chenalopex
        - Family 2. Anatidae
            - Genus Tadorna, Cairina, Rhynchaspis, Chauliodus, Dafila, Anas, Mareca
    - Subtribe 2. Membranipollices
        - Family 3. Fuligulidae
            - Genus Clangula, Undina, Harelda, Fuligula, Oedemia, Somateria
- Tribe 2. Serrirostres
    - Family 1. Mergidae
        - Genus Mergus, Merganser
    - Subtribe 3. Totipalmae
        - Family 2. Fregatidae
            - Genus Fregata
        - Family 3. Carbonidae
            - Genus Carbo, Sula
- Tribe 3. Sacculirostres
    - Family Pelecanidae
        - Genus Pelecanus
- Tribe 4. Tubinarirostres
    - Subtribe 4. Longipennes
        - Family Procellariadae
            - Genus Diomedea, Procellaria, Puffinus, Thalassidroma

Tribe 5. Medionarirostres

Family Laridae

Genus Cataracta, Lestris, Larus, Rissa, Xema

Tribe 6. Subulirostres

Family Sternidae

Genus Anous, Viralva, Pontochelidon, Sterna

Tribe 7. Cuspidirostres

Subtribe 5. Brevipennes

Family 1. Podicipidae

Genus Podiceps

Family 2. Colymbidae

Genus Colymbus, Uria

Tribe 8. Sulcirostres

Family 1. Mormonidae

Genus Mergulus, Mormon, Ultamania

Family 2. Alcidae

Genus Alca

Appendix 25

Huxley

Order 1 Saururae: Archaeopteryx

Order 2 Ratitae: Ratites

Order 3 Carinatae:

Dromaeognathae: Tinamous

Schizognathae:

Charadriomorphae: Plovers and Scolopacids

Geranomorphae: Bustards, Cranes, Trumpeters, Kagu, Rails.

Cecomorphae: Gulls & Terns, Tube-nosed Swimmers, Divers, Auks.

Spheniscomorphae: Penguins

Alectoromorphae: Game Birds

Peristeromorphae: Pigeons, Dodos.

Heteromorphae: Hoatzin

Desmognathae:

Chenomorphae: Geese, Swans & Ducks, Screamers.

Amphimorphae: Flamingoes

Pelargomorphae: Herons, Storks

Dysporomorphae: Totipalmes

Aetomorphae: Birds of Prey and Owls

Psittacomorphae: Parrots.

Coccygomorphae: Colies, Cuckoos, Bee-eaters, Hornbills, Hoopoes, Rollers, Trogons.

Aegithognathae:

Celeomorphae: Woodpeckers & Wryneck

Cypselomorphae: Hummingbirds, Swifts, Nightjars & allies.

Coracomorphae: Passerines; divided into

(a) Lyrebirds, which have no maxillo-palatine processes, and

(b) all the rest.

Appendix 26
Sundevall's Tentamen

[Translated, with notes by Francis Nicholson]

Agmen 1 Psilopaedes Altricial Birds

Order 1 Oscines

1st series Laminiplantares

Cohort 1 Cichlomorphae

Phalanx 1 Ocreatae

Family 1 Lusciniinae (contains Erithacus, Luscinia, Petroica, Chaimarrornis, Sialis etc.)
Family 2. Saxicolinae (Saxicola & Myrmecocichla)
Family 3. Turdinae (Turdus, Catharus, Geocichla)
Family 4. Cinclidae (Cinclus)
Family 5. Henicurinae (Henicurus only)
Family 6. Myiophoninae (Zoothera, Thamnocichla, Myiophoneus)
Family 7. Eucichlinae (Pittas)

Phalanx 2. Brevipennes

1. Warbler-like species of Old World

Family 8. Acanthizinae [Sundevall here includes the South African genus *Eremomela* in the Acanthizinae]
Family 9. Cisticolinae (some of the Sylvidae)
Family 10. Malurinae (contains a variety of Sylvid genera. "With the wing more convex, small, the second quill shorter than the secondaries". Nicholson adds a note that Sundevall has overlooked the difference in the number of tail feathers).
Family 10b Bradypterinae. (contains Sylvid & Turdid genera)
Family 11. Aegithininae

2. Short-winged thrush-like birds of the Old World.

Family 12. Copsychinae (Copsychus, Napothera etc.)
Family 13. Crateropodinae (some Turdidae and Babblers)
Family 14. Brachypteryginae (Thrushes & Babblers)
Family 15. Eupetinae (Eupetes, a Babbler)
Family 16. Malaconotinae. (Malaconotus shrikes, Scimiter Babblers, etc)

3. American short-wings.

Family 17. Hylophilinae. (Hylophilus, Dulus, Cyclorhis)
Family 18. Troglodytinae.
Family 19. Toxostominae. (some Mimidae & Chamaea, the Wrentit)

Phalanx 3. Aequiparatae.

Family 20. Miminae (some & Polioptila, Gnatcatchers)
Family 21. Vireoninae.
Family 22. Phyllopseustinae. (Regulus, Phylloscopus & Ephthianura!)
Family 23. Sylviinae. (some & some flycatchers)
Family 24. Calamodytinae. (some Sylvids)
Family 25. Ptenoedinae. [Not clear what this is. 2 genera, Ptenoedus & Cinclorhamphus are given, both Australian. See Sharpe, Cat. Bds, 7, p. 498.)
Family 26. Pachycephalinae. (Pachycephala, some Babblers & Lalage)
Family 27. Parinae. (inc. Paradoxornis & Catamblyrhamphus)[Nicholson correctly notes that this last-named is a finch].

Family 28. Laniinae. (part)
Family 29. Ptilorhynchinae. (Bower Birds & some Australian Babblers)
Phalanx 4. Brachypodes.
Family 30. Ampelidinae. (Ampelis [Waxwing], Hypocolius, Phainopepla, etc., & some Tyrannids)
Family 31. Pycnonotinae.
Family 32. Phyllornithinae. (Chloropsis)
Family 33. Oriolinae.
Family 34. Artaminae.
Family 35. Campophaginae. (except Lalage)
Family 36. Prionopinae. (Laniidae, part; Irena)
Family 37. Dicrourinae.
Phalanx 5. Latirostres
Family 38. Ficedulinae. (Muscicapidae, part)
Family 39. Platystirinae. (Sylvidae, Muscicapidae, pt.)
Family 40. Myiagrinae. (Monarchinae, part)
Family 41. Muscipetinae. ditto.
Family 42. Rhipidurinae. (part)
Family 43. Muscicapinae. (part)
Family 44. Pericrocotinae (Pericrocotus).
Phalanx 6. Novempennatae.
Family 45. Motacillinae.
Family 46. Dendroecinae. (this family is a mixture of Flowerpeckers, Zosterops, Parulids, and slenderer-billed Tanagers)
Family 47. Setophaginae. (Parulids, part)
Family 48. Icteriinae. (part)
Family 49. Hemithraupinae. (Thraupidae, part)
Family 50. Pardalotinae. (Pardalotus)
Cohort 2. Conirostres.
Phalanx 1. Decempennatae.
Family 1. Plocenae.
Family 2. Viduinae. (Vidua, Estrildinae, Panurus, Aegithalus, Euplectes)
Family 3. Accentorinae. (Accentor)
Phalanx 2. Amplipalatales.
Family 4. Chloridinae. (Fringillidae. part)
Family 5. Fringillinae. (Coccothraustes, Fringilla, Passer, Petronia)
Phalanx 3. Arctipalatales.
Family 6. Loxiinae. (Loxia)
Family 7. Emberizinae. (part)
Family 8. Zonotrichinae. (Emberizidae, part)
Family 9. Pitylinae. (Emberizidae, part)
Family 10. Arremoninae (Emberizidae, part)
Family 11. Cissopinae. (Cissopis) (Thraupidae, part)
Phalanx 4. Simplicirostres.
Family 12. Tachyphoninae. (Thraupidae, part)
Family 13. Thraupinae (part)
Family 14. Tanagrinae. (Thraupidae, part)
Cohort 3. Colipmorphae.
Phalanx 1. Novempennatae.

Family 1. Chalcophaninae. (Icteridae, part)
Family 2. Agelaeinae. (Ictertidae, pat)
Family 3. Icterinae. (part)
Phalanx 2. Humilinares
Family 4. Callaeadinae.
Family 5. Sturninae.
Family 6. Buphaginae.
Family 7. Fregilinae (Pyrrhocorax, Fregilus, Corcorax)
Phalanx 3. Altinares.
Family 8. Nucifraginae. (Nucifraga, Podoces)
Family 9. Garrulinae. (Corvidae, except those placed in other groups)
Family 10. Corvinae (Corvus)
Phalanx 4. Idiodactylae
Family 11. Subgarrulinae. (Jays & Magpies (part: gen: Kitta, Chlorosoma, Glenargus, [Glaucopis])
Family 12. Gymnorhininae. (Grallina, Gymnorhina & Cracticus)
Family 13. Paradiseinae. (Birds of Paradise, part)
Family 14. Epimachinae. (Birds of Paradise, part)
Family 15. Irrisorinae. (Wood hoopoes)
Cohort 4. Certhiomorphae.
Family 1. Certhiinae. (Mniotilta, Certhia, Salpornis, Caulodromus, Climacteris & Tichodroma)
Family 2. Sittinae. (Sitta, Sittella)
Family 3. Acanthisittinae. (New Zealand Wrens)
Cohort 5. Cinnyrimorphae.
Family 1. Arbelorhininae. (Chlorophanes & Arbelorhina)
Family 2. Drepanidinae.
Family 3. Nectariniinae.
Family 4. Meliphaginae. (inc. Promerops)
Family 5. Philedoninae. (Meliphagidae, part)
Cohort 6. Chelidonomorphae.
Family 6. Hirundininae.
2nd series. Scutelliplantares.
Cohort 1. Holaspideae.
Family 1. Alaudinae.
Family 2. Upupinae. (Upupa)
Cohort 2. Endaspideae.
Family 1. Furnariinae.
Family 2. Synallaxinae.
Family 3. Dendrocolaptinae.
Cohort 3. Exaspidaea.
Family 1. Oxyrhynchinae.
Family 2. Tyranninae.
Family 3. Todinae.
Family 4. Piprinae.
Cohort 4. Pycnaspideae.
Family 1. Rupicolinae. (Cotingas, part, Cocks-of-the-Rock, Broadbills)
Family 2. Ampelionidae. (Cotingas, part)
Family 3. Tityrinae. (Cotingas, part)

Cohort 5. Taxaspideae.
Family 1. Paictinae (Philepitta)
Family 2. Thamnophilinae. (Formicariidae, part)
Family 3. Myrmornithinae. (Formicariidae, part)
Family 4. Hypsibaemoninae. (Formaricariidae, part)
Family 5. Scytalopodinae. (Formicariidae, part & Menura)
Order 2. Volucres.
1st series. Volucres Zygodactylae.
Cohort 1. Psittaci.
Family 1. Camptolophini.
Family 2. Androglossini.
Family 3. Conurini.
Family 4. Platycercini.
Family 5. Stringopini.
Family 6. Trichoglossini.
Cohort 2. Pici
Family 1. Pici Angusticolles.
Family 2. Pici Securirostres.
Family 3. Pici Ligonirostres.
Family 4. Pici Nudinares.
Family 5. Picumnini.
Family 6. Iynginae.
Cohort 3. Coccyges.
Family 1. Indicatorinae.
Family 2. Megalaeminae.
Family 3. Ramphastinae.
Family 4. Galbulinae.
Family 5. Bucconinae.
Family 6. Leptosominae. (Cuckoo Roller)
Family 7. Centropodinae. (Coucals)
Family 8. Zanclostominae. (Cuckoos, part)
Family 9. Cuculinae. (Cuckoos, part)
Family 10. Coccystinae. (Cuckoos, part)
Family 11. Diplopterinae. (Cuckoos, part)
Family 12. Crotophaginae. (Guira & Crotophaga).
2nd series.
Cohort 4. Coenomorphae.
Family 1. Musophaginae.
Family 2. Coliinae.
Family 3. Coraciinae. (inc. Brachypterus) [Ground Roller]
Cohort 5. Ampligulares.
Family 1. Trogoninae.
Family 2. Podarginae. (Steatornis, Nyctibius, Podargus, Batrachostomus, Aegotheles)
Family 3. Caprimulginae. (inc. Glareola!)
Family 4. Cypselinae. (Swifts)
Cohort 6. Volucres Longilingues. Hummingbirds
Family 1. Rhamphodontinae.
Family 2. Phaethornithinae.

Family 3. Lampornithinae.
Family 4. Hylocharinae.
Family 5. Lesbiinae.
Family 6. Prymnacanthinae.
Family 7. Mellisuginae.
Family 8. Heliangelinae.
Family 9. Orthorhynchinae.
Family 10. Heliothrichinae.
Family 11. Heliantheinae.
Family 12. Hypermetrinae (Patagona)

Cohort 7. Volucres Syndactylae
Family 1. Meropinae.
Family 2. Prionitinae. (Motmots)
Family 3. Alcedininae.
Family 4. Bucerotinae.

Cohort 8. Peristeriodeae.
Family 1. Didinae. (Didus* & Didunculus)
[*Sundevall states it was Reinhardt the younger who first determined that the Dodo was a pigeon]
Family 2. Columbinae.
Family 3. Megapeliinae.(Goura)

Agmen 2. Ptilopaedes.

Order 3. Accipitres.

Cohort 1. Nyctharpages Owls
Family 1. Glaucinae.
Family 2. Ululinae.
Family 3. Buboninae.
Family 4. Noctuinae.

Cohort 2. Hemeroharpages.
Family 1. Asturinae.
Family 2. Buteoninae.
Family 3. Falconinae.
Family 4. Circaetinae.
Family 5. Herpetotherinae.
Family 6. Aquilinae.
Family 7. Milvinae.
Family 8. Pandioninae. (Pandion & Icthyophaga)

Cohort 3. Saproharpages
Family 1. Gypaetinae.
Family 2. Vulturinae.

Cohort 4. Necroharpages.
Family 1. Cathartinae.
Family 2. Polyborinae.

Order 4. Gallinae.

Cohort 1. Tetraonomorphae.
Family 1. Pteroclinae.
Family 2. Tetraoninae.

Cohort 2. Phasianomorphae.
Family 3. Phasianinae.

Family 4. Pavoninae (inc. Turkeys)
Family 5. Perdicinae. (inc. Guineafowl)
Family 6. Hemipodiinae. (Turnix, the Buttonquails)
Cohort 3. Macronyches.
Family 7. Catheturinae. (Megapodes except Megapodius)
Family 8. Megapodiinae. (Megapodius)
Cohort 4. Duodecimpennatae.
Family 9. Cracinae.
Family 10. Penelopinae.
Cohort 5. Struthioniformes.
Family 11. Crypturinae. (Tinamous)
Cohort 6. Subgrallatores.
Family 12. Thinocorinae.
Family 13. Chionidinae.
Order 5. Grallatores.
Cohort 1. Herodii.
Family 1. Ardeidae.
Cohort 2. Pelargi.
Family 1. Plataleinae (Platalea)
Family 2. Ciconiinae.
Family 3. Ibidinae.
Family 4. Scopinae (Scopus & Balaeniceps)
Cohort 3. Limicolae.
Family 1. Totaninae. (Scolopacids, inc. Ibidorhyncha & Phalaropes)
Family 2. Himantopodinae. (Himantopus & Recurvirostra)
Cohort 4. Cursores.
Family 1. Charadriinae.
Family 2. Otidinae. (inc. Cursorius & Burhinus)
Family 3. Rhinochaetinae. (Kagu & Plains Wanderer)
Family 4. Eurypyginae.
Family 5. Gruinae.
Family 6. Rallinae. (inc. Jacanas)
Family 7. Heliornithnae. (Podica & Heliornis, no mention of Heliopais)
Family 8. Palamedeinae. (Screamers)
Order 6. Natatores
Cohort 1. Longipennes.
Family 1. Sterninae. (including the Crab Plover)
Family 2. Rhynchopinae.
Family 3. Larinae. (including Skuas)
Cohort 2. Pygopodes.
Family 1. Alcariae. Auks
Family 2. Eudytinae. (Divers)
Family 3. Colymbinae. Grebes
Cohort 3. Totipalmatae.
Family 1. Pelecaninae (including Pelicans, Cormorants, Gannets, Frigatebirds, Anhinga, Tropicbirds)
Cohort 4. Tubinares.
Family 1. Halodrominae. (Diving Petrels)
Family 2. Procellarinae. (including Storm Petrels)

Family 3. Diomedinae.
Cohort 5. Impennes.
Family 1. Spheniscinae.
Cohort 6. Lamellirostres
Family 1. Phoenicopterinae.
Family 2. Anatinae. (Geese, Swans & Ducks)
Order 7. Proceres
Cohort 1. Proceres Veri.
Family 1. Struthioninae. (Struthio & Rhea)
Family 2. Dromaeinae. (Emu & Cassowaries)
Cohort 2. Subnobilis
Family 3. Apteryginae.
Order 8. Saururae.
Archaeopteryx.

Appendix 27
Lilljeborg 1866

Subclass 1. Natatores
Group 1. Simplicirostres
Order 1. Pygopodes
Family 1. Aptenodytidae
Family 2. Alcidae
Family 3. Colymbidae
Family 4. Podicipidae
Order 2. Longipennes
Family 5. Procellaridae
Family 6. Laridae
Order 3. Steganopodes
Family 7. Dysporidae
Family 8. Pelecanidae
Group 2. Lamellirostres
Order 4. Lamellirostres
Family 9. Mergidae
Family 10. Anatidae
Subclass 2. Cursores
Order 5. Grallae
Family 11. Phoenicopteridae
Family 12. Rallidae
Family 13. Palamedeidae
Family 14. Psophidae
Family 15. Ardeidae
Family 16. Ciconidae
Family 17. Gruidae
Family 18. Totanidae
Family 19. Scolopacidae
Family 20. Charadriidae
Family 21. Otididae
Order 6. Brevipennes
Family 22. Struthionidae
Family 23. Apterygidae
Order 7. Gallinae
Family 24. Crypturidae
Family 25. Tetraonidae
Family 26. Phasianidae
Family 27. Pteroclididae
Subclass 3. Insessores
Order 8. Pullastrae
Family 28. Megapodidae
Family 29. Penelopidae
Family 30. Columbidae
Family 31. Didunculidae
Order 9. Accipitres
Family 32. Vulturidae
Family 33. Falconidae
Family 34. Strigidae
Order 10. Strisores
Family 35. Caprimulgidae
Family 36. Cypselidae
Family 37. Trochilidae
Family 38. Coracidae
Family 39. Meropidae

Family 40. Alcedinidae
Family 41. Bucerotidae
Order 11. Zygodactyli
Family 42. Musophagidae
Family 43. Trogonidae
Family 44. Galbulidae
Family 45. Bucconidae
Family 46. Ramphastidae
Family 47. Cuculidae
Family 48. Picidae
Family 49. Psittacidae
Order 12. Passeres

Appendix 28

Reichenow

GRALLATORES
Calamicolae
Rallidae
Aramidae
Jacanidae
Eurypigidae
Mesitidae
Arvicolae
Otididae
Charadridae (Chioninae, Charadriinae, Haematopdinae, Cursorinae, Oednicmeminae)
Dromadidae
Scolopacidae (Himantopodinae, Totaninae, Scolopacinae)
Thinocoridae
Gressores
Ibididae
Ciconidae
Scopidae
Balaenicipidae
Ardeidae
NATATORES
Lamellirostres
Mergidae
Anatidae
Anseridae
Cygnidae
Palamedeidae
Phoenicopteridae
Longipennes
Procellariidae (Pelecanoidinae, Diomedeinae, Procellariinae, Hydrobatinae)
Laridae (Larinae, Sterninae)
Steganopodes
Phaethontidae
Sulidae
Fregatidae
Phalacrocoracidae
Pelecanidae
Pygopodes
Alcidae
Colymbidae (Divers & Grebes)
Spheniscidae
RATITAE
Struthionidae
Rheidae
Casuariidae
Dromaeidae
Apterigidae
Dinornithes (Moas)
GALLINACEI
Turnicidae
Pteroclidae
Tinamidae
Megapodidae
Cracidae
Opisthocomidae
Phasianidae
Odontophoridae
Tetraonidae
Columbae
Didunculidae
Geotrygonidae
Columbidae (Phabinae, Peristerinae, Turturinae, Columbinae)
Carpophagidae (Carpophaginae, Ptilopodinae, Treroninae)
FIBULATORES
Psittaci
Nestoridae

Lorriidae
Cyclopsittacidae
Nasiternidae
Cacatuidae
Strigopodidae
Platycercidae
Psittacidae
Scansores
Rhamphastidae
Capitonidae
Bucconidae
Galbulidae
Indicatoridae
Picidae
Coliidae
Trogondae
Musophagidae
Cuculidae
PASSERES
Oscines
Menuridae
Hirundinidae
Muscicapidae
Campephagidae
Laniidae (Laniinae, Prionopinae, Malaconotinae, Vanginae, Pachycephalinae, Vireoninae, Cracticinae)
Corvidae
Paradisaeidae (inc. Ptilonorhynchinae)
Dicruridae
Oriolidae
Artamidae
Sturnidae
Paramythidae
Icteridae
Ploceidae
Fringillidae (inc. Emberizinae)
Tanagridae
Mniotiltidae (= Parulidae)
Motacillidae
Alaudidae
Pycnonotidae
Zosteropidae
Meliphagidae
Nectariniidae
Dacnididae
Dicaeidae (Dicaeinae, Drepanidinae)
Certhiidae (Certhiinae, Sittinae)
Paridae (Parinae, Paradoxornithinae, Polioptilinae)
Sylviidae (Timaliinae, Cisticolinae, Sylviinae, Troglodytinae, Miminae, Crateropodinae, Turdinae, Erithacinae)
Climactores
Eurylaemidae
Cotingidae (inc.Phytotomidae)
Tyrannidae (inc. Piprinae, Oxyruncinae)
Conopophagidae
Dendrocolaptidae (inc. Furnariinae)
Formicariidae
Pteroptochidae
Philepittidae
Pittidae
Strisores
Caprimulgidae
Macropteridae
Trochilidae
Insessores
Bucerotidae
Alcedinidae
Meropidae
Upupidae
Coraciidae (inc. Podarginae)
Momotidae
Todidae
RAPTATORES
Accipitres
Cathartidae
Vulturidae
Serpentariidae
Falconidae
Accipitridae
Striges
Ketupinae
Buboninae
Nycteinae
Syrninae
Striginae

APPENDIX 29

FÜRBRINGER

Subclass 1. SAURURAE
- Order Archornithes
  - Suborder Archaeopterygiformes
    - Family Archaeopterygidae

Subclass 2. ORNITHURAE
- Order Struthiornithes
  - Suborder Struthioniformes
    - Family Struthionidae
- Order Rheornithes
  - Suborder Rheiformes
    - Family Rheidae
- Order Hippalectryornithes
  - Suborder Casuariiformes
    - Family Casuariidae
  - Intermediate suborder Aepyornithes
    - Family Aepyornithidae
    - Intermediate suborder Palamedeiformes
    - Family Palamedeidae
- Order Pelargornithes
  - Suborder Anseriformes
    - Family Gastornithidae
    - Family Anatidae
  - Suborder Podicipitiformes
    - Family Enalornithidae
    - Family Hesperornithidae
    - Family Colymbidae
    - Family Podicipidae
  - Suborder Ciconiiformes
    - Family Palaeolodidae
    - Family Phoenicopteridae
    - Family Plataleidae
    - Family Ciconiidae
    - Family Scopidae
    - Family Ardeidae
    - Family Balaenicipitidae
    - Family Gypogeranidae
    - Family Falconidae
    - Family Phaethontidae
    - Family Phalacrocoracidae
    - Family Pelecanidae
    - Family Fregatidae
  - Intermediate suborder Procellariiformes
    - Family Procellariidae
  - Intermediate suborder Aptenodytiformes

Family Aptenodytidae
Order Charadriornithes
Suborder Ichthyornithiformes
Family Ichthyornithidae
Family Apatornithidae
Suborder Charadriiformes
Family Charadriidae
Family Glareolidae
Family Dromadidae
Family Chionididae
Family Laridae
Family Alcidae
Family Thinocoridae
Family Paridae [from Parra, not Parus]
Family Oedicnemidae
Family Otididae
Suborder Gruiformes
Family Eurypygidae
Family Rhinochetidae
Family Aptornithidae
Family Gruidae
Family Psophidae
Family Cariamidae
Suborder Ralliformes
Family Heliornithidae
Family Rallidae
Family Mesitidae
Family Hemipodidae
Order Alectrornithes
Suborder Apterygiformes
Family Apterygidae
Family Dinornithidae
Suborder Crypturiformes
Family Crypturidae
Suborder Galliformes
Family Megapodidae
Family Cracidae
Family Gallidae
Intermediate suborder Columbiformes
Family Pteroclidae
Family Dididae
Family Columbidae
Intermediate suborder Psittaciformes
Family Psittacidae
Order Coracornithes
Suborder Coccygiformes
Family Musophagidae
Family Cuculidae
Family Bucconidae

Family Galbulidae
Suborder Pico-Passeriformes
Family Capitonidae
Family Rhamphastidae
Family Indicatoridae
Family Picidae
Family Pseudoscines
Family Passeridae
Family Cypselidae
Family Trochilidae
Family Coliidae
Family Trogonidae
Suborder Halcyoniformes
Family Halcyonidae
Family Alcedinidae
Family Upupidae
Family Bucerotidae
Family Meropidae
Family Momotidae
Family Todidae
Suborder Coraciiformes
Family Coraciidae
Family Leptosomidae
Family Caprimulgidae
Family Steatornithidae
Family Podargidae
amily Strigidae

Appendix 30
Gadow

Class AVES
Subclass Archornithes
1. Archaeopterygiformes
I. Archaeopteryges
1. Archaeopterygidae
Subclass Neornithes
Division 1. Neornithes Ratitae
1. Struthiones
2. Rheae
3. Casuarii
4. Apteryges
5. Dinornithes (Moas)
6. Aepyornithes
Division 2. Neornithes Carinatae
7. Colymbiformes
I. Colymbi
II. Podicipedes
8. Sphenisciformes
9. Procellariiformes
10. Ardeiformes
I. Steganopodes
1. Phaethontidae
2. Phalacrocoracidae (inc. Suliinae, Plotinae, Phalacrocoracinae)
3. Pelecanidae
4. Fregatidae
II. Herodii
1. Ardeidae
2. Scopidae
III. Pelargi

1. Ciconiidae
2. Phoenicopteridae
11. Falconiformes
I. Cathartae
II. Accipitres
1. Vulturidae
2. Gypogeranidae
3. Pandionidae
4. Falconidae
12. Anseriformes
I. Palamedeae
II. Anseres
13. Crypturiformes
I. Crypturae (Tinamous)
14. Galliformes
I. Turnices
II. Galli
1. Gallidae
2. Opisthocomidae
15. Gruiformes
I. Eurypygae
1. Eurypygidae
2. Rhynochetidae
3. Mesitidae
II. Ralli
III. Grues
IV. Dicholophi (Cariamas)
V. Otides
16. Charadriiformes
I. Limicolae
1. Chionididae
2. Charadriidae
a. Charadriinae
b. Scolopacinae
3. Glareolidae
4. Thinocoridae (inc. Attagis)
5. Oedicnemidae
6. Parridae
II. Gaviae
1. Alcidae
2. Laridae
17. Columbiformes
I. Pterocles
II. Columbae
1. Dididae
2. Columbidae
18. Cuculiformes
I. Coccyges
1. Cuculidae
2. Musophagidae
II. Psittaci
1. Psittacidae
19. Coraciiformes
I. Striges
1. Strigidae
II. Macrochires
1. Caprimulgidae
2. Cypselidae
3. Trochilidae
III. Colii
1. Colidae
IV. Trogones
1. Trogonidae
V. Coraciae
1. Coraciidae
2. Momotidae
3. Alcedinidae
4. Meropidae
5. Upupidae
a. Upupinae
b. Bucerotinae
c. Irresorinae
20. Passeriformes
I. Pici
1. Galbulidae
2. Picidae
3. Capitonidae
4. Rhamphastidae
II. Passeres
1. Eurylaemidae
2. Menuridae
a. Menurinae
b. Atrichiinae
3. Passeridae

Appendix A

Birds from Emperor Rudolph II's collection. "Le Bestiare de Rudolphe II" by H. Haupt, T. Vignau-Wilberg and E. Irblich (?ed) M. Staudinger. Éditions Citadelles-Éditio, 33 Rue de Naples, Paris 1990.
[The first 89 plates are not of birds]

90-91 Aquila chrysaetos
92 Aegypius monachus
93 Gyps fulvus
94 Gypaetus barbatus
95 Pandion haliaetus
96 Falco tinnunculus (albino)
97 Falco rusticolus white morph
98 ditto, grey morph
99 Falco femoralis
100 Falco vespertinus
101 Paradisaea apoda
102 Paradisaea minor: Cicinnurus regius
103 Paradisaea minor: crickets
104 Seleucides melanolauca [a very tatty specimen]: Eudocimus ruber
105 Ara ararauna
106 Ara macao
107 Ara ararauna
108 Cacatua moluccensis
109 Cacatua galerita
110 Turdus merulus [partial albino]: Pionus sordidus
111 Lorius domicellus
112 Lorius lory jobiensis: Lorius garrulus flavopallidus
113 Lorius domicellus
114 Agapornis pullaria: Ara sp. [This *Ara* resembles *macao* except that it has a broad ring of small red feathers round the eye. The wing coverts and secondaries are red, with just a few green and yellow feathers, the quills are dark. The tail is missing].
115 Larus sp.: Trichoglossus haematodus haematodus
116 Paradisaea apoda
117 Struthio camelus
118-9 Casuarius casuarius
120 Raphus cucullatus [a thin bird with feathers like an Emu, but for the bill, it could almost pass for an Emu]
121 Aphanapteryx bonasia [much thinner and scruffier than Rothschild's plate]
122 Phoenicopterus ruber
123 Balearica pavonina
124-5 Anthropoides virgo
126 Ciconia nigra
127 Platalea leucorodia
128 Himantopus himantopus
129 xobrychus minutus
130 Nycticorax nycticorax
131 Ixobrychus minutus
132 Philomachus pugnax ["Chevallier Combattant"]

133 Cardinalis cardinalis: Recurvirostra avosetta
134 Pavo cristatus (partial albino)
135 Phasianus colchicus (partial albino)
136-7 Phasianus colchicus (albinos)
138 Tetrao urogallus
139 Otis tarda
140 Lagopus mutus: Nycticorax nycticorax
141 Perdix perdix (two varieties)
142 Perdix perdix (partial albino): Lanius excubitor
143 Crax pauxi
144 Gallus gallus (domestic variety)
145 Tetrao urogallus
146-8 Numida meleagris
149 Columba livia (with two heads): Plectrophenax nivalis
150 Columba livia (domestic variety)
151 Starnoenas cyanocephala
152 Gallinula chloropus
153 Branta bernicla (with barnacles!!)
154 Anser anser (with three legs)
155 Anser anser (with a growth of feather fans on the neck)
156 Branta leucopsis
157 ?Luscinia megarhynchos [Doubtful, probably a South American sub-oscine]
158 nest of Remiz pendulina
159 Alauda arvensis (partial abino): Plectrophenax nivalis
160 Turdus pilaris (partial albino)
161 Pica pica: Corvus frugilegus (?) (partial albinos, the "Rook" has a fully feathered face)
162 Pica pica (partial albino)
163 Perdix perdix: Corvus monedula (partial albinos)
164 Merops apiaster: Coturnix coturnis (two, each with three legs)
165 Merops apiaster
166 Picus viridis
167 Garrulus glandarius
168 Alcedo atthis: Loxia curvirostra
169 Ramphastos tucanus: Ramphastos toco
170 [bat]
171 Pelecanus crispus
172 Platalea leucorodia
173 Phalacrocorax carbo: Anas clypeata
174 Geronticus eremita: Mergus albellus
175 Podiceps cristatus: Netta rufina
176 Cairina moschata (white domestic variety)
177 Cairina moschata (domestic variety)
178 Spheniscus demersus
179-80 Merops apiaster

Appendix B

Birds described by Quoy & Gaimard on Freycinet's Voyage 1817-1820.
Itinerary: (places where landings were made?)

1. Brazil & Rio de la Plata
2. Cape of Good Hope
3. Timor, Rawak & Waigiou
4. Mariannes Islands
5. Hawaiian Islands
6. Australia
7. Falklands

p. 91- 141.
p. 91. Falco leucorrhous Brazil = Buteo leucorrhous see Peters, vol. 1, ed. 2, p. 364.
p. 92. Falco polyosoma Cayenne = Buteo polysoma see Peters, vol. 1, ed. 2, p. 367.
p. 93. Falco histrionicus Falkland Islands = synonym of Circus cinereus Vieillot (Peters 1(2), p. 318.
p. 96. Lanius ferrugineus Cape of Good Hope = Laniarius f. ferrugineus (Gmelin) Peters 9, p. 330.
p. 98. Vanga striata Brazil = Batara cinerea (Vieillot)
p. 100. Barita tibicen Australia = Gymnorhina tibicen (Latham)
p. 103. Graucalus viridis Timor = Specotheres viridis (Vieillot)
p. 104. Turdus falcklandii Falklands = Turdus falcklandii see Peters 10, p. 215.
p. 105. Oriolus regens Australia = Sericulus chrysocephalus (Lewin)
p. 107. Malurus textilis Australia = Amytornis textilis (Dumont) see Peters 11, p. 405.
p. 108. Malurus leucopterus Isle Dirk-Hatichs = Malurus leucopterus Dumont see Peters 11, p 395.
p. 109. Emberiza melanodera Falklands = Melanodera melanodera see Peters 13, p. 109.
p. 110. Xanthornus gasquet Rio de la Plata = Pseudoleistes guirahuro (Vieillot)
p. 112. Dacelo gaudichaud Gueba = Dacelo gaudichaud see Peters, 5, p. 191.
[NB: Mayr (1941, List of New Guinea Birds, p. 89) restricted the type loclity from Papuan Islands to Waigeu, but Quoy & Gaimard give Gueba (=Gebe) as locality. Mayr gave no reason for his action.
p. 114. Cuculus guira [not a new description] = Guira guira (Gmelin)
p. 116. Psittacus erythropterus [not a new description] = Aprosmictus erythropterus (Gmelin)
p. 118. Columba pinon Rawak = Ducula pinon pinon see Peters, 3, p. 52.
p. 119. Columba aenea [not a new description] = Ducula aenea (Linn).
p. 121. Columba pampusan Mariannes = Gallicolumba xanthonura (Temminck 1823)
[NB: Peters, 3, p. 134, spells Quoy & Gaimard's name *pampusana*, but it is *pampusan* in the original]
p. 122. Columba macquarie Australia = Geopelia cuneata (Latham)
p. 125. Megapodius freycinet Guebe, Waigiou = Megapodius freycinet (Gaimard 1823) See Peters 2, p. 6.
p. 127. Megapodius la Perouse Mariannes = Megapodius La Perouse (Gamard 1823) see Peters 2, p. 7.
p. 129. Haematopus niger Falklands = Haematopus ater (Vieillot & Oudard)
p. 131. Chionis alba [not a new description] = Chionis alba (Gmelin)
p. 133. Podiceps rolland Falklands = Rollandia rolland see Peters vol. 1, ed. 2, p. 141.
p. 135. Procellaria berard = Pelecanoides urinator berard
p. 137. Lestris catarractes = Catharacta antarctica (Lesson)
p. 139. Anas brachyptera [not a new description] = Tachyeres brachypterus (Latham)

Bibliography

This bibliography includes sources used in the preparation of this work, and works on ornithology written by authors discussed in the text. Works by them on other subjects, which are mentioned in the text in the course of biographical discussion, but which are not of an ornithological nature, are not included here.

Anon. 1884. Zoological Nomenclature [report of the meeting held in the Lecture Room of the Natural History Museum], *Nature*, 10 July 1884, p. 256-259, 17 July 1884, p. 277-279.

Anon. 1823. "Ornithology" in Encyclopaedia Britannica, Sixth Edition, vol. 15, p. 467-565.

Adanson, M. 1757. Histoire Naturelle de Senegal. Paris. Eng. trans. 1759.

Aelian (Claudius Aelianus). On the characteristics of Animals. [Published editions 1554, 1744, etc.]

Albin, E. 1731, 1734, 1738. A Natural History of Birds. 3 vols. London.

Albin, E. 1737. A Natural History of English Song Birds, and such of the foreign as are usually brought over and esteemed for their singing. London.

Aldrovandus, Ulysse. 1599, 1600, 1603. Ornithologiae hoc est de Avibus historiae libri. 3 vols. Bononiae.

Aldrovandus, Ulysse. Memoir of. in The Naturalists Library, 1838, vol. 13. Edinburgh.

Allen, E.G. 1951. A Sixteenth Century Classification of Birds. Proceedings of the Tenth Ornithological Congress. Uppsala, June 1950.

Allen, E.G. 1951. The History of American Ornithology before Audubon, *Transactions of the American Philosophical Society*, vol. 41, pt. 3.

Allen, J.A. 1871. On the Mammals and Winter Birds of East Florida, *Bulletin of the Museum of Comparative Zoology*, vol. 2, p. 161-451.

Allen, J.A. 1876. The Availability of Certain Bartramian names in Ornithology, *American Naturalist*, vol. 10, p. 21-29.

Allen, J.A. 1876. Geographical Variation among North American Mammals, especially in respect to Size, *Bulletin of the United States Geographical Survey*, vol. 2, p. 309-344.

Allen, J.A. 1877. The Influence of Physical Conditions in the Genesis of Species, *Radical Review*, vol. 1, p. 108-140. [not in NHM].

Allen, J.A. 1883. On Trinomial Nomenclature, *Zoologist*, vol. 7, p. 97-100.

Allen, J.A. 1907. Linnaeus as Zoologist, *Annals of the New York Academy of Sciences*, p. 9-19.

Allen, J.A. 1908. Pennant's Indian Zoology, *Bulletin of the American Museum of Natural History*, vol. 24, p. 111-116.

Allen, J.A. 1910. Collation of Brisson's Genera of Birds with those of Linnaeus. *Bulletin of the American Museum of Natural History*, vol. 28, p. 317-335.

Altum, B. 1868. Die Vögel und sein Leben [The bird and its life]. Münster.

American Ornithologists Union. 1886. The Code of Nomenclature and Check-List of North American Birds.

Anderson, J. 1878. Anatomical and Zoological Researches: comprising an account of the Zoological Results of the Two Expeditions to Western Yunnan in 1868 and 1875, vol. 1, introduction. London.

André, E. 1904. A Naturalist in the Guianas. London.

Aristotle. The History of Animals. various editions.

Aristotle. The Parts of Animals. various editions.

Atsunobu Kaibara. 1709. Yamato Honzo (The Natural History of Japan).

Audebert (see Vieillot)

Audubon, J.J. 1827-38. The Birds of America, 4 vols. London.

Audubon, J.J. 1831-39. Ornithological Biography, 5 vols. Edinburgh.

Azara, F. Don de. 1802-5. Apuntamientos para la natural, etc., 3 vols. Madrid.

Baird, S.F., Brewer. T.M. and Ridgway, R. 1874. A History of North American Birds, 3 vols. Boston.

Bancroft, E. 1769. An Essay on the Natural History of Guiana. London.

Barrère, P. 1741. [checklist of birds of Cayenne] in Essai sur l'histoire naturelle de la France quinoxiale, etc. Paris.

Barrère, P. 1745. Ornithologiae Specimen novum. Perpignan.

Bartholomew de Glanville. app.1470. De Proprietatibus Rerum.

Bartram, W. 1791. Travels through North and South Carolina, Georgia, East and West Florida ... Philadelphia.

Bechstein, J.M. 1789-95. Gemeinnützige Naturgeschichte Deutschlands nach allen drey Reichen [General Natural History of Germany], 4 vols. Leipzig.
Beddall, B.G. 1983. The Isolated Spanish Genius - Myth or Reality? Felix de Azara and the Birds of Paraguay, *Journal of the History of Biology*, vol. 16, no. 2, p. 225-258.
Beebe, W. 1907. Geographical variation in birds, *Zoologia*, p. 3-41.
Belon, P. 1555. L'Histoire de la nature des Oyseaux. Paris.
Belovacensis, V. app.1264. Speculum Naturae.
Bergtold, W.H. 1917. A study of the incubation periods of birds. Denver.
Berkenhaut, J. 1769. Outlines of the Natural History of Great Britain. 3 vols. London.
Berthold, A.A. 1831. Beiträge zur Anatomie, Zootomie und Physiologie. Göttingen.
Bewick, Thomas. 1797, 1804. History of British Birds, 2 vols. Newcastle upon Tyne.
Blainville, H.D. de. 1821. Mémoire sur l'emploi de la forme du Sternum et de ses annexes, pour l'établishment ou la confirmation des familles naturelles parmi les oiseaux. *Journal de Physique de Chimie et d'Histoire Naturelle*, vol. 92, p. 185-216. [paper read at the Academy of Sciences, 6 December 1815].
Blakiston, T.W. and Pryer, H. 1878. A Catalogue of the Birds of Japan, *Ibis*, p. 209-250.
Blakiston, T.W. and Pryer, H. 1882. Birds of Japan, *Transactions of the Asiatic Society of Japan*, vol. 10, p. 88-186.
Blanchard, C.E. 1859. Recherches sur les caractères ostéologiques des Oiseaux appliquées à la Classification naturelle de ces animaux, *Ann. Sc. Nat. Zoologie*, vol. 11, pp. 11-145, pls. 2-4.
Blumenbach, J.F. 1779-80. Handbuch der Naturgeschichte, 2 vols. Göttingen. English translation by R. Gore. 1825. A Manual of the Elements of Natural History. London.
Blyth, E. 1838. Outlines of a new arrangement of Insessorial birds, *Annals & Magazine of Natural History*, vol. 2, p. 256-268, 314-319, 351-361, 420-426, 589-601, vol. 3, p. 76-84.
Boddaert, P. 1783. Table des Planches enlumines. Reprinted 1876 with an introduction by W.B. Tegetmeir. London.
Bodkin, D. B. and Miller, R. S. 1974. Mortality rates and survival of birds. *American Naturalist*, vol. 108, p. 181–192.
Boeseman, M. 1970. The Vicissitudes & dispersal of Albertus Seba's zoological specimens, *Zoologische Mededelingen*, vol. 44, p. 177-206.
Boie, F. 1826. [classification of birds] *Isis von Oken*, p. 970-982.
Bonaparte, C.L. 1825-33. American Ornithology, 4 vols. Philadelphia.
Bonaparte, C.L. 1841. New Systematic Arrangement of Vertebrated Animals: Aves. *Transactions of the Linnean Society of London*, vol. 18, p. 258-278.
Bonaparte, C. L. 1845. Observations on the State of Zoology in Europe, *in* Reports on the Progress of Zoology and Botany for 1841 and 1842. pub. Ray Society; trans. by H.E. Strickland.
Bonaparte, C.L. 1850, 1857. Conspectus genera avium, 2 vols, unfinished. Lugduni Batavorum.
Bonelli, F.A. 1811. Catalogue des Oiseaux du Piedmont, *Annales de l'Observatoire de l'Académie de Turin*, 24pp. Journal not seen, separate of paper catalogued as book in Rothschild Library. Birds are arranged alphabetically.
Bonnaterre, J.P. Abbé de. 1790-92. Tableau encyclopédique et méthodique: Ornithologie. Paris.
Bonnet, C. 1764. Contemplation de la Nature, 2 vols. Amsterdam. Translated: The Contemplation of Nature, 1766, 2 vols. London.
Bosman, W. 1704. Nauwkeurige beschryving van de Guinese Goud- Tand- en Slave-Kust, etc. 2 vols. Utrecht. English translation. 1705. A new ... description of the Coast of Guinea ... Written originally in Dutch ... and now faithfully done into English. London. Another edition in Pinkerton's General Collection of Voyages, 1814.
Bowdich, T. E. 1821. An Introduction to the Ornithology of Cuvier. J. Smith: Paris.
Brandt, J.F. von. 1836-1839. [various papers] *Mémoires de l'Académie Impériale Sciences de St. Petersbourg*. 3 vols.
Brehm, C.L. 1820-22. Beiträge zur Vögelkunde. 3 vols. Neustadt an der Orla.
Brehm, C.L. 1831. Handbuch der Naturgeschicte aller Vögel Deutschlands. Ilmenau.
Brisson, M.T. 1760. Ornithologie, 6 vols. Paris.
Bruch, C.F. 1828. Ornithologische Beyträge, *Isis von Oken*, vol. 21 (1828), col. 725.
Brünnich, M.T. 1763. Die natürliche Historie des Eider-Vogels [Natural History of the Eider Duck]. Copenhagen.
Brünnich, M.T. 1764. Ornithologia Borealis. Hafniae.
Buffon, G.L. Comte de. 1750. Histoire et Thèorie de la Terre [Theory of the Earth] in 1749-1804. Histoire Naturelle ... avec la description du Cabinet du Roi. 44 volumes. Paris. English translation, 1781, by W. Smellie. London.
Buffon, G.L. Comte de. 1770-83. Histore Naturelle des Oiseaux, 10 vols. Paris.
Bullock, W. 1809. A Companion to the Liverpool Museum. Bath. [various subsequent editions after it had moved

to London and become the London Museum].
Bullock, W. 1819. Catalogue ... of the Roman Gallery ... and the London Museum of Natural history ... which will be sold ... by Mr. Bullock. London.
Burke, T. and Bruford, M.W. 1987. DNA fingerprinting in birds. *Nature*, vol. 327, p. 149–152.
Burns, F.L. 1908-9. Alexander Wilson, pts. 1-8. *Wilson Bulletin*, vols. 20-21,
Burns, F.L. 1917. Miss Lawson's Recollections of Ornithologists, *Auk*, p. 275-282.
Butterfield, W.R. 1906, A Plea for further recognition of Subspecies in Ornithology, *Zoologist*, 4th series, vol.10, p. 62-64.
Bruce, J. 1790. Travels to discover the source of the Nile in 1768-73. Edinburgh & London.
Cabanis, J.L. & Heine, F. 1863. Museum Heineanum, 4 pts. Halberstadt.
Call, R.E. 1895. The Life and Writings of Rafinesque. Pub. for the Filson Club by Morton & Co., Louisville, Kentucky.
Cassin, J. 1858. United States Exploring Expedition: Mammalogy & Ornithology (second edition). Philadelphia.
Cassin, J. 1864. Fasti Ornithologiae, no. 1. "Philipp Ludwig Statius Muller", *Proceedings of the Academy of Natural Sciences of Philadelphia*, p. 234-257.
Castlenau, F.L.N. de C, Comte de, 1850-9. Expédition dans les parties centrales de l'Amerique du Sud, de Rio de Janeiro à Lima, et de Lima au Para; executee ... pendant ... 1843 à 1847, sous la direction de F. de Castelnau. Paris
Catesby, M. 1731-43. Natural History of Carolina. London.
Cetti, F. 1776. Gli Ucelli di Sardegna. [Natural History of Sardinia]. Sassari.
Chalmers Mitchell, P. 1929. Centenary History of the Zoological Society of London. Published by the Society.
Chapman, F. 1922. In Memoriam: Joel Asaph Allen. *Auk*, vol. 39, p. 1-14.
Chapman, F. 1926. A Review of the History of Ecuadorean Ornithology, in Distribution of Bird-life in Ecuador, *Bulletin of the American Museum of Natural History*, vol. 55, p. 7-14, 723-735.
Chapman, F. 1917. A Review of Colombian Ornithology in Distribution of Bird-life in Colombia, *Bulletin of the American Museum of Natural History*, vol. 36, p. 11-19.
Charleton, W. 1668. Onomasticon Zoicon. London.
Christy, B.H. 1933. Topsell's 'Fowles of Heauen', *Auk*, vol. 50, p. 275-283.
Clusius, C. 1605. Exoticorum libri decum. Leyden.
Coiter, V. 1572. Tables of the Principal External and Internal Parts of the Human Body and Various Anatomical Exercises and Observations Illustrated with New, Diverse and Very Ingenious Figures, Extremely Useful Especially to Those Devoted to Anatomical Study. Nürnberg.
Coiter, V. 1575. De Differentiis Avium.
Condamine, C.M. de la. 1751. Journal du Voyage fait ... à l'Equateur etc., Paris.
Condamine, C.M. de la. 1813. Abridged narrative of travels through the interior of South America; in Pinkerton, J. A general collection of Voyages., vol. 14. London.
Coues, E. Progress of American Ornithology, *American Naturalist*, vol. 5, p. 364-373. [Review of Allen's paper above].
Coues, E. 1872. Key to North American Birds. Salem.
Coward, T.A. 1920-26. The Birds of the British Isles, 3 vols. London. Many subsequent editions.
Cramp, S. *et al.* 1977-94. The Birds of the Western Palaearctic. 9 vols. Oxford University Press.
Cretzschmar. P.J. (see Rüppell).
Crichton, A. 1835. Memoir of Pliny, in The Naturalists' Library, vol. 19. Edinburgh.
Crichton, A. 1837. Memoir of Le Vaillant, in The Naturalists' Library, vol. 23. Edinburgh.
Cuba, Johannes de. 1475. Hortus Sanitatis. Mainz.
Cunningham, R.O. 1866. On the Solan Goose or Gannet (*Sula bassana* Linn.), *Ibis*, p. 5. [Comments on Jonston].
Cuvier, G.L.C.F.D. Baron. 1817. Le Règne Animal, 4 vols. Paris.
Dagger, John H.K. 1988. The Dartford Warbler: Historical Notes: Copies of the Relevant Letters from John Latham, Dartford. Privately printed, pp. 14.
Darwin, C. and A.R.Wallace 1858. On the Tendency of Species to form Varieties, Read before the Zoological Society on 1 July 1858. *Proceedings of the Zoological Society of London*.
Darwin, C. 1859. On the Origin of Species by Natural Selection. London.
D'Aubenton, E.L. 1771-86. Planches enluminées, 10 vols. Paris.
Daudin, F.M. 1800. Trait élémentaire et complet d'Ornithologie, 2 vols. Paris.
David, A. & Oustalet, E.A. 1877. Les Oiseaux de Chine, 2 vols. Paris.
Des Murs, O. 1860. Traité général d'Oologie ornithologique au point de vue de la Classification. Paris.

Dickey & van Rossem, 1938. Summary of Ornithological work *in* Birds of El Salvador, *Field Museum of Natural History: Zoology*, vol. 23, p. 11-13.
Dictionary of National Biography.
Dieffenbach, E. 1843. Travels in New Zealand, 2 vols. London.
Dixon, C. 1902. Birds' Nests: An Introduction to the Science of Caliology. London.
Donati, V. 1750. Della storia naturale marina dell' Adriatico. Venice.
Donovan, E. 1794. Natural History of British Birds, 10 vols. London.
d'Orbigny, A.C.V. 1835-1847. Voyage dans l'Amerique meridionale etc., 7 vols plus 2 vols Atlas, and maps. Paris.
Dunmore, J. 1965. French Explorers in the Pacific, 2 vols. Clarendon Press, Oxford.
Edwards, G. 1743-51. Natural History of Birds, 4 vols. London.
Edwards, G. 1758-64. Gleanings of Natural History, 3 vols. London.
Ehrenberg (see Hemprich)
Elton, C. 1927. *Animal Ecology*. Sidgwick and Jackson, London.
Evans, A.H. 1903. Turner on Birds. Cambridge.
Evans, W. 1891. On the Periods occupied by Birds in the Incubation of their Eggs, *Ibis*, p. 52-93.
Evans, W. 1892. Some further Notes on the Periods occupied by Birds in the Incubation of their Eggs, *Ibis*, p. 55-58.
Eyton, T.C. 1859-1875. Osteologia Avium. Wellington.
Faber, F. 1821. Prodromus der Isländischen Ornithologie. Copenhagen.
Faber, F. 1825-6. Ueber das Leben der hochnordischen Vögel. Leipzig.
Fermin, P. 1765. Histoire naturelle de la Hollande équinoxiale: ou description des Animaux, etc. 4 parts. Amsterdam.
Fermin, P. 1769. Description ... de la Colonie de Srinam, etc. 2 vols in one. Amsterdam.
Finsch, F.H.O. 1867-8. Die Papageien, 2 vols. Leyden.
Finsch, F.H.O. & Hartlaub, G. 1867. Ornithologie der Samoa und Tonga-Inseln. Halle.
Finsch, F.H.O. & Hartlaub, G. 1870. Die Vögel Ost-Africas. Leipzig & Heidelberg.
Forbes, V S. 1965. Pioneer Travellers in South Africa. Cape Town. [Sections on Carl Peter Thunberg, p. 25-36, Anders Sparrmann, p. 46-58 and Francois Levaillant p. 117-127.]
Forster, J.R. 1844. Descriptiones animalium. Berlin.
Franklin, Maj. 1830-2. [various papers on Indian birds] *Proceedings of the Zoological Society*, parts1-2.
Frederick II, Emperor of Germany. 1596. De Arte Venandi cum Avibus [A useful English translation is "The Art of Falconry" trans. Casey A. Wood & F. Marjorie Fyfe 1943 (reprinted 1969). Stanford University Press].
Fürbringer, M. 1888. Untersuchungen zur Morphologie und Systematik der Vögel. Amsterdam.
Gadow, Hans. 1892. On the Classification of Birds. *Proceedings of the Zoological Society of London*, p. 229-256.
Gadow, Hans, 1893. Vögel, in Dr. H.G. Bronn's *Klassen und Ordnungen des Thier-Reichs, wissenschaftlich dargestellt in Wort und Bild*. Sechster Band. Vierte Abtheilung.
Garnot, P. in Duperry, M.L.I. 1826-30. Voyage autour du monde ... sur ... la Coquille, etc., Paris.
Gätke, H. 1891. Die Vogelwarte Helgoland. Braunschweig. English translation by R. Rosenstock. 1895. Heligoland as an Ornithological observatory. Edinburgh.
Gay, C. 1844-71. Historia fisica y politica de Chile, 28 vols. [Ornithology in vol. 1, 1844, by O. Des Murs.] Paris & Santiago.
Gesner, C. 1555. Historia Animalium, vol. 3. Tiguri.
Giraldus Cambrensis. Topography of Ireland [written 1187, published 1587].
Gloger, C.W.L. 1829. Concerning the colours of Birds' Eggs, etc., *Verhandlungen der Gesselschaft Naturforschender Freunde*, Berlin, vol. 1, p. 332-347.
Gloger, C.W.L. 1833. Schlesiens Wirbelthier-Fauna. Breslau.
Gloger, C.W.L. 1833. Das Abändern der Vögel durch Einflus des Klima's. [Change in Birds produced by Climatic Influence] Breslau.
Gmelin, J.F. 1788-93. Systema Naturae, 13th edition. Lipsiae.
Gould, J. 1831. A Century of Birds from the Himalaya Mountains. London.
Gould, J. 1832-37. The Birds of Europe, 5 vols. London.
Gould, J. 1840-48. The Birds of Australia, 7 vols. London.
Gould, J. 1849-1861. Monograph of the Trochilidae or Family of Hummingbirds, 5 vols. London.
Gould, J. 1862-73. The Birds of Great Britain, 5 vols. London.
Grant, C.H.B. and Mackworth-Praed, C.W. 1956. Moehring's Genera, etc., *Annals and Magazine of Natural History*, ser. 12, vol. 9, p. 774-8.

Grant, C.H.B. 1957. Levaillant's travels in South Africa, *Ostrich*, p. 83-97.
Gray, G.R. 1844-9. The Genera of Birds, 3 vols. London.
Gray, G.R. 1855. A Catalogue of the Genera and subgenera of Birds. London.
Gray, G.R. 1869-71. Hand-list of genera and species of Birds, 3 vols. London.
Grew, N. 1681. Museum Regialis Societatis: Catalogue & description of the Natural and Artificial Rarities belonging to the Royal Society and preserved at Gresham College. London.
Griffith, E. 1827-35. The Animal Kingdom of Baron Cuvier, 16 vols. [Aves, published 1829] London.
Griscom, L. 1935. History of Panama Ornithology *in* The Ornithology of the Republic of Panama, *Bulletin of the Museum of Comparative Zoology*, vol. 78, p. 262-268.
Grote, A. 1875. "Introduction" [biography and bibliography of Edward Blyth] in "Catalogue of Mammals of Birds of Burma", by the late Edward Blyth, *Journal of the Asiatic Society of Bengal*, pt. 2, extra number, August 1875, pp. iii-xxiv.
Güldenstädt, J.A. 1787, 1791. Reisen durch Russland und im Caucasischen Gebürge, 2 vols. St. Petersburg.
Hamilton, R. 1839. Memoir of M. le Comte de Lacépède, in The Naturalist's Library, vol. 16. Edinburgh.
Hargitt, E. [Obituary] *Ibis*, 1895 p. 302.
Harris, H. 1928. Robert Ridgway, with a bibliography of his published writings and fifty illustrations, *Condor*, vol. 30, p. 5-118.
Hartert, E. 1903-22. Die Vögel der Paläarktischen Fauna. Berlin.
Hartlaub, C.J.G. 1847. Systematischer Index zu Don Felix de Azara's Apuntamientos para la historia natural de los Pxaros del Paraguay y Rio de la Plata. Bremen.
Hartlaub, C.J.G. 1857. System der Ornithologie Westafrica's. Bremen.
Hartlaub, C.J.G. 1901. [Obituary] *Ibis*, p. 348-351.
Hayes, W. 1775. A natural history of British Birds. London.
Helbig, A.J., Matens, J., Seibold, I., Heening, F., Schottler, B. and Wink, M. 1996. Phylogeny and species limits in the Palaearctic chiffchaff *Phylloscopus collybita* complex; mitochondrial genetic differentiation and bioacustic evidence. *Ibis*, vol. 138, p. 650–666.
Hellmayr, C.E. 1926. Historical Sketch in A Contribution to the Ornithology of Northeastern Brazil, *Field Museum of Natural History: Zoology*, vol. 12, p. 236-240.
Hellmayr, C.E. 1932. Historical Sketch of Chilean Ornithology in Birds of Chile, *Field Museum of Natural History: Zoology*, vol. 19, p. 6-12.
Hemprich, F.W. & Ehrenberg, C.G. 1828-45. Symbolae Physicae, seu Icones et descriptiones Corporum Naturalium. Berlin.
Hernandez, F. 1651. Rerum medicarum Novae Hispaniae Thesaurus, seu Plantarum etc. Rome. An earlier edition, said to have been published in 1635, could not be traced.
Herrick, H. 1901. The Home Life of Wild Birds: A New Method of the Study and Photography of Birds. New York & London.
Hershkovitz, P. 1987. A history of Recent Mammalology of the Neotropic Region from 1492 to 1850. *Fieldiana: Zoology*, new series, no. 39, p. 11-98.
Hetherington, W.M. 1843. Memoir of Alexander Wilson, in The Naturalist's Library, vol. 40. Edinburgh.
Hildebrandt, Hugo. 1932. Christian Ludwig Brehm: a German Ornithologist. [translated by Rev. M.M. Vischer] *Ibis*, p. 308-316.
Hindwood, K.A. 1950. A note on William Swainson, *Emu*, vol. 49, p. 208-210.
Hoare, M. E. 1982. The Resolution Journal of Johann Reinhold Forster 1772-1775, 4 vols. Edited and with an introduction by M.E. Hoare. The Hakluyt Society: London.
Hogg, J. 1846. On the Classification of Birds, and particularly of the Genera of European Birds, *The Edinburgh New Philosophical Journal*, vol. 41, p. 50-71.
Holthius, L.B. 1969. Albertus Seba's "Locupletissimi Rerum Naturalium Thesauri..." (1734-1756) and the "Planches de Seba" (1827-1831), *Zoologische Mededelingen*, vol. 43, p. 239-253.
Hooke, R. 1665. Micrographia Restaurata, or the Copper-Plates of Dr. Hooke's Wonderful Discoveries by the Microscope, p. 32-33, and pl. 19. London.
Hopkinson, E. 1940. Latham as a bird fancier, *Avicultural Magazine*, 5th series, vol. 5, pp. 128-134, 176-182, 227-230, 259-264, 276-281.
Howard, H.E. 1907-14. The British Warblers: A History with Problems of Their Lives. 2 vols. London.

Hudson, W.H. and Sclater, P.L. 1888-9. Argentine Ornithology. A descriptive catalogue of the Birds of the Argentine Republic. 2 vols. London.

Humboldt, F.H.A. von, Baron & Bonpland, A.J.A. 1805-37. Voyage aux Régions Equinoxiales du Nouveau Continent, fait en 1799-1804. Paris.

Hume, A.O. 1874. [review of] "Die Papageien" [of Otto Finsch], *Stray Feathers*, vol. 2, p. 1-28.

Huxley, T.H. 1867. On the Classification of Birds, *Proceedings of the Zoological Society of London*, pp. 415-472.

I-Ching, or Book of Changes. Ancient Chinese Classic. Many modern editions and translations. Quotation used taken from translation by James Legge, 1899, Clarendon Press; vol. 16 of *The Sacred Books of the East*. Reprinted 1963, Dover Publications. New York, p. 178-9.

Ihering, H von & R. von. 1907. As Aves do Brazil etc. São Paulo – Museu Paulista. Catalogos da Fauna Braziliera, etc., vol. 1.

Illiger, J.C.W. 1811. Prodromus systematis Mammalium et Avium. Berlin.

Isidore, Bishop of Seville (560-636) Etymologus. Published 1833 as "Isidori Hispalensis episcopi Etymologiarum libri", edit. F.V. Otto.

Jannequin, C. Sieur de Rochefort. 1643. Voyage de Lybie au royaume de Senegra, le long du Niger. Paris.

Jardine, W. 1833. Memoir of Pennant, The Naturalist's Library, vol. 3. Edinburgh.

Jardine, W. 1843. Memoir of Francis Willughby, The Naturalist's Library, vol. 36. Edinburgh.

Jefferies, A.J., Wilson, V. and Thein, S.L. 1985. Individual-specific 'finger-prints' of human DNA. *Nature*, vol. 316. p. 76–79.

Jefferies, A.J., Wilson, V. and Thein, S.L. 1985a. Hypervariable 'minisatellite' regions in human DNA. *Nature*, vol. 314, p. 67–73.

Jefferies, A.J., Wilson, V. and Thein S.L. 1985b. Individual-specific 'finger-prints' of human DNA. *Nature*, vol. 316, p. 76–79.

Jefferies, A. J., Brookfield, J. F. Y. and Semeonoff, R.1985c. Positive identification of an immigration test-case using human DNA fingerprints. *Nature*, vol. 317, p. 818–819.

Jerdon, T.C. 1862-4. The Birds of India, 2 vols. Calcutta.

Johnsell, B. 1981 (1982). Linnaeus and his two Circumnavigating Apostles, *Proceedings of the Linnean Society of New South Wales*, vol. 106, p. 1-19.

Johnston, A.K. 1848. The Physical Atlas of Natural Phenomena. Edinburgh. Second edition, 1856, Edinburgh & London.

Jonston, J. [Johannes Jonstonus] 1650-53. Historia Naturalis, 6 parts. Frankfurt. Anonymous English translation 1657.

Kalm, P. 1753-61. En resa til Norra America, 3 vols. Stockholm. English translation in Pinkerton's Collection of Voyages, vol. 13, 1812. London.

Kämpfer, E. 1727. Historia Imperii Japonici. Translated by J.G. Scheuchzer as The History of Japan. London.

Kaup, J.J. 1854. Einige Worte über die systematische Stellung der Familie der Raben, Corvìdae, *Journal für Ornithologie*, vol. 2, p. xlvii-lvi.

Kaup, J.J. [various monographs on Birds of Prey] in Jardine's Contributions to Ornithology. Edinburgh.

Kaup, J.J. 1862. Monograph of the Strigidae, *Transactions of the Zoological Society of London*, vol. 4, p. 201-260.

Keyserling, A. Graf von. & Blasius, J.H. 1839. Ueber ein Zoologisches Kennzeichen der Ordnung der Sperlingsartigen- oder Singvögel, *Archiv für Naturgeschichte*, vol. 5, p. 332-4.

Kirkman, F.B., Hartert, E., Jourdain, F.C.R., Pycraft, W.P. and Selous, E. 1910-1913. The British Bird Book. London & Edinburgh.

Kittlitz, F.H. von, Baron. 1832-3. Kupfertafeln zur Naturgeschichte der Vögel. 3 parts. Frankfurt.

Klein, J.T. 1750. Historiae Avium Prodromus. Lubecae.

Kolbe, P. 1719. Caput Bonae Spei Hodiernum. [Eng. trans. 1731].

Kosai Naoumi. 1755. Ko Yamato Honzo (Enlarged Natural History of Japan).

Kuhl, H. 1820. Conspectus psittacorum. Bonn.

Kuhl, H. 1820. Buffoni et Daubentoni figuararum Avium coloratarum nomina systematica. Gronigen.

Lacaille, Abbé de. 1763. Journal historique du voyage fait au Cap de Bon-Espérance.[not seen]

Lacépède, B-G-E, de la Ville, Comte de. 1799 [1802]. Tableaux Méthodiques des Mammifres et des Oiseaux. [Dated 1799, but apparently not issued till 1802]. Paris.

Lack, D. 1943a. The life of the Robin. Witherby, London.

Lack, D. 1943b. The age of the Blackbird. *British Birds* vol. 36, p. 166–175.

Lamarck, J.B. 1802. Recherches sur l'Organisation des Corps Vivant. Paris.
Lamarck, J.B. 1809. Philosophie Zoologique. Paris.
Lamarck, J.B. 1815-1822. Histoire Naturelle des Anmaux sans Vertèbres. Paris.
Landbeck, L. [1850s] papers on birds of Chile.
Latham, John. 1781-1785. General Synopsis of Birds, 3 vols. Supplement 2, 1802. London.
Latham, John. 1790. Index Ornithologicus. London.
Latham, John. 1821-1828. General History of Birds, 10 vols. Winchester.
LaTouche, J.D. 1925-37. A Handbook of the Birds of Eastern China. 2 vols. London.
Lear, E. 1832. Illustrations of the Family of Psittacidae or Parrots. London.
Lesson, R.P. 1826-30. Voyage autour du monde sur La Coquille, 2 vols and Atlas. Paris.
Lesson, R.P. 1829-30. Histoire naturelle des Oiseaux-Mouches. Paris.
Lesson, R.P. 1830-32. Histoire naturelle des Colibris. Paris.
Lesson, R.P. 1832-33. Les Trochilidées, ou les Colibris et les Oiseaux-Mouches. Paris.
Lesson, R.P. 1837. Journal de la Navigation autour du Globe de la frègate Thetis, 2 vols. Paris.
Levaillant, F. 1790. Travels from the Cape of Good Hope, into the interior parts of Africa. Trans. by Elizabeth Helme, 2 vols. London
Levaillant, F. 1796. New travels into the interior parts of Africa. Translated from the French, 3 vols. London.
Levaillant, F. 1796-1808. Histoire naturelle des oiseaux d'Afrique, 6 vols. Paris.
Levaillant, F. 1801-2. Histoire naturelle d'une partie d'Oiseaux nouveaux et rares de l'Amerique et des Indes. Paris & Amsterdam.
Levaillant, F. 1801-5. Histoire naturelles des perroquets, 2 vols. Paris.
Levaillant, F. 1801-6. Histoire naturelles des Oiseaux des Paradis et des Rolliers, 2 vols. Paris.
Lewin, J.W. 1813. Birds of New South Wales. Sydney.
Lewin, W. 1789. Birds of Great Britain, 7 vols. London.
Lilljeborg, W. 1866. Outlines of a Systematic Review of the Class of Birds. *Proceedings of the Zoological Society of London*, p. 5-20.
Linnaeus, C. 1748. Systema Naturae, 6th edition. Stockholm.
Linnaeus, C. 1758. Systema Naturae, 10th edition. Holmiae.
Linnaeus, C. 1766. Systema Naturae, 12th edition. Holmiae.
L'Herminier, F.J. 1827. Recherches sur l'appareil sternal des Oiseaux, "*Actes Linnean Society of Paris*", vol. 6, p. 3-93. [So cited by Newton].
Lord, T. 1791. Entire New System of Ornithology; or Æcumenical History of British Birds. London.
Lorenz, K. 1927. Beobachtungen an Dohlen [diary of a Jackdaw] *Journal für Ornithologie*, vol. 75, p. 511-519.
Lorenz, K. 1952. King Solomon's Ring. Methuen, London.
Lyell, C. 1830-2. Principles of Geology, being an attempt to explain the former changes of the Earth's surface. London.
Lysaght, A. 1952. Manchots de l'Antarctique en Nouvelle-Guinee, *L'Oiseau*, vol. 22, p. 120-124.
Macartney. 1819. "Feathers" article in *Rees's Cyclopaedia*, vol. 14.
Macdonald, J.D. and Grant, C.H.B. 1951. *Bulletin of the British Ornithologist's Club*, vol. V, 71, p. 30.
MacGillivray, W. 1836. Descriptions of the Rapacious Birds of Great Britain. Edinburgh.
MacGillivray, W. 1837-52. A History of British Birds, 5 vols. London.
Macintyre, B. 16 July 1993 "An artistic license to kill", *The Times*.
MacLeay, W.S. 1819. Horae Entomologicae. London.
Magnus, Albertus. 1478. De Animalibus. 26 books.
Marcgraf, G. 1648. Historiae rerum naturalium Brasiliae. Amsterdam.
Marcgraf, G. 1658. De Indiae utriusque re naturali et medica libri quatuordecim. Amsterdam.
Mason, A. S. 1992. George Edwards, the Bedell and the Birds. London: Royal College of Surgeons.
Mathews, G.M. 1912. [bibliographical note on Shaw's Zoology of New Holland]. *Emu*, vol. 11, p. 255-7.
Mathews, G.M. 1931. "John Latham (1740-1837): an early English Ornithologist", *Ibis*, p. 466-475.
Mathews, G.M. & Iredale, T. 1915. On the "Table des Planches Enlum." of Boddaert, *Australian Avian Record*, vol. 3, no. 2, p. 31-51.
Mearns, B. & R. 1988. Biographies for Birdwatchers; The Lives of Those Commemorated in Western Palaearctic Bird Names. Academic Press.
Mearns, B. & R. 1992. Audubon to Xántus; The Lives of Those Commemorated in North American Bird Names. Academic Press.

Medway, D. G. 1976. Extant types of New Zealand birds from Cook's Voyages. Pt. 1. Historical, and type paintings. *Notornis*, vol. 23, p. 44-60, Pt. 2. The Type specimens. pp. 120-137.

Merrem, B. 1813. An attempt at an outline of the general history and natural classification of birds, *Abhandlungen* Academy of Sciences of Berlin.

Merrett, C. 1666. Pinax Rerum Naturalium Britannicarum. London.

Merriam, C. H. Suggestions for a new method of discriminating between species and subspecies, *Science*, 14 May 1897.

Meyer, A.B. 1879-97. Abbildungen von Vogel-Skeletten, 2 vols. Dresden.

Meyer de Schauensee, R. 1984. The History of Ornithology in China, in The Birds of China, p. 23-36. Oxford University Press.

Möhring, P.H.G. 1752. Avium Genera. Bremen.

[Möhring, P.H.G. 1758] Geslachten der Volelen: Proposed suppression under the Plenary Powers. *Bulletin of Zoological Nomenclature*, vol. 21, part 5, p. 368-369, Nov. 1964.

Molina, J.I. 1782. Saggio sulla Storia Naturale di Chile. Bologna. English translation: 1802, 1809. The Geographical, Natural, and Civil History of Chili. 2 volumes. London.

Montagu, G. 1802. Ornithological Dictionary, 2 vols. London.

Mullens, W.H. 1908-9. Some Early British Ornithologists and their Work, *British Birds*, vol. 2; William Turner, p. 5-13; Richard Carew, p. 42-50; Christopher Merrett, p. 109-118, 151-164; Martin Martin, p. 173-182; Robert Plot, p. 218-225; Thomas Pennant, p. 259-266; John Ray and Francis Willughby, p. 290-300; Thomas Bewick and George Montagu, p. 351-361; William MacGillivray and William Yarrell, p. 389-399. [NB: These are listed in full for convenience, though not all were used in the text of the present work].

Mullens, W.H. 1911. Walter Charleton and his "Onomasticon Zoicon", *British Birds*, vol. 5, p. 64-71.

Mullens, W.H. 1915. Some Museums of Old London, 1. The Leverian Museum, *Museums Journal*, vol. 15, pp. 123-129, 162-172.

Mullens, W.H. 1917. Some Museums of Old London, 2. William Bullock's London Museum, *Museums Journal*, vol. 17, pp. 51-56, 132-137, 180-187.

Mullens, W.H. and Kirke Swann, H. 1917. Bibliography of British Ornithology. London.

Müller, P.L.S. 1773-76. Des Ritters C. von Linné [German edition of Systema Naturae]. Nürnberg.

Naumann, J.A. 1795-1803. Naturgeschicte der Land und Wasser-Vögel etc., [Detailed description of all birds of wood, field, and stream that reside in or pass through the principality of Anhalt and some of its neighbouring districts]. 4 vols. Köthen.

Naumann, J.F. 1795-1817. Naturgeschichte der Land- und Wasser-Vögel des nördlichen Deutschlands [Natural History of Birds in Northern Germany], 8 vols. Köthen.

Newman, E. 1850. First thoughts on a physiological arrangement of birds, *Proceedings of the Zoological Society of London*, p. 46-48.

Newton, A. 1864, 1902-7. Ootheca Wolleyana; an illustrated Catalogue of the Collection of Birds' Eggs, formed by the late John Wolley, Jun., MA., F.L.S. Edited from the original notes by Alfred Newton, M.A., F.L.S., 4 vols. London.

Newton, A. 1885 [1875-89] Ornithology, in Encyclopaedia Britannica, Ninth Edition, vol. 18, p. 2-50.

Newton, A. 1896. A Dictionary of Birds. London.

Newton, A. and "P.C.M.". 1911. Ornithology, in Encyclopaedia Britannica, Eleventh edition, vol. 20, p. 299-326.

Newton, I. 1989. Lifetime reproduction in birds. Academic Press, London.

Nice, M.M. 1937. Studies in the life history of the Song Sparrow. *Transactions of the Linnean Society*, New York, vol. 4, p. 1–247.

Nice, M.M. 1953. Incubation Periods of Birds of Prey, *Die Vogelwarte*, band 16, heft 4, p. 154-157. [in English].

Nice, M.M. 1954. Problems of Incubation Periods in North American Birds, *Condor*, vol. 56, p. 173-197.

Nichol, M.J. 1930. Nichol's Birds of Egypt, 2 vols. edited by R. Meinertzhagen. London.

Nicholson, F. (translator). 1889. Sundevall's Tentamen. London.

Nitzsch, C.L. 1820. [treatise on nasal glands of birds] in Meckel's Deutsches Archiv für dei Physiologie, 6, pp. 251-269. The gist of this system was given by Owen in "Aves", Todd's Cyclopaedia Anat., 1, p. 226.

Nitzsch, C.L. 1829. Observationes de Avium arteria carotide communi. Halae.

Nitzsch, C.L. 1833. *Pterlyographiae Avium pars prior.* Halae. Eng. Trans. 1867 as "Nitzsch's Pterylography". Ray Society.

Nobutoshi Okada. 1891. Catalogues of the Animals of Japan: Birds.

Nordenskiöld, E. 1928. The History of Biology, translated from the Swedish by Leonard Bucknell Eyre. New York.

Oehser, P H. 1948. Louis Jean Pierre Vieillot (1748-1831) [biography], *Auk*, vol. 65, p. 568-576.

O'Hara, R J. 1991. Representations of the Natural System in the Nineteenth Century, *Biology and Philosophy*, vol. 6, p. 255-274.

O'Hara, R J. 1988 [1989]. Diagrammatic Classifications of Birds, 1819-1901: Views of the Natural System in 19th-century British Ornithology, in *Acts XIX Congressus Internationalis Ornithologici*, p. 2746-2759.

Oken, L. [Entry in Encyclopaedia Britannica]

Oken, L. 1809-11. Lehrbuch der Naturphilosophy. 2nd ed. 1831. 3rd ed. 1843. Leipzig & Jena. Eng. trans. 1847 as Elements of Physiophilosophy. Ray Society.

Oliver, S. P. 1909. The Life of Philibert Commerson. London.

Olson, S.L., and James, H.F. 1994. A chronology of ornithological exploration in the Hawaiian Islands, from Cook to Perkins. In: Jehl Jr, J. R., and Johnson, N. K., (eds.). *A century of avifaunal change in western North America*. Studies in Avian Biology No. 15. Cooper Ornithological Society, California.

Pallas, P.S. 1811. Zoographia Rosso-Asiatica, 3 vols. Petropoli.

Pander, C. 1817. Beiträge zur Entwickelungsgeschichte des Hühnchens im Eye. Würzburg.

Pasquet, E. 1998. Phylogeny of the nuthatches of the *Sitta canadensis* group and its evolutionary and biogeographic implications. *Ibis*, vol. 140, p. 150–156.

Peale, T. 1848. United States Exploring Expedition: Mammalology & Ornithology. Philadelphia.

Penard, T.E. 1924. Historical Sketch of the Ornithology of Surinam, *De West-Indische Gids*, August 1924, pp. 145-168: Reprinted as separate pamphlet, p. 1-24.

Pennant, T. 1761-1766. British Zoology, 4 vols. London & Chester.

Pennant, T. 1769. Indian Zoology. London.

Pennant, T. 1784-5. Arctic Zoology, 2 vols. London.

Pernau, J.F.A. von. 1702. Lesson on what can be done, for pleasure and amusement, with those delightful creatures, birds, in addition to catching them, only by inquiring into their qualities and taming or otherwise training them. Revised editions, 1707, 1716.

Pernau, J.F.A. von. 1720. Agreeable Country Pleasures.

Phillip, Capt. A. 1789. The Voyage of Governor Phillip to Botany Bay. London.

Philippi, R.A. 1850s. papers on birds of Chile.

Pliny. Historia Naturalis, various editions.

Poeppig, E.F. 1835-36. Reise in Chile, Peru und auf dem Amazonenstrome während ... 1827-32. 2 vols & Atlas. Leipzig.

Pontoppidan, E.L. 1755. The Natural History of Norway (translated). London.

Pontoppidan, E. L. 1763. Dansk Atlas. Copenhagen.

Quoy, J.C.R. & Gaimard, J.P. 1824-26. Voyage de l'Uranie: Zoologie. Paris.

Quoy, J.C.R. & Gaimard, J.P. 1830-35. Voyage de l'Astrolabe: Zoologie, 4 vols. Paris.

Rafinesque, C.S. 1836. A Life of Travels and researches in North America and South Europe. Philadelphia, privately printed.

Rafinesque. See also Call, and Rhoads.

Ralph, R. 1993. William MacGillivray. NHM, London.

Ranzan Ono. 1803. Honzokomuku Keimo.

Ray, J. 1713. Synopsis Methodica Avium et Piscium. London.

Reichenow, A. 1913-4. Die Vögel, Handbuch der systematischen Ornithologie. 2 vols. Stuttgart.

Rees's Cyclopaedia, vol. 14. article: "feathers". London.

Rhoads, S N. 1911. Constantine S. Rafinesque as an Ornithologist, *Cassinia*, no. 15, p. 1-12.

Rhoads, S N. 1912. Additions to the known ornithological publications of C.S. Rafinesque, *Auk*, vol. 29, p. 191-198.

Richardson, Sir J. & W. Swainson. 1831. Fauna Boreali-Americana, or the Zoology of the northern parts of British America, vol. 2, Birds. London.

Richdale, L.E. 1949. A study of a group of penguins of known age. *Biological Monographs* (Dunedin, New Zealand) vol 1. p. 1–88.

Richmond, C W. 1909. A reprint of the Ornithological writings of C.S. Rafinesque, *Auk*, vol. 26, pp. 37-55, 248-262.

Ridgway, R. 1881. Nomenclature of North American Birds, *Bulletin of the United States National Museum*, no. 21.

Ridgway, R. 1901-1919. The Birds of North and Middle America, 8 vols. 1941-50, 2 further vols, edited by H. Friedmann. *Bulletin of the United States National Museum*, no. 50.

Ridgway, R. 1912. Color Standards and Color Nomenclature. Washington, D.C.

Rookmaaker, L.C. 1989. Zoological exploration of Southern Africa 1650-1790. Rotterdam. [Section on Francois Levaillant, p. 177-271].
Rosenberg, C.B.H. von. 1878. Der Malayische Archipelago. Leipzig.
Rüppell, W.P. E. S 1826-1830. Atlas zu der Reise im nördlichen Afrika [Atlas of Rüppell's Travels in Northern Africa]. Ornithology compiled by P. Cretzschmar. Frankfurt.
Ruysch, F. 1710. Thesaurus Animalium primus, etc. Amsterdam.
Ryoan Terashima. 1713. Wakan Sansai Zuye (Illustrations of the Natural History of Japan and China).
Sangster, G. 2000. Genetic distance as a test of species boundaries in the Citril Finch *Serinus citrinella*: a critique and taxonomic reinterpretation. *Ibis*, vol. 142. p. 487–490.
Savi, P. 1827-31. Ornitologia Toscana, 3 vols. Pisa.
Savi, P. 1873-76. Ornitologia Italiana, 3 vols. Firenze.
Schaeffer, J.C. 1774. Elementa Ornithologica. Ratisbon.
Schaeffer, J.C. 1789. Museum Ornithologicum. Ratisbon.
Schlegel, H. 1859-1861. [papers in] *Journal für Ornithologie*.
Schlegel, H. 1862-80. Muséum d'histoire naturelle des Pays-Bas. 14 vols. Leyden.
Schomburgh, R.H. 1840. A description of British Guiana. London.
Schomburgk, R.H. 1848. The History of Barbados. London.
Schwenckfeld, C. 1603. Theriotropheum Silesiae. Lignicii.
Sclater, P.L. 1858. On the general Geographical Distribution of the Members of the Class Aves, *Journal of the Proceedings of the Linnean Society*, vol. 2, p. 130-145.
Sclater, P.L. 1880. Remarks on the present State of the Systema Avium, *Ibis*, p. 340-350, 399-411.
Sclater, W.L. 1905. South African Ornithologists, *Journal of the South African Ornithologists' Union*, vol. 1, p. 3-8.
Sclater, W.L. 1929. Notes on the early sources of our knowledge of African Ornithology, *Journal für Ornithologie*, Ernst Hartert Supplement, p. 184-196 [in English].
Scopoli, G.A. 1769-72. Anni Historico-Naturales, 5 parts. Lipsiae.
Scopoli, G.A. 1786-88. Deliciae Flora et Fauna Insubricae, 3 parts. Ticini.
Seba, A. 1734-65. [Accurate Description of the Most Richly Endowed Treasury of Nature, and an Illustration with the most Skilful Pictures, for a Universal History of the Physical World]. Amsterdam
Seebohm, H. 1880. Siberia in Europe. London.
Seebohm, H. 1882. Siberia in Asia. London.
Seebohm, H. 1882. [comments on Pallas's "Zoographia Rosso-Asiatica" and trinomial nomenclature] in Ornithology of Siberia, *Ibis*, p. 425-428.
Seebohm, H. 1882. Letter to editor [further comments on Pallas's "Zoographia Rosso-Asiatica"], *Ibis*, p. 611-612.
Seebohm, H. 1887. The Geographical Distribution of the Charadriidae, [introductory chapters]. London & Manchester.
Seebohm, H. 1890. The Birds of the Japanese Empire. London.
Seebohm, H. 1901. The Birds of Siberia. London.
Sharpe, R.B. 1899-1909. Hand List of Genera and Species of Birds. 5 vols. London.
Shaw, G. 1784. Zoology of New Holland, pp. 33. [see Mathews 1912].
Sherborn, C.D. 1934. On the dates of Pallas's "Zoographia Rosso-Asiatica", *Ibis*, p. 164-167.
Sibley, C.G. and Ahlquist, J.E. 1986. Reconstructing bird phylogeny by comparing DNAs. *Scientific American*. February 82–92.
Sibley, C.G. and Ahlquist, J.E. 1990. *Phylogeny and classification of birds*. Yale University Press, New Haven.
Sibley, C.G. and Monroe, B. 1990. *Distribution and taxonomy of birds of the world*. Yale University Press, New Haven.
Siebold, P.F. von. 1845-50. Fauna Japonica: Oiseaux. Lugduni.
Sloane, Sir Hans. 1707, 1725. A Voyage to the Islands Madera, Barbados, Nieves, S. Christophers and Jamaica, 2 vols. London.
Smith, C.H. 1839. Memoir of Pallas, in The Naturalist's Library, vol. 25. Edinburgh.
Smith, C.H. 1841. Memoir of Gesner, in The Naturalist's Library, vol. 31. Edinburgh.
Smith, C.H. 1840. Memoir of Felix de Azara, in The Naturalist's Library, vol. 28. Edinburgh.
Sonnerat, P. 1776. Voyage à Nouvelle Guinea. Paris. English translation of part: An account of a Voyage to New Guinea. Bury St. Edmunds.
Sparrman, A. 1786. A Voyage to the Cape of Good Hope, towards the Antarctic Polar Circle, and round the

World with Captain Cook: but chiefly into the country of the Hottentots and Caffres, from 1772 to 1776. Translated by J.R. Forster. London.
Sparrman, A. 1786-89. Museum Carlsonianum. Holmiae.
Sparrman, A. 1806. Svensk Ornithologie. Stockholm.
Spix, J.B. von. 1824-25. Avium Species Novae. 2 vols. Monachii.
Spix, J.B. von. & Martius, K.F.P. von. 1823-31. Reise in Brasilien, 3 vols. München.
Steller, G. 1751. De Bestiis Marinis.
Stone, W. 1899. Some Philadelphia Ornithological Collections & Collectors, *Auk*, vol. 16, p. 166-177.
Stone, W. 1912. Vroeg's Catalogue, *Auk*, vol. 29, p. 205-207.
Stone, W. 1915. Titian Ramsey Peale, *Cassinia*, no. 19, p. 1-13.
Stowell Rounds, R. 1990. Men and Birds in South America 1492-1900. Q.E.D. Press, Fort Bragg, California.
Stresemann, E. 1951. Date of publication of Pallas's "Zoographia Rosso-Asiatica", *Ibis*, p. 316-318.
Stresemann, E. 1952. On the Birds collected by Pierre Poivre in Canton, Manilla, India and Madagascar (1741-1756). *Ibis*, vol. 94, p. 499-523.
Stresemann, E. 1947. Baron von Pernau, Pioneer Student of Bird Behaviour, *Auk*, vol. 64, p. 35-52.
Stresemann, E. 1951. Die Entwicklung der Ornithologie von Aristoteles bis zur Gegenwart. Berlin.
Stresemann, E. 1975. Ornithology from Aristotle to the Present. Translated by Hans J. and Cathleen Epstein, with a Foreword and an Epilogue on American Ornithology by Ernst Mayr. Cambridge, Mass. & London.
Steullet, A.B. and Deautier, E.A. 1935. Catálogo Sistemático de las Aves de la República Argentina, tomo 1. *Obra del Cincuentenario del Museo de La Plata, Buenos Aires*. [Section of history of ornithology in Argentina kindly translated by Alejandro Mouchard].
Strickland, H.E. (*et al.*) 1843. Report of a Committee appointed to consider of the rules by which the Nomenclature of Zoology may be established on a Uniform and Permanent Basis. ["The Strickland Code"] in Report of 12th Meeting of the British Association for the Advancement of Science. 1842 (1843) p. 105-121.
Strickland, H.E. 1835. On the arbitrary Alteration of established Terms in Natural History, *Magazine of Natural History*, p. 36-40.
Strickland, H.E. 1844 (1845). Report on the Recent Progress and Present State of Ornithology, in Report of the Fourteenth Meeting of the British Association for the Advancement of Science. vol. 13, p. 247-315. London.
Strickland, H.E. and Melville, A.G. 1848. The Dodo and its Kindred. London.
Sundevall, C.J. 1857. Om le Vaillant's Oiseaux d'Afrique, *Kritish Framställning*, p. 23-60.
Sundevall, C.J. 1865. Les Oiseaux d'Afrique de Levaillant, critique de cet ouvrage, *Revue et Magazin de Zoologie*, p. 282-285, 323-329, 408-414, (1866) p. 42-46, 189-197, 227-233.
Sundevall, C.J. 1872-3. Methodi naturalis Avium disponendarum Tentamen. Stockholm. [Translated by Francis Nicholson (q,v.)].
Swainson, W. 1808. Instructions for Collecting & preserving Subjects of Natural History. Liverpool.
Swainson, W. 1820-3. Zoological Illustrations. London.
Swainson, W. 1836-7. Natural History & Classification of Birds. London.
Swainson, W. 1840. Taxidermy, with the biography of zoologists, [p. 105. Comments on.Albin]. London
Swinhoe, R. 1863. Catalogue of the Birds of China, *Proceedings of the Zoological Society of London*, p. 259-339; id. 1871. A Revised catalogue of the Birds of China, p. 337-423.
Taczanowski, L. 1884-6. Ornithologie du Pérou, 4 vols. Rennes.
Taka-Tsukasa, Prince. 1935-1943. History of Japanese Ornithology, in The Birds of Nippon, vol. 1. [only one volume issued]. London & Tokyo.
Temminck, C.J. & Madame Knip. 1808-11. Les Pigeons. Paris.
Temminck, C.J. 1813-5. Histoire générale des pigeons et des gallinacées. 3 vols. Amsterdam & Paris.
Temminck, C.J. 1815. Manuel d'Ornithologie. Amsterdam & Paris.
Temminck, C.J. & Laugier de Chartrouse M. Baron de. 1820-39. Planches coloriées, 5 vols. Paris.
Thomas de Cantimpré. c1233-1247. Die Naturis Rerum.
Tinbergen, N. 1951. The Study of Instinct. Oxford University Press, Oxford.
Tinbergen, N. 1953. The Herring Gull's world. Collins, London.
Topsell, E. The Fowles of Heauen. MS in Henry E. Huntingdon Library; edited and published with introduction by T.P. Harrison and F.D. Hoeniger, 1972. University of Texas Press, Austin.
Tree, I. 1991. The Ruling Passion of John Gould. A biography of the bird man. London.
Tristram, H.B. 1859. On the Ornithology of Northern Africa, pt. 3, *Ibis*, p. 429-33 [415-435].

Tristram, H.B. 1867. Natural History of the Bible. London.
Tschudi, J.J. 1844. Avium Conspectus quae in Republica Peruana, *Archiv für Naturgeschichte*, p. 262-317.
Tunstall, M. 1771. Ornithologia Britannica. London.
Turner, W. 1544. A Short and succinct account of the principal birds mentioned by Pliny and Aristotle. Cologne.
Ulloa, A. 1760. A voyage to South America., translated from the Spanish. 2 vols. London.
Valentijn, F. 1726. Verhandeling der Vogelen van Amboina, in *Amboina*, p. 297-329. [The East Indies Old and New] Amsterdam.
Vieillot, L.P.J. & Audebert, J.B. 1802 Oiseaux Dorés. Paris.
Vieillot, L.P.J. 1816-1819. [contributions to] Nouveau Dictionaire d'Histoire Naturelle. Paris.
Vieillot, L.P.J. 1816. Analyse d'une nouvelle Ornithologie élémentaire. Paris.
Vigors, N.A. 1824. [description of Psittacula Kuhlii] *Zoological Journal*, vol. 1, p. 412, pl. 16.
Vigors, N.A. 1825. Observations on the Natural Affinities that connect the Orders and Families of Birds. *Transactions of the Linnean Society of London*, vol. 14. p. 395-517.
Vigors, N.A. 1840. [notice of his death]. *Proceedings of the Zoological Society of London*, p. 129.
Vigors, N.A. [Entry in Dictionary of National Biography].
Vincent, Levin. 1719. Description Abrege des Planches qui representent les cabinets & quelques-unes des Curiositis, contenus dans le Theatre des merveilles de la nature. [Descriptions and Plates of cabinets of Curiosities and Marvels of Nature].
Violani, C. [no date]. MS notes on Scopoli.
Wagler, J.G. 1827. Systema Avium. Stuttgart & Tubingen.
Wagler, J.G. 1830. Natürliches System der Amphibien mit vorangehender Classification der Säugthiere und Vögel. München.
Wagler, J.G. 1832. Monographia Psittacorum, *Denkschriften der Königlichen Academie der Wissenschaften in München*, vol. 1, p. 469 et. seq. Reprinted 1835 as a separate work.
Wagstaffe, R. and G. Rutherford. 1954-5. "Letters from Knowsley Hall", *Northwestern Naturalist*, vol. 2, p. 173-183.
Walcott J. 1789. Synopsis of British birds, 2 vols. London.
Wallace, A.R. 1876. The Geographic Distribution of Animals. London.
Wallace, A.R. 1869. The Malay Archipelago, 2 vols. London.
Wedderburn, W. 1913. Allan Octavian Hume ... 1829 to 1912. London.
Wendt, H. 1968. Before the Deluge. Translated from the German by Richard and Clare Winston. London.
White, G. 1789. Natural History and Antiquities of Selbourne. London. Many subsequent editions.
White, J. 1790. Journal of a Voyage to New South Wales. London.
Whitehead, P.J.P., van Vliet, G. and Stearn, W.T. 1989. The Clusius and other natural history pictures in the Jagiellon Library, Kraków, *Archives of Natural History*, vol. 16, p. 15-32.
Wied-Neuwied, Prince A.P. Maximilian zu. 1820-21. Reise nach Brasilien. Frankfurt.
Wied-Neuwied, Prince A.P. Maximilian zu. 1825-33. Beiträge zur Naturgeschichte von Brasilien. Weimar.
Wied-Neuwied, Prince A.P. Maximilian zu. Notice of his death in *Ibis*. 1867. p. 472.
Willughby, F. 1676. Ornithologiae. London.
Willughby, F. 1678. Ornithology (translated into English ... with ... additions). London.
Wilson, A. 1808-14. American Ornithology, 9 vols. Philadelphia.
Wilson, J. 1853-60. Ornithology, in Encyclopaedia Britannica, Eighth edition, vol. 16, p. 725-831.
Witherby, H.F., Jourdain, F.C.R. and Ticehurst, H.F. 1912. A Handlist of British Birds. London.
Witherby, H.F. et. al. 1940-41. The Handbook of British Birds. 5 vols. London.
Wollaston, A.F.R. 1921. Life of Alfred Newton. John Murray: London.
Wood, N. 1836. The Ornithologist's Text-Book, being reviews of ornithological works; with an appendix, containing discussons on various topics of interest. London.
Worm, Olao. [Ole Wormius] 1655. Museum Wormianum. Lugduni Batavorum.
Yarrell, W. 1829. On the Organs of Voice in Birds, *Transactions of the Linnean Society*, vol. 16, p. 305-321, pl. 17, 18.
Yarrell, W. 1837-43. History of British Birds, 3 vols. 2nd ed. 1845, 3rd. ed. 1856. London.
Zorn, J.H. 1742-3. Ornithotheology, or an attempt to encourage men, through closer observation of birds, to the admiration, love, and reverence of their most powerful, wise, and good Creator.

INDEX